PENGUIN ENGLISH LIBRARY

VOYAGES AND DISCOVERIES

RICHARD HAKLUYT

Jack Beeching was brought up in England and began to publish poetry there just before the outbreak of war. He has travelled extensively in both Europe and Latin America, and his publications include novels and volumes of poetry and history. His recent books include *The Chinese Opium Wars* (1975), *An Open Path: Christian Missionaries 1515–1914* (1980), and *The Galleys at Lepanto* (1982). His novel, *Death of a Terrorist*, was published in 1981, and *Twenty-five Short Poems* in 1982.

HAKLUYT

Voyages and Discoveries

The Principal Navigations
Voyages, Traffiques and Discoveries
of the English Nation

EDITED, ABRIDGED
AND INTRODUCED BY
JACK BEECHING

PENGUIN BOOKS

Penguin Books Ltd, Harmondsworth, Middlesex, England
Penguin Books, 625 Madison Avenue, New York, New York 10022, U.S.A.
Penguin Books Australia Ltd, Ringwood, Victoria, Australia
Penguin Books Canada Ltd, 2801 John Street, Markham, Ontario, Canada L3R 1B4
Penguin Books (N.Z.) Ltd, 182–190 Wairau Road, Auckland 10, New Zealand

—

This edition first published in Penguin English Library 1972
Reprinted 1982

—

Made and printed in Singapore by
Richard Clay (S.E.Asia) Pte Ltd
Set in Linotype Juliana

CONTENTS

Contents

Contents

INTRODUCTION

1

The Principal Navigations, Voyages, Traffiques and Discoveries of the English Nation is a collection of documents made in the sixteenth century, extending to a million and a half words. By rigorous excision and abridgement, about a tenth of Hakluyt's matter has been compressed into this Penguin edition, and put into chronological order. The many volumes of the complete Hakluyt, usually unopened, were often enough to be found on private bookshelves years ago when England thought of herself as imperial. Since those days Hakluyt has been used mainly as a source of rattling good yarns. He is very much more; and this selection has been made to whet the appetite of anyone who ever wanted to try the serious pleasure of bookworming through the entire collection.

To the title of his second and enlarged edition, 1598–1600, Hakluyt added the one significant word, *Traffiques*. The English nation in the previous fifty years had travelled and discovered, not for fun, but in order to do business. Nearly all these documents – ships' logs, salesmen's reports to head office, secret economic intelligence, captured enemy papers – relate, in the last analysis, to trade. Hakluyt does his best to prove that the English were travellers even before the Crusades; perhaps because any tradition is a comfort, even a fake one. But England had, in fact, arrived laggardly on the world's commercial scene.

In the previous century the carrying trade around the European littoral had largely been in Venetian or German hands. England was a kingdom on the margin of affairs: a purveyor of raw materials, still selling abroad some raw wool that might have been woven for the market. But soon afterwards the Portuguese, another small people on Europe's outer edge, found a vastly profitable detour to the Spice Islands around the Cape of Good Hope: on the cargo he brought back to Lisbon, Vasco da Gama is said to have made a profit of 10,000 per cent. The discovery of America decisively shifted the centre of gravity in world trade. And there was England, an island with a conveniently indented coastline, already launched forth expectantly into the Atlantic to confront the New World. The opportunity was immense.

By the 1550s the extravagant output of bullion from Potosí and other American mines was beginning to agitate European trade. Already the Tudor dynasty had endowed England with a new dominant class: men who had risen by the law and mopped up their share of church lands, men to whom enterprise and profitability were congenial ideas. England began to challenge the Portuguese and Spanish monopoly of sea-routes to the Indies, East and West. The French were some little way ahead, the Dutch would soon be coming up on their heels, when belatedly but with their usual obstinacy, shrewdness and courage, English seamen began to try out one way after another of getting at the gold and the spices. Many of the ways they chose were next to impossible.

Since the Portuguese-dominated route round the Cape and thence to India and Cathay could not be forced, they were obliged to try heroic outflanking manoeuvres through the ice-bound seas north of Canada and Siberia – routes that have become viable only in our lifetimes. For these Tudor seamen the Northeast and Northwest Passages were never more than hope's graveyard.

The Venetian hold over trade to the Levant was successfully if wearisomely bypassed when an Arctic route was found into Russia – northward around Scandinavia and thence into the White Sea. Since Tsar Ivan by conquering Kazan and Astrakhan 1552–6 had forced a way south to the Caspian Sea, goods could be worked up the Dvina and down the Volga, across the Caspian and actually into Persia; there to be traded or bartered for Oriental goods coming overland – though at a diminishing rate of profit as distance and risk and political friction increased.

Others like John Hawkins tried running the Spanish blockade, and trading directly with America – where African slaves to work the plantations were discovered to be a prime article of commerce. Others again, before and during the long drawn out naval war with Spain, experimented with more or less glossed over piracy: the readiest means of diverting commercial profits from one pocket to another, without the usual need first to strike a bargain.

The years from 1575 to 1620, observed J. M. Keynes, were the 'palmy days of profit'.* Poor people in town and country might suffer from the headlong rise in prices which followed the dissemination through Europe of Spanish silver, but price inflation is good for business. For merchants trading abroad this was a time of exhilarating opportunity and they grabbed at it with both hands.

Organized mercantile exploration may be said to have begun in earnest in 1553, with Sebastian Cabot's instructions to Sir Hugh Willoughby, whose expedition, though intended for Cathay, pioneered the northward route to Russia. This veteran explorer of North America, tempted back to England after long years in the service of Spanish navigation, is here setting down from experience a series of practical rules for those who sail under discipline into strange seas. Any reader who happens to have served on the lower deck of the Royal Navy will recognize at once the distinctive tone of voice. Cabot's set of imperatives for shipboard life, derived from Spanish experience, were to serve the English well for the next four centuries.

2

Hakluyt meant his documents to be useful. He accumulated for instance a mass of navigational fact and economic information – of which only small samples are here given. Important pieces of knowledge, formerly held as trade secrets, had been yielded up at Hakluyt's urging by merchants and mariners, and put on display for the common good. But Hakluyt's intention as archivist and editor was also consciously historical. He deliberately arranged his documents to throw light on recent national experience. Devoid of vanity and pedantry, he conceals this historical intention behind arrays of plain fact; but the real function of the book is to promote national confidence. By displaying the facts intelligibly he hopes

*Keynes, J. M., A *Treatise on Money*, London, 1930, p. 149 ff.

to give policy, whether private or national, a better sense of direction. Later, as more specialist works took the place of this first agglomeration, its practical usefulness to merchants and mariners may have been exhausted. But in history and literature, in geography, and even in the natural and social sciences, Hakluyt's remoter implications reverberate underground to this day. As Bacon observed in the *Advancement of Learning*, 'This proficience in navigation and discoveries may plant also an expectation of the further proficience and augmentation of all sciences.'

Hakluyt is our first serious geographer – the Englishman who decisively parts company with medieval cosmography, weighed down as it was by the authority of classical texts, and the astronomy of Ptolemy. Hakluyt's informants are practical, hard-headed and fearless men, who root their lives in the written certainties of logbook and ledger and *Book of Common Prayer*. His rough-and-ready shipmasters can have had no inbuilt prejudice against a Copernican world. They knew that circumnavigation was possible. Sun and stars overhead are not symbols of perfection, but useful signals of whereabouts; the stars control a navigator's fate by telling him his latitude. Hakluyt himself is a scholar who has learned to get on terms with such men as these. He is the human link in Elizabethan London between ship's captain, merchant, map-maker, courtier, statesman, speculator and foreign geographer.

Because the method of Hakluyt's collection, informed by the mood of the new learning, is accurate, generous, sensible, free of dogma and based upon living practice, information he was the first to organize underpins a number of our contemporary academic disciplines. His lists of words in rum native lingos may have been provided chiefly so that a ship's captain could make himself understood when he came to a strange place – but here, or hereabouts, is the germ of comparative linguistics. Shipmen breaking in upon more primitive, hitherto untouched societies for the purposes of trade had a faculty of observing and recording curious customs with the lack of prejudice which distinguishes the anthropologist, who is their historical legatee. A diplomat might get a better grip on his profession from reading

Hakluyt's book; a cloth merchant could plan his export drive. There is also much descriptive botany and natural history and a quantity and variety of significant economic information that in an abridgement can only be suggested. All sciences begin with collections – and in this early, multiplex collection the cast of mind is proto-scientific.

Probably the last thing Hakluyt intended was to make an anthology of recent English prose. But as the dates appended to his documents jog on from the fifties to the eighties and his informants grow more aware of the size and strangeness of their world, there is a perceptible enrichment of language. Cadence and vocabulary begin to reflect the intoxicating taste of success in complex endeavour. The English language is, visibly, becoming equivalent to Shakespeare's mind.

Earlier merchants' letters, repetitive, eccentrically punctuated, languid with law terms, grip one's attention only when some novelty is described in words which happen to glow from personal experience. Compare specimens of the earliest prose – index of a society incompletely broken out from the static values of feudalism – with the authentic eloquence that arrives within one generation. Raleigh's account of the *Revenge*'s last fight off the Azores is a deliberate work of art – at moments so vivid that one feels almost physically the throb in the writer's throat as he meditates the action. What officer of the Crown writes like this nowadays?

3

By origin the Hakluyts were border gentry from Herefordshire. In the 1540s the antiquary John Leland, making notes for his *Itinerary*, was entertained by them at Eaton Manor near Leominster, and wrote, 'The Hackluits have been gentlemen in times out of memory.' Sir Walter Hakelutel, the family's first notability, had been Justice of West Wales in the time of Edward I, and died in 1315. By Tudor times the name was written Hakluyt and evidence indicates that it was pronounced *Haklet*.

Younger sons had as usual gone into law, trade and the church. Take for example Hakluyt's cousin and namesake – a lawyer, whose name too has a place in the history of travel and trade – Richard Hakluyt of the Middle Temple, an older man than Richard the preacher and geographer. His father, also a lawyer, was for a long time Clerk to the Council for Wales; an uncle had been admitted apprentice to the Skinners' Company in 1510; another uncle took an Oxford degree in Canon Law, and held a Suffolk living: thoroughly typical Tudor gentry. Richard the geographer was the protégé of Richard of the Middle Temple, and may well have been his ward.

'I do remember that being a youth,' wrote Hakluyt in the dedication to Sir Francis Walsingham of his first edition, '... it was my hap to visit the chamber of Mr Richard Hakluyt my cousin, a gentleman of the Middle Temple, well known unto you, at a time when I found lying open upon his board certain books of Cosmography, with an universal Map: he, seeing me somewhat curious ... began to instruct my ignorance ... he pointed with his wand to all the known Seas, Gulfs, Bays ... Kingdoms, Dukedoms and territories of each part, with declaration also of their special commodities, and particular wants, which by the benefit of traffic, and intercourse of merchants, are plentifully supplied ... my cousin's discourse ... took in me so deep an impression, that I constantly resolved ... I would by God's assistance prosecute that knowledge and kind of literature ...'

Sir Francis Walsingham, the Secretary of State responsible for Queen Elizabeth's foreign policy, was also the organizer of her elaborate secret service. Hakluyt's cousin, the lawyer who kept maps in his chambers, a man familiar in the City of London as a projector of new discoveries, was perhaps 'well known unto' Walsingham as a source of intelligence. In February 1570–71, for example, when there was a threat of Ireland's being invaded, we find him giving confidential information to Burghley about a Spanish mobilization order, acquired by his informant, he cheerfully admits, by 'harkening and listening at a lock hole'. More usually he seems to have been an occasional adviser to London merchants on matters of economic geography. In 1578 he was involved in Gilbert's first attempt to plant a colony in America; and in 1580 advised the Russia Company on its Northeast Passage attempt. He must evidently have made accessible to the younger and now more famous Richard Hakluyt many of the documents relating voyages to Russia and the Levant; not only did he direct his young cousin's early enthusiasm, he would have been able also to effect the right introductions. Hakluyt's book is so comprehensive because this elder cousin of his broke the ground.

Our Richard Hakluyt was born in 1551 or 1552, and would have been about sixteen when he was shown those fascinating maps in his cousin's chambers. In the late sixties he was elected from Westminster School, where he had been Queen's Scholar, to Christ Church, Oxford. He was a contemporary there of Sir Philip Sidney, later to become Walsingham's son-in-law. Between 1577 when he proceeded M.A., and 1583 when Student of Christ Church, Hakluyt took Holy Orders.

Interest in geographical studies in the University was by no means dead, though it still kept a medieval cast of mind and was tied to classical texts. The picture of our universe that most educated men held by was Ptolemaic – sun and planets making complicated gear-wheel revolutions round an Earth at the centre of everything. Even among learned men, academic knowledge of a world now known, practically, to include whole new continents would probably not extend beyond information conveyed from antiquity by geographers like Pliny, Strabo and Ptolemy, and by subsequent more or less fanciful travellers' tales. Hakluyt, in his first edition, reprints the Latin text of Mandeville, whose delightful though often far-fetched medieval collection of travels was still thought of as geographically informative, if not authoritative. Martin Frobisher, in his ship's library, actually took a copy of Mandeville with him as a geographical guide, on his first attempt to find the Northwest Passage.

Just before Hakluyt's time at Oxford, the continental mapmakers Mercator and Ortelius had begun to arouse intellectual interest by presenting an augmented picture of our world in the light of the new discoveries. Hakluyt gave geographical lectures on the use of these new maps and globes, but much of his time must have been spent learning languages: a decisive study in his chosen career. The Spaniards were beginning to publish descriptions of the lands they had discovered; so were the French. In 1580 Hakluyt is found writing to Mercator for information about Siberian coasts and currents which might be of use to an English expedition setting out once more to find a Northeast Passage. He spoke with Sir Francis Drake about a plan for subsidizing a chair of navigation. The Portuguese ambassador in London, his country now swallowed up by Spain, might be of mind to purvey information: Hakluyt had with him 'a great conference in matters of cosmography'. He speaks of having mastered Greek, Latin, Italian, Portuguese, Spanish and French, but evidently his Spanish was weakest, for later on he prefers the French or Italian translation of a Spanish original, whenever he can lay his hands on one.

In 1582 Hakluyt published his *Divers Voyages touching the discovery of America*, dedicated to Sir Philip Sidney. To furnish timely argument for Walsingham's policy of planting an English colony on the American mainland, Hakluyt had brought together, from written records, an account of much it might be well to know about America. From this response to a need of practical politics, his documentary method was beginning to emerge.

As early in the century as 1527 Robert Thorne in Seville, alive to the implications for England of Spanish discoveries, had written encouragingly, 'There is no land unhabitable, nor sea innavigable.' The voyages of Cabot and Frobisher, as set down by Hakluyt, gave point to this saying. He baits the argument for the City of London with a somewhat overstated list of American products available for trade, and gives again the arguments for a Northwest Passage to Cathay – that recurring dream of making an immensely profitable detour by going to the Indies northward around Canada.

In his later *Particular Discourse on the Western Planting* – presented to Queen Elizabeth in Walter Raleigh's name, a state paper not printed in the *Principal Navigations* – Hakluyt was at liberty to give more thoroughly the argument for colonial expansion. Strategic reasons were given emphasis. Spain, the dominant European power – financing the Counter-Reformation with American gold, trying by force to impose Catholicism on France and the Low Countries – is identified as the enemy of an England already hoping to prosper by the inroads that may be made upon the Spanish colonial monopoly. An English colony midway along the North American coast could serve as a naval base for attacks on the West Indies and the Spanish Main, not to mention on the Plate Fleet – Spain's annual convoy bringing gold and silver in bulk across the Atlantic to Seville.

Hakluyt's report on the state of the Spanish possessions, procured, he says, from an 'excellent French captain', describes them as under-manned and surrounded by Indians made hostile by maltreatment. A plantation, in Hakluyt's view, must also be a mission, where religion is employed to win the natives over. (He made several attempts to go to America as a chaplain; once he spoke of following Gilbert to Newfoundland; later he sought leave of absence from a post in Paris to go on the 1584 voyage to Virginia. The chance finally came in middle age, when his enthusiasm had waned.)

He goes on, in his *Particular Discourse*, to suggest how America could ultimately make England economically independent of Europe. America produced all those sought-after articles of commerce for which the English had been obliged to pay highly elsewhere – Hakluyt gives a list of fifty-four, all of which came to light eventually in the course of trade. He quotes with satisfaction the case of a Rouen merchant who sold in France for 440 crowns a parcel of furs he had bought in Maine for forty. The real source of wealth in the New World, Hakluyt insists, will be not those imaginary gold mines that had blinded Gilbert's and Frobisher's backers, but the fur trade, not to mention the timber and naval stores and potash that might make England independent of the Baltic. Hakluyt for the moment drops the modest mask of archivist to show uncommon powers of historical insight into the problems, then entirely novel, of colonization.

Hakluyt about this time was sent to Bristol, the second port of the Kingdom, on a mission as Walsingham's agent and was successful in gaining the support of the Corporation there for

Gilbert's Newfoundland colony. Hakluyt was rewarded for these recent services by preferment in the chapter of Bristol Cathedral.

4

As interlopers into the Spanish American Empire, the French were the pace-makers. They had been established on the banks of the St Lawrence River since Cartier's discoveries in the 1530s, and had planted short-lived colonies in Florida as early as the 1560s. For the time being Queen Elizabeth had no envoy in Spain itself; so Paris, then a focus of anti-Spanish sentiment, was the next-best listening-post to unattainable Madrid. In the autumn of 1583 the thirty-one year old Oxford scholar Richard Hakluyt made his way there as chaplain to Sir Edward Stafford, the new English ambassador.

Hakluyt described his prime duty in Paris as 'diligent inquiry of such things as may yield any light unto our western discovery.' This was a time when exact geographical knowledge of America was politically vital. As late as 1583 there had been no certain knowledge in England of the coast of Newfoundland, despite the long experience offshore of our deep-sea fishermen, not to mention John Cabot's landfall made there, or thereabouts, three generations earlier. In 1585 the first colonizing expedition to Virginia had sailed ill-prepared, to be brought back in distress by Drake in '86. French knowledge of America was wider and better. The fur trade had already taken the French down the New England coast, perhaps as far as the Hudson River. Moreover Spanish ships from America sometimes called at French ports. Paris, therefore, was as close as Walsingham could reasonably get to the organized secrets of Spanish navigation, and the threatening, covert motions of Spanish policy.

Incidental intelligence that Hakluyt gained in the course of his geographical researches would never come amiss. 'I have talked twice with Don Antonio of Portugal, and with five or six of his best captains', Hakluyt reported. He procured a survey of the West Indies, and a plan of how the Portuguese colonies were organized. Disappointed in his hope of accompanying the Virginia colonists to Roanoke, Hakluyt found himself sent instead on a dull but useful tour of nothern French ports to get comparative figures about the profitability of French fishery and trade across the Atlantic. He also served as a courier. Said the ambassador, 'I have taken his oath upon a book for secrecy; for his honesty I will answer.'

But Hakluyt was only a part-time spy. He was primarily a scholar in the magnificent Renaissance tradition – a man for whom the gratuitous pursuit of knowledge for its own sake was life's most important end. In 1587 he spent half a year's income from his Bristol sinecure on financing the Paris publication, in Latin, the cosmopolitan language of learning, of the complete *Decades* of Peter Martyr – a collection in seven volumes of New World discoveries, not available in full for nearly sixty years. 'It will cost me', wrote Hakluyt to Raleigh, '40 French crowns and five months travail.'

In 1588, Armada year, when his own *Principal Navigations* was well advanced, Hakluyt was employed carrying confidential despatches from his ambassador to Lord Burleigh. The first edition in quarto of his life-work came out in 1589–90, after the Spanish invasion fleet had miraculously been dispersed, and the English, though still in an exalted mood, could breathe more freely. Printing his work, about half as long as the later three-volume edition, may have cost the author as much as fifty Elizabethan pounds. Hakluyt has by now settled on his method – the surprisingly modern one of giving validity to history by assembling chosen documents. He has accepted the roles of archivist and editor, but does both superlatively.

Hakluyt prefers, when he can get them, eye-witness accounts and authentic business correspondence. (Perhaps the mental attitudes of social scientist and professional spy are not so very far apart.) He will go to great lengths to get at the facts: to take down an account of the 1536 Newfoundland expedition from the last survivor, Hakluyt had been willing to journey 200 miles. He had over the years won the confidence of those institutions holding stores of valuable documents – from chartered companies of merchant venturers to the Privy Council itself. From sources he was no doubt helped to by his cousin, he gives a continuous account of the Russian voyages since 1553. He documents the progress made by the Turkey Company since 1579, but omits ventures for instance to the African coast which broke in upon the Portuguese monopoly and might perhaps spoil the good terms he had established with Dom Antonio. He tends to avoid foreign accounts; the patriotic note is struck.

This quarto edition of Hakluyt's *Principal Navigations* emerged into the afterglow of Armada year. At once it took place amid that cluster of works – Spenser's *Faerie Queen*, Holinshed's *History*, Camden's *Britannia*, Stow's *Antiquities*, which together helped to give nascent English patriotism a tone of voice, indeed a life-style, it was not to discard until very recent times. Once the Protestant Englishman had been accorded his distinctive profile, the conquest of Empire had from then on a subtle moral sanction. The world was to be mastered for its own good. This new type – imperialist as godfearing knight-errant – had already been marvellously delineated throughout hundreds of Spenserian stanzas, written down, many of them, in a lonely castle amid bog and forest, surrounded by the fierce papist denizens of a temporarily conquered Ireland. To Hakluyt just then, imperialism must have seemed a worthy national destiny.

Tennyson, three centuries later, was to put that report of Sir Richard Grenville's wrong-headed, but sublimely heroic, single-ship engagement off the Azores directly into rousing and memorable verse. Of the many hundreds of young English subalterns who fell in the hopeless trench-warfare offensives on the Somme, one may safely assume that most would have been made to learn lines of that poem by heart.

The counterpart of Spenser's knight-errant was the gallant English sea-dog – another representative national hero; even though, like John Hawkins, he may have condemned the world that came after him to the horrid and interminable consequences of the slave trade. Respected, pillaged, but seldom read critically – 'the prose epic of our modern English nation', as one enthusiastic Victorian imperialist described his *Principal Navigations* –

Richard Hakluyt's value, not merely as a geographer but as an originator across a broad spectrum of social sciences, has too long been obscured by the glowing patriotic penumbra diffused about his work since the Armada.

5

Before 1590 Hakluyt gave up the Christ Church Studentship he had kept while in Paris, together with his chaplaincy, and in that year became rector of Wetheringsett in Suffolk, a parish with two hundred communicants – a spy and a scholar thriftily rewarded, in Tudor fashion, with a cure of souls. Presumably this was when he married. For the next ten years of his life Richard Hakluyt acted as an unofficial but acknowledged geographical consultant, sometimes working for fees.

Concurrent with their own long, national fight against Spain, the Dutch were henceforth to make the running in world trade. By 1595 Dutch ships were sailing round the Cape to India and pressing onward to trade in Malaysia; by 1601 the Dutch East Indiamen comprised sixty-five ships, in fifteen squadrons. In 1599, under the provocative stimulus of this growing competition, an East India Company was organized in London, the Privy Council was petitioned, and £30,000 subscribed. Hakluyt, for the part he played in these negotiations, received a fee of ten pounds 'and thirty shillings for three maps by him provided'. As early as 1594 merchants in Zeeland were planning an attempt of their own on the Northeast Passage by sailing northward round the island of Novaya Zemlya, and they approached Hakluyt for his professional advice.

'Touching the reward of my labour,' he replied, 'I believe I shall well deserve twenty pounds sterling. But it may so happen that my advices will give your friends such light and courage that it shall profit them many thousand pounds ...' At this point Hakluyt faltered. More scholar than businessman, he went on to cut his own price by a third. 'I will not communicate all my secrets in that matter of so great import,' he added, 'under the sum of twenty marks at the least.' From the hard-headed Dutch twenty marks was what he got. 'It seems to me well worth it', remarked the go-between.

Barents, the Dutch explorer, also failed to find the hoped-for passage northward through the ice to Cathay. A century ago, among the objects preserved when his abandoned camp was unexpectedly discovered on Novaya Zemlya, were a manuscript version in Dutch of the English voyage of 1580 and two maps, perhaps those supplied by Hakluyt for his fee.

Sir Walter Raleigh, having expended £40,000 on his unsuccessful Virginia plantations, began to advocate the colonization of Guiana, casting the somewhat out-dated lure (for which a certain, mystifying justification could be found in Spanish histories) of *Eldorado*: a treasure in gold comparable with those looted by the conquistadores from Mexico and Peru. Raleigh, in his account given here of Guiana exploration, can be detected using his literary faculty to put the best complexion on the evidence. Hakluyt's expert opinion in the matter was sought by Sir Robert Cecil, the statesman obliged to appraise the Guiana project. For, later, trying to re-live in Guiana his country's earlier, more robust and more romantic history, Raleigh – the most intellectual of Hakluyt's patrons – was to end his career on the block.

The world was changing. Commerce with Russia, as well as becoming more competitive, had declined. Spain dominated the entrance to the Mediterranean, inhibiting the Levant trade, and giving point to James Lancaster's voyage in 1591. He had gone east by the Cape route, to return in 1594 with an impressively profitable cargo, and proof that the easiest way to India was wide open. Cavendish, possibly related to Hakluyt by marriage, had followed Drake's example in one circumnavigation of the globe and failed in a second; but apart from Arctic exploration the spectacular voyages were over until the days of Captain Cook. Inexorably the world was to become one market, disintegrating and laying waste local and particular cultures. In the centuries ahead European powers would virtually divide up the world: the plunder from an eighteenth century conquest of Bengal was to finance an Industrial Revolution. Even when at last the pendulum swung the other way, those countries menaced or subordinated by European commerce would be able to retrieve their independence only by some kind of deliberate imitation of Europe. Hereabouts then the modern world begins.

6

During the decades of maritime war, declared or undeclared, between England and Spain, the more spectacular profits had come from what the Spaniards spoke of as piracy – run as a speculative business, financed by syndicates. An English fleet would linger off the Azores waiting hopefully for a carrack or plate ship; or cross the Atlantic to raid some unfortunate town on the Spanish Main. Though some privateering ventures failed dismally, Drake's famous voyage had paid its backers a dividend of 4,700 per cent. All this captured treasure went eventually to prime the pump of trade. J. M. Keynes describes how 'those countries which had established trading relations with the Levant and with Asia were then able to export the treasure thus received on terms which were quite immensely profitable.'*

Keynes goes on to make a fascinating and significant calculation: 'the booty brought back by Drake in the *Golden Hind* may fairly be considered the fountain and origin of British Foreign Investment. Elizabeth paid off out of the proceeds the whole of her foreign debt, and invested a part of the balance (about £42,000) in the Levant Company. Largely out of the profits of the Levant Company, there was formed the East India Company, the profits of which, during the seventeenth and eighteenth centuries, were the main foundation of England's foreign connections, and so on ... the £42,000 invested by Elizabeth out of Drake's booty in 1580 would have accumulated by 1930 to approximately the actual aggregate of our present foreign investments, namely, £4,200,000,000 – or say 100,000 times greater than the original investment.' Piracy and sober, patriotic compound interest here work hand in hand.

*op. cit.

7

In the second and greatly enlarged edition of Hakluyt's *Principal Navigations* – from which our selections are taken – the emphasis has shifted from America to the entire world. Published in three successive volumes, 1598–1600, Hakluyt's master-work reflects the emerging situation: England, a naval power, will seek trade and colonies everywhere. English achievement in trade and travel is now self-evident; consequently Hakluyt uses non-English narratives more freely: they account for a quarter of the book. He has gone to Camden for yet more medieval records of travel – as if to give these recently enriched merchant venturers a decent genealogy. From the new edition, Mandeville has been dropped altogether. Of 350,000 words of translation from foreign tongues in the three folio volumes, a quarter of a million have been translated by Hakluyt himself. Nor is there any need now to go abroad for a decent map: the one included is taken from the terrestrial globe made in 1592 by Emeric Molineux of Lambeth for a merchant called Sanderson who had backed Davis in his Northwest Passage attempts. This is Shakespeare's 'new map, with the augmentation of the Indies', immortalized in *Twelfth Night* (Act III, Sc. 2).*

Hakluyt had lost his wife in 1597. In 1602 he was installed prebendary in Westminster Abbey. In 1604, when he married again, he was Archdeacon of Westminster and (by Sir Robert Cecil's patronage) chaplain at the Savoy. In 1612 he was presented to the living of Gedney in Lincolnshire. Hakluyt had prospered with the rest of them: a pluralist parson, rewarded somewhat incongruously out of Church property for services that had benefited not religion, but business and the State.

In 1616 Hakluyt died, after friendly attempts had been made

***Twelfth Night*, believed to have been written about the time when the three volumes of *Principal Navigations* were appearing successively, is full of curious minor echoes of Hakluyt's phraseology. Might a more meticulous scrutiny enable the date of composition of the play to be bracketed with precision?

by the men he had served to bestow on him a distinctive kind of immortality. In 1608 Henry Hudson in *Hopewell*, some of whose navigational information had been provided by Hakluyt, gave to a promontory on an island north of Spitzbergen the name Hakluyt's Headland. The name has not survived: the island, now called Amsterdam Island, is in about 80°N. – the last bit of Europe seen by the polar explorer who follows the Gulf Stream Drift to the northernmost limit of open water. In 1616 Baffin broke through the west Greenland ice north of 70°N. and, finding an island in 77°30′N., named it Hakluyt's Island. Headland and Island indicated as if symbolically the northern limits of English Arctic enterprise from Hakluyt's lifetime until the nineteenth century.

Richard Hakluyt's last years had been spent in gathering additional materials for yet another edition. He was buried on 26 November 1616 in an unmarked grave in Westminster Abbey, leaving a son reputed to have run through his estate. Hakluyt's manuscript collection came into the hands of the Reverend Samuel Purchas (pronounced, apparently, *Purkas*), chaplain to the Archbishop of Canterbury. Purchas, with the stay-at-home reader in mind, began to cut away at Hakluyt's conservatively edited documents. Hakluyt's *Principal Navigations* had established a taste for closet travel: his literary legacy provided two-fifths of the 1625 edition of *Purchas His Pilgrims*, a vast and popular compilation extending to four million words. The documents curtailed by Purchas may make more lively reading, but they are prone also to mislead. By suppressing, for example, Baffin's log of his important 1616 voyage to Baffin Bay and Smith Sound, Purchas contrived to distort the history of navigation. Travel and exploration were being turned at his busy hands into national myth. Purchas was awarded £100 from the East India Company, and in the two centuries that followed, being readable and romantic, his book was more highly regarded than Hakluyt's work of sober scholarship. Purchas began a different tradition: one that ultimately flowered splendidly with Coleridge in *The Rime of the Ancient Mariner* and *Kubla Khan*.

8

John Donne sailed in the expedition to the Azores of 1597 as a volunteer: through a few lines of his *Satyre III* flickers the mood of those Englishmen who went out in small wooden ships to discover an unexplored world:

> dar'st thou lay
> Thee in ships woodden Sepulchers, a prey
> To leaders rage, to stormes, to shot, to dearth?
> Dar'st thou dive seas, and dungeons of the earth?
> Hast thou couragious fire to thaw the ice
> Of frozen North discoveries?

The syndicates of merchants gratefully accepted their profits; the gallant sea-dogs and the gentlemen adventurers were endowed by Hakluyt and Purchas with enduring celebrity. But the usually nameless seamen in their wooden sepulchres died by the dozen – of scurvy and calenture, washed overboard, blown up, cut down. They were often enough the unconscious victims of Tudor prosperity – displaced from their traditional lives by price inflation and enclosure. They were the first of many generations of long-suffering English countrymen to break from the cage of humiliating poverty by taking the Queen's shilling or going for a sailor. Their manly endurance and their good nature are perhaps best of all conveyed in the tale told by those of Hawkins's men set ashore in Central America, who fell later into the hands of the Spanish Inquisition, until at last the most obstinate among them won a way home.

Hakluyt's collection also affords several glimpses of the social connections that from its very genesis have linked modern science with the practical needs and discoveries of expanding trade. Problems of magnetism are matters of deadly earnest to men who will need to sail their small wooden ships close to the Magnetic Pole: the Dr Gilbert who is famous as an investigator of magnetism was also interested in polar navigation. Thomas Harriot, the one-time tutor in Sir Walter Raleigh's family, the first surveyor of

Virginia, and indeed the man who took Hakluyt's place there, is now best known for having discovered the Factor Theorem, a fundamental simplification in algebraical procedures.

Travel too is no longer an incidental aspect of pilgrimage, war or trade, but has become professional. Anthony Jenkinson – who opened up Russia and Persia for the merchant venturers, and reached Bokhara, and provided Ortelius with geographical data for his great atlas – lists the many lands through which he has journeyed with a deliberate, resonant pride.

The extent to which Hakluyt's *Principal Navigations* entered the popular mind can also be noticed in literature.

'Her husband's to Aleppo gone, master o' th' *Tiger*' remarks the witch in Act I of *Macbeth*; and turning Hakluyt's page we read, with a secret shock, of a ship called *Tiger*, her destination Aleppo. Hard-headed Ben Jonson (a pupil of Camden) turns it another way in Act V, Scene 2 of *Volpone*:

SECOND MERCHANT: If you could ship him away, 'twere excellent.
THIRD MERCHANT: To Zant, or to Aleppo?
PEREGRINE: Yes, and have his
Adventures put i' the Book of Voyages,
And his gull'd story registered for truth.

Though the twin corrosives of puritanism and trade are evidently going to finish off Merry England pretty soon, and eat away also at the tissue of primitive societies abroad, Hakluyt's great narratives are still permeated by a sense of national unity. Prosperity, empire, industrialism, the production in public schools of synthetic gentlemen, and the simultaneous pauperizing of the common people, have not yet created Disraeli's *Two Nations*. The schoolmaster has not yet had time to brand the upper class on the tongue; the lower orders have not so far been struck dumb. Seaman and gentleman and poet use a virtually identical language: it would be hard at times to say which has the most admirable powers of expression. Gentleman and foremast hand are in the same wooden sepulchre and take the same risks. The original identity of interest linking trade and science, literature and knowledge – the need for national survival – is still vividly apparent, though it will not last long. Culture is soon to split between religion and science, reason and faith, poetry and prose, as English society begins its open and covert civil war of opposing classes, inexorably confronting each other down the centuries.

9

To sailors of the time, the *Principal Navigations* were simply their own account of themselves. Hakluyt's documents, in Ben Jonson's words, were 'registered for truth'. They were reliable. They would help a wayfarer in distress. The East India fleet on its third voyage was in trouble off Sierra Leone in 1607. The master sent for 'the Book', and by using the facts about the coast that Hakluyt provided he brought the fleet to a safe anchorage, saving the Company £20,000. Admiralty charts and *Admiralty Pilots* may have systematized what Hakluyt clumsily began, and seamen everywhere have learned to rely on them. But someone, in a truthful and methodical if elementary way, had to begin all this: in a number of different directions others of us nowadays are simply carrying on.

10

The text of Hakluyt's second and best edition has been reprinted twice in this century: reverently by the Hakluyt Society (Glasgow, 1903–5), and somewhat more cursorily by Everyman's Library (London, 1908). The spelling of this edition has been modernized for Penguin, but the punctuation retained, except for the suppression of the confusing Elizabethan full point after numbers, and the insertion of a possessive apostrophe. On the very rare occasions when, for clarity, a word is here added to Hakluyt's text, it has been put in square brackets. Much use has been made in this Introduction of facts contained in G. B. Parks's *Richard Hakluyt and the English Voyages* (New York, 1928).

A satisfactorily annotated edition of Hakluyt would be the work of a lifetime well spent: the additional information provided for this condensed version in notes and glossary is little more than what the alert Penguin reader could have looked up for himself, were he sitting within reach of such common works of reference as the *Encyclopaedia Britannica*, the *Dictionary of National Biography*, the *Oxford Dictionary*, *The Times Atlas*, *Shepherd's Historical Atlas*, and good French and Spanish dictionaries.

DEDICATIONS

The Epistle Dedicatory in the first edition, 1589. To the Right Honourable Sir Francis Walsingham Knight, Principal Secretary to Her Majesty, and one of Her Majesty's most honourable Privy Council.

Right Honourable, I do remember that being a youth, and one of Her Majesty's scholars at Westminster that fruitful nursery, it was my hap to visit the chamber of Mr Richard Hakluyt my cousin, a gentleman of the Middle Temple, well known unto you, at a time when I found lying open upon his board certain books of Cosmography, with a universal Map: he seeing me somewhat curious in the view thereof, began to instruct my ignorance, by showing me the division of the earth into three parts after the old account: he pointed with his wand to all the known seas, gulfs, bays, straits, capes, rivers, empires, kingdoms, dukedoms and territories of each part, with declaration also of their special commodities, and particular wants, which by the benefit of traffic, and intercourse of merchants, are plentifully supplied. From the map he brought me to the Bible, and turning to the 107 Psalm, directed me to the 23 and 24 verses, where I read, that they which go down to the sea in ships, and occupy by the great waters, they see the works of the Lord and his wonders in the deep, &c. Which words of the prophet together with my cousin's discourse (things of high and rare delight to my young nature) took in me so deep an impression, that I constantly resolved, if ever I were preferred to the University, where better time, and more convenient place might be ministered for these studies, I would by God's assistance prosecute that knowledge and kind of literature, the doors whereof (after a sort) were so happily opened before me.

According to which my resolution, when, not long after, I was removed to Christ Church in Oxford, my exercises of duty first performed, I fell to my intended course, and by degrees read over whatsoever printed or written discoveries and voyages I found extant either in the Greek, Latin, Italian, Spanish, Portugal, French or English languages, and in my public lectures was the first, that produced and showed both the old imperfectly composed, and the new lately reformed maps, globes, spheres, and other instruments of this art for demonstration in the common schools, to the singular pleasure, and general contentment of my auditory. I grew familiarly acquainted with the chiefest captains at sea, the greatest merchants, and the best mariners of our nation: by which means having gotten somewhat more than common knowledge, I passed at length the narrow seas into France with Sir Edward Stafford, where during my five years abroad with him in his dangerous and chargeable residency in Her Highness's service, I both heard in speech and read in books other nations miraculously extolled for their discoveries and notable enterprises by sea, but the English of all others for their sluggish security, and continual neglect of the like attempts especially in so long and happy a time of peace, either ignominiously reported, or exceedingly condemned.

Thus both hearing, and reading the obloquy of our nation, and finding few or none of our own men able to reply herein: and further, not seeing any man to have care to recommend to the world, the industrious labours, and painful travels of our countrymen: for stopping the mouths of the reproachers, myself being the last winter returned from France determined notwithstanding all difficulties, to undertake the burden of that work wherein all others pretended either ignorance, or lack of leisure, or want of sufficient argument, whereas (to speak truly) the huge toil, and the small profit to ensue, were the chief causes of the refusal. I call the work a burden, in consideration that these voyages lay so dispersed, scattered and hidden in several hucksters' hands, that I now wonder at myself, to see how I was able to endure the delays, curiosity, and backwardness of many from whom I was to receive the originals.

To harp no longer upon this string, and to speak a word of that just commendation which our nation does indeed deserve: it cannot be denied, but as in all former ages, they have been men full of activity, stirrers abroad, and searchers of the most remote parts of the world, so in this most famous and peerless government of Her most excellent Majesty, her subjects, in compassing the vast globe of the world more than once, have excelled all the nations and people of the earth. For, which of the kings of this land before Her Majesty, had their banners ever seen in the Caspian Sea? Which of them hath ever dealt with the Emperor of Persia, as Her Majesty hath done, and obtained for her merchants large and loving privileges? Who ever found English Consuls and agents at Tripolis in Syria, at Aleppo, at Babylon, at Basra, and which is more, whoever heard of Englishmen at Goa before now? What English ships did heretofore ever anchor in the mighty river of Plate? Pass and re-pass the unpassable (in former opinion) strait of Magellan, range along the coast of Chile, Peru, and all the backside of Nova Hispania,[1] further than any Christian ever passed, traverse the mighty breadth of the South Sea, land upon the Luzones[2] in despite of the enemy, enter into alliance, amity and traffic with the princes of the Moluccas, and the isle of Java, double the famous Cape of Bona Speranza, arrive at the isle of St Helena, and last of all return home most richly laden with the commodities of China, as the subjects of this now flourishing monarchy have done?

[*Preface to the first edition*]
Richard Hakluyt to the favourable Reader.

I have thought it very requisite for thy further instruction and direction in this history (good reader) to acquaint thee briefly with the method and order which I have used in the whole course thereof: and by the way also to let thee understand by whose friendly aid in this my travel I have been furthered.

I meddle in this work with the navigations only of our own nation: and albeit I allege in a few places (as the manner and occasion required) some strangers as witnesses of the things done, yet are they none but such as either faithfully remember or sufficiently confirm the travels of our own people.

I must crave thy further patience (friendly reader) in acquainting thee with those virtuous gentlemen, which have yielded me their several good assistances. Master Richard Staper merchant of London, hath furnished me with divers things touching the trade of Turkey and other places in the east. Master William Burrough, Clerk of Her Majesty's Navy, and Master Anthony Jenkinson, both gentlemen of great experience and observation in the north regions, have much pleasured me. Beside mine own extreme travail in the histories of the Spaniards, my chiefest light hath been received from Sir John Hawkins, Sir Walter Raleigh, and my kinsman Master Richard Hakluyt of the Middle Temple.

I have contented myself with inserting into the work one of the best general maps of the world only, until the coming out of a very large and most exact terrestrial globe, collected and reformed according to the newest, secretest and latest discoveries, both Spanish, Portugal and English, composed by Mr Emery Molineux of Lambeth, a rare gentleman in his profession.

This being the sum of those things which I thought good to admonish thee of (good reader) it remaineth that thou take the profit and pleasure of the work: which I wish to be as great to thee, as my pains and labour have been in bringing these raw fruits unto this ripeness, and in reducing these loose papers into this order. Farewell.

[*Preface to the second edition, 1598*]

A preface to the reader as touching the principal voyages.

Having for the benefit and honour of my country zealously bestowed so many years, so much travail and cost, to bring antiquities smothered and buried in dark silence, to light, and to preserve certain memorable exploits of late years by our English nation achieved, from the greedy and devouring jaws of oblivion: to gather likewise, and as it were to incorporate into one body the torn and scattered limbs of our ancient and late navigations by sea, our voyages by land, and traffics of merchandise by both: I do this second time (friendly reader) presume to offer unto thy view this discourse. For the bringing of which into this homely and rough-hewn shape, which here thou seest, what restless nights, what painful days, what heat, what cold I have endured; how many long and chargeable journeys I have travelled; how many famous libraries I have searched into; what variety of ancient and modern writers I have perused, what expenses I have not spared; and yet what fair opportunities of private gain, preferment, and ease I have neglected. Howbeit, the honour and benefit of this Commonweal wherein I live and breathe hath made all difficulties seem easy.

Be it granted that the renowned Portugal Vasco da Gama traversed the main ocean southward of Africa. Did not Richard Chancellor and his mates perform the like, northward of Europe? Suppose that Columbus that noble and high-spirited Genoese escried unknown lands to the westward of Europe and Africa: did not the valiant English knight Sir Hugh Willoughby; did not the famous pilots Stephen Burrough, Arthur Pet, and Charles Jackman accost Novaya Zemlya[3] to the north of Europe and Asia? Howbeit you will say perhaps, not with the like golden success, not with such deductions of colonies, nor attaining of conquests. True it is, that our success hath not been correspondent unto theirs: yet in this our attempt the uncertainty of finding was far greater, and the difficulty and danger of searching was no whit less.

Into what dangers and difficulties they plunged themselves, I tremble to recount. For first they were to expose themselves unto the rigour of the stern and uncouth northern seas, then they were to sail by the ragged and perilous coast of Norway, to frequent the unhaunted shores of Finmark, to double the dreadful and misty North Cape, and as it were to open and unlock the sevenfold mouth of Dvina.[4] Unto what drifts of snow and mountains of ice even in June, July, and August, unto what hideous overfalls, uncertain currents, dark mists and fogs, and divers other fearful inconveniences they were subject and in danger of, I wish you rather to learn out of the voyages of Sir Hugh Willoughby, Stephen Burrough, Arthur Pet and the rest.

To return to our voyages performed within the bounds of Russia, neither hath our nation been contented only thoroughly to search into all parts, but also to visit Kazan[5] and Astrakhan.[6] They have adventured their persons, ships, and goods, homewards and outwards, fourteen times over the unknown and dangerous Caspian Sea; that valiant, wise and personable gentleman Mr Anthony Jenkinson being their first ring-leader.

But that no man should imagine that our foreign trades of merchandise have been comprised within some few years, there may he plainly see in an ancient testimony translated out of the Saxon tongue, how our merchants were often wont for traffic's sake, so many hundred years since, to cross the wide seas. Yea, there mayest thou observe (friendly reader) what privileges the Danish king Canute obtained at Rome of Pope John for our English merchant adventurers of those times. There may you partly see what our state of merchandise was in the time of King Stephen, and how the city of Bristol was then greatly resorted unto with ships from Norway and from Ireland.

King John passed into Ireland with a fleet of 500 sails; so great were our sea-forces even in his time. Neither did our shipping for the wars first begin to flourish with King John, but long before his days in the reign of King Edward the Confessor, of William the Conqueror, of William Rufus and the rest, there were divers men of war which did valiant service at sea. Then have you the frank and bountiful charter granted by King Edward the First upon the Cinque Ports: and next thereunto a roll of the mighty fleet of seven hundred ships which King Edward the Third had with him unto the siege of Calais.

The Epistle Dedicatory in the second volume of the Second Edition, 1599. To the Right Honourable Sir Robert Cecil, Principal Secretary to Her Majesty.

Right honourable. After the coming in of the Normans, and so downward for a space of above 300 years, such was the ardent desire of our nation to visit the Holy Land, and to expel the Saracens and Mahometans, that not only great numbers of Earls, Bishops, Barons and Knights, but even Kings, Princes, and peers of the Blood Royal, with incredible devotion, courage and alacrity intruded themselves into this glorious expedition. All these, either Kings, Kings' sons, or Kings' brothers, exposed themselves with invincible courage to the manifest hazard of their persons, leaving their ease, their countries, wives and children, induced with a zealous devotion and ardent desire to protect and dilate the Christian faith. These memorable enterprises I have brought together in the best method and brevity that I could devise.

And here by the way if any man shall think, that an universal peace with our Christian neighbours will cut off the employment of the courageous increasing youth of this realm, he is much deceived. There is under our noses the great and ample country of Virginia; the inland whereof is found of late to be so sweet and wholesome a climate, so rich and abundant in silver mines, so apt and capable of all commodities, and in a secret map of those parts made in Mexico for the King of Spain (which original with many others is in the custody of the excellent mathematician Mr Thomas Harriot), as also in their intercepted letters come unto my hand, bearing date 1595 they acknowledge the inland to be a better and richer country than Mexico itself. If upon a good and godly peace obtained, it shall please the Almighty to stir up Her Majesty's heart to continue with transporting one or two thousand of her people, and such others as upon mine own knowledge will most willingly at their own charges become adventurers in good numbers with their bodies and goods; she shall by God's assistance, in short space, work many great and unlooked for effects, increase her dominions, enrich her coffers, and reduce many pagans to the faith of Christ.

This treatise containeth our ancient trade and traffic with English shipping to the islands of Sicily, Candia[7] and Chios,[8] which I find to have been begun in the year 1511 and to have continued until the year 1552 and somewhat longer. But shortly after it was given over, by occasion of the Turk's expelling of the four and twenty governors of the Genoese out of the island of Chios, and by taking the island wholly into his own hand; and afterward by his growing over-mighty and troublesome in those seas.

Lastly, I have here put down the happy renewing and much increasing of our interrupted trade in all the Levant: the traffic of our nation in all the chief havens of Africa and Egypt: the voyages over land and by river through Aleppo, Babylon and Basra, and down the Persian Gulf, to Ormuz,[9] and then by the ocean sea to Goa, and even to the frontiers of the Empire of China. I have likewise set in order such voyages as our nation, and especially the worthy inhabitants of this city of London, have painfully performed without the Strait of Gibraltar, upon the coasts of Africa, about the Cape of Buona Speranza, to and beyond the East Indies. I have here set down the whole course of the Portugal carracks from Lisbon to the bar of Goa in India, with the sundry and infallible marks of approaching unto and doubling the Cape of Good Hope.

Because our chief desire is to find out ample vent for our woollen cloth, the natural commodity of this our realm, the fittest place I find for that purpose are the manifold islands of Japan and the northern parts of China, and the regions of the Tartars next adjoining. Therefore I have here inserted treatises of the said countries, one of which was printed in Latin in Macao, a city of China, in China-paper, in the year 1590, and was intercepted in the great carrack called *Madre de Dios* two years after, enclosed in a case of sweet cedar wood, and lapped up almost an hundred fold in fine calicut-cloth, as though it had been some incomparable jewel.

When it pleased your honour to have some conference with me, and to demand mine opinion touching the state of the country of Guiana, and whether it were fit to be planted by the English: I then (to my small joy) did admire the exact knowledge which you had gotten of those matters of Indian navigations: and how careful you were, not to be overtaken with any partial affection for the action, also, by the sound arguments which you made of the likelihood and reason of good or ill success, before the state and commonwealth (wherein you have an extraordinary voice) should be farther engaged. I think myself thrice happy to have these my travels censured by your honour's so well approved judgement. I think it my bounden duty in all humility and with much bashfulness to recommend myself and them unto your right honourable and favourable protection.

The Epistle Dedicatory in the third volume of the second edition, 1600. To the Right Honourable Sir Robert Cecil.

The subject matter herein contained is the fourth part of the world, more commonly than properly called America: but by the chiefest authors The New World. New, in regard to the new and late discovery thereof made by Christopher Colon, alias Columbus, a Genoese by nation, in the year of grace 1492. And world, in respect of the huge extension thereof, which to this day is not thoroughly discovered. Albeit my work do carry the title of the English voyages, where our own men's experience is defective, there I have been careful to supply the same with the best and chiefest relations of strangers.

Moreover, because since our wars with Spain, by the taking of their ships, and sacking of their towns and cities, their secrets of the West Indies are fallen into our people's hands, I have used my best endeavour to translate out of Spanish, and here in this present volume to publish such secrets of theirs, as may any way avail us or annoy them, if they drive and urge us by their sullen insolencies, to continue our course of hostility against them, and shall cease to seek a good and Christian peace. There is no chief river, no port, no town, no city, no province of any reckoning in the West Indies that hath not here some good description, as well for the inland as the sea coast.

I have brought to light certain new advertisements of the late alteration of the mighty monarchy of Japan, and the new conquest of the Kingdom of Korea, as also of the Tartars adjoining on the east and northeast parts of Korea, where I think the best utterance of our natural and chief commodity of cloth is like to be, if it please God hereafter to reveal unto us the passage thither by the northwest.

Thus, sir, I have portrayed out in rude lineaments my Western Atlantis, or America: I humbly desire you to receive her with your wonted and accustomed favour at my hands.

Your honour's most humble to be commanded,
RICHARD HAKLUYT, *Preacher*

II

The voyage of Octher made to the northeast parts beyond Norway, reported by himself unto Alfred the famous King of England, about the year 890.

Octher said that the country wherein he dwelt was called Heligoland.[10] Octher told his lord King Alfred that he dwelt furthest north of any other Norman: and affirmed that the land, notwithstanding it stretched marvellous far towards the north, yet it is all desert and not inhabited, unless it be very few places, here and there, where certain Finns dwell upon the coast, who live by hunting all the winter, and by fishing in the summer. He fell into a fantasy and desire to prove and know how far that land stretched northward, and whether there were any habitation of men. Whereupon he took his voyage directly north along the coast, having upon his starboard always the desert land, and upon the larboard the main ocean: and continued his course for three days. In which space he was come as far towards the north, as commonly the whale hunters use to travel. He proceeded in his course still towards the north so far as he was able to sail in other three days. He perceived that the coast turned towards the east, or else the sea opened with a main gulf into the land, he knew not how far. He was fain to stay until he had a western wind and somewhat northerly: and thence he sailed plain east along the coast still so far as he was able in the space of four days. At the end of which time he was compelled again to stay until he had a full northerly wind, forasmuch as the coast bowed thence directly towards the south, or leastwise the sea opened into the land he could not tell how far: at the fifth day's end he discovered a mighty river which opened very far into the land. At the entry of which river he stayed his course, and in conclusion turned back again, for he durst not enter thereinto for fear of the inhabitants of the land: perceiving that on the other side of the river the country was thoroughly inhabited: which was the first peopled land that he had found since his departure from his own dwelling: except that in some places, he saw a few fishers, fowlers, and hunters, which were all Finns: and all the way upon his larboard was the main ocean. The principal purpose of his travel this way, was to increase the knowledge and discovery of these coasts and countries, for the more commodity of fishing of horse-whales, which have in their teeth bones of great price and excellency: whereof he brought some at his return unto the King. Their skins are also very good to make cables for ships, and so used.

III

Certain privileges obtained for the English and Danish merchants of Conrad the Emperor and John the Bishop of Rome by Canute the King of England in his journey to Rome. Extracted out of a letter of his.

At the feast of Easter there was a great company of nobles with Pope John and Conrad the Emperor assembled at Rome. Wherefore I conferred with the Emperor himself and the Pope, and with the other princes who were there present, concerning the necessities of all my subjects both Englishmen and Danes; that a more favourable law and secure peace in their way to Rome might be granted unto them, and that they might not be hindered by so many stops and impediments in their journey, and wearied by reason of unjust exactions. And the Emperor condescended unto my request, and King Rodolphus also, who hath greatest authority over the foresaid stops and straits, and all the other princes confirmed by their edicts, that my subjects, as well merchants, as others who travelled for devotion's sake, should without all hindrance and restraint of the foresaid stops and customers, go unto Rome in peace, and return from thence in safety.

IV

The state of the shipping of the Cinque ports from Edward the Confessor and William the Conqueror, and so down to Edward the First, by the learned gentleman Mr William Lambert, in his Perambulation of Kent.

I find in the book called Doomsday, that Dover, Sandwich and Romney, were in the time of King Edward the Confessor, discharged almost of all manner of impositions and burdens (which other towns did bear) in consideration of such service to be done by them upon the sea.

Sheppey was before King Edward the First's time the place of assembly for the pleas of the Five Ports. Hastings and Hythe, Dover, Romney and Sandwich were accounted the five principal havens or ports, which were endowed with privilege, and had the same ratified by the Great Charter of England. Soon after, Winchelsea and Rye might be added to the number.

The Book of Doomsday chargeth Dover with twenty vessels at the sea, whereof each to be furnished with one and twenty men for fifteen days together: Romney and Sandwich answered the like service. At each time that the King passeth over the sea, the ports ought to rig up fifty and seven ships, (whereof every one to have twenty armed soldiers) and to maintain them at their own costs, by the space of fifteen days together.

This duty of attendance (being devised for the honourable transportation, and safe conduct of the King's own person or his army over the narrow seas) the ports have not only most diligently ever since that time performed, but furthermore also valiantly behaved themselves against the enemy in sundry exploits by water.

V

The voyage of Ingulphus Abbot of Croyland unto Jerusalem in the year of our Lord, 1064.

I Ingulphus a humble servant of reverend Guthlac and of his monastery of Croyland, born in England, and of English parents, at the beautiful city of London, was in my youth, for the attaining of good letters, placed first at Westminster, and afterward sent to the University of Oxford. And as I grew in age, disdaining my parents' mean estate, and forsaking mine own native soil, I affected the courts of kings and princes, and was desirous to be clad in silk, and to wear brave and costly attire. And lo, at the same time, William our sovereign King now, but then Earl of Normandy, with a great troop of followers and attendants came unto London, to confer with King Edward the Confessor his kinsman. Into whose company intruding myself, and proffering my service for the performance of any speedy or weighty affairs, in short time I was known and most entirely beloved by the victorious Earl himself, and with him I sailed into Normandy. And there being made his secretary, I governed the Earl's court. Being carried with a youthful heat and lusty humour, I began to weary even of this place, wherein I was advanced so high above my parentage. There went a report throughout all Normandy, that divers archbishops of the Empire, and secular princes were desirous for their soul's health, and for devotion sake, to go on pilgrimage to Jerusalem. Wherefore out of the family of our lord the Earl, sundry of us, both gentlemen and clerks (principal of whom was myself) with the licence and goodwill of our said lord the earl, sped us on that voyage, and travelling thirty horses of us into High Germany, we joined ourselves unto the Archbishop of Mainz. And being with the companies of the bishops seven thousand persons sufficiently provided for such an expedition, we passed prosperously through many provinces, and at length attained unto Constantinople. Where doing reverence unto the Emperor Alexius, we saw the church of Santa Sophia, and kissed divers sacred relics. Departing thence through Lycia,[11] we fell into the hands of the Arabian thieves: and after we had been robbed of infinite sums of money, and had lost many of our people, hardly escaping with extreme danger of our lives, at length we joyfully entered into the most wished city of Jerusalem. Where we were received by the most reverend, aged, and holy patriarch Sophronius, with great melody of cymbals and with torchlight, and were accompanied unto the most divine church of Our Saviour His Sepulchre with a solemn procession. Here, how many prayers we uttered, what abundance of tears we shed, what deep sighs we breathed forth, our Lord Jesus Christ only knoweth. Wherefore being conducted from the most glorious sepulchre of Christ to visit other sacred monuments of the city, we saw with weeping eyes a great number of holy churches and oratories, which Achim the Sultan of Egypt had lately destroyed. And so having bewailed with sad tears, and most sorrowful and bleeding affections, all the ruins of that most holy city both within and without, and having bestowed money for the re-edifying of some, we desired with most ardent devotion to go forth into the country, to wash ourselves in the most sacred river of Jordan, and to kiss all the steps of Christ. Howbeit the thievish Arabians lurking upon every way, would not suffer us to travel far from the city, by reason of their huge and furious multitudes. Wherefore about the Spring there arrived at the port of Joppa[12] a fleet of ships from Genoa. In which fleet (when the Christian merchants had exchanged all their wares at the coast towns, and had likewise visited the holy places) we all of us embarked, committing ourselves to the seas: and being tossed with many storms and tempests, at length we arrived at Brindisi: and so with a prosperous journey travelling through Apulia towards Rome, we there visited the habitations of the holy apostles Peter and Paul, and did reverence unto divers monuments of holy martyrs in all places throughout the city. From thence the archbishops and other princes of the Empire travelling towards the right hand for Germany, and we declining towards the left hand for France, departing asunder, taking our leaves with unspeakable thanks and courtesies. And so at length, of thirty horsemen which went out of Normandy fat, lusty, and frolic, we returned thither scarce twenty poor pilgrims of us, being all footmen, and consumed with leanness to the bare bones.

VI

The flourishing state of merchandise in the days of William of Malmesbury, who died in the year 1142 in the reign of King Stephen.

Not far from Rochester, about the distance of five and twenty miles, standeth the noble city of London, abounding with the riches of the inhabitants, and being frequented with the traffic of merchants resorting thither out of all nations, and especially out of Germany. Whereupon it cometh to pass, that when any general dearth of victuals falleth out in England, by reason of the scarcity of corn, things necessary may there be provided and bought with less gain unto the sellers, and with less hindrance and loss unto the buyers, than in any other place in the realm. Outlandish wares are conveyed into the same city by the famous river of Thames.

The famous town of Bristol, with a haven belonging thereunto, is a commodious and safe receptacle for all ships directing their course for the same, from Ireland, Norway, and other outlandish and foreign countries.

VII

The voyage of Macham an Englishman, wherein he first of any man discovered the Island of Madeira.

In the year 1344 King Peter the fourth of that name reigning in Aragon, about this time the Island of Madeira, standing in 32 degrees, was discovered by an Englishman, which was named Macham, who sailing out of England into Spain, with a woman that he had stolen, arrived by tempest in that island, and did cast anchor in that haven or bay, which now is called Machico,[13] after the name of Macham. And because his lover was sea-sick, he went on land with some of his company, and the ship with a good wind made sail away, and the woman died for thought. Macham, which loved her dearly, built a chapel or hermitage, to bury her in, calling it by the name of Jesus, and caused his name and hers to be written or graven upon the stone of her tomb, and the occasion of their arrival there: and afterwards he ordained a boat made of one tree (for there be trees of great compass about), and went to sea in it, with those men that he had, and were left behind with him, and came upon the coast of Africa, without sail or oar. And the Moors which saw it, took it to be a marvellous thing, and presented him unto the King of that country for a wonder, and that King also sent him and his companions for a miracle unto the King of Castile.

In the year 1395 King Henry the Third reigning in Castile, by the information which Macham gave of this island, moved many of France and Castile to go and to discover it, and also Grand Canary.

VIII

A brief relation of William Esturmy and John Kington concerning their embassies into Prussia, and the Hans towns.

In the month of July, and in the year of Our Lord 1403, there came into England the ambassadors of the mighty lord Fr Conradus de Jungingen, being then master-general of Prussia,[14] with his letters directing unto our sovereign lord the King, requiring amends and recompense for certain injuries unjustly offered by Englishmen unto the subjects of the said master-general.

Our King sent William Esturmy, knight, Mr John Kington, clerk, and William Brampton citizen of London, from his court of parliament holden at Coventry, very slightly informed, as his ambassadors into Prussia.

Before the arrival of the ambassadors in Prussia, all intercourse of traffic between the English and the Prussians, in the realm of England, and in the land of Prussia was altogether restrained and prohibited: and in the same land it was ordained and put in practice, that in whatsoever port of the land of Prussia any English merchant had arrived with his goods, he was not permitted to convey the said goods, out of that port, unto any other place in the land of Prussia, either by water, or by land, under the pain of the forfeiting of the same: but was enjoined to sell them in the very same port, unto the Prussians only and to none other, to the great prejudice of our English merchants.

After the arrival of the English ambassadors in the land of Prussia, it was ordained, that from the eighth day of October, in the year of Our Lord 1405, all English merchants whatsoever should have free liberty to arrive with all kinds of their merchandise in whatsoever port of the land of Prussia, and to make sale of them, as hath heretofore from ancient times been accustomed.

Also sundry other commodious privileges unto the realm of England were then ordained and established.

The English ambassadors being arrived in the land of Prussia, demanded of the master-general, a reformation and amends, for the damages and injuries offered by the Prussians unto the liege people of our sovereign lord and king, which losses amounted to the sum of 4535 nobles.

It was afterwards concluded, that upon the first of May next, namely in the year of Our Lord 1406 or within the space of one year immediately following there should be made a convenient, just, and reasonable satisfaction, for all molestations unjustly offered on both parts. Which satisfaction not being performed, the Prussians with their goods and merchandises were without molestation or impediment, enjoined to depart out of the realm of England with their ships and goods, and the English men likewise, out of the territories and dominions of the said master-general.

William Esturmy and John Kington in their return homewards from Prussia towards England passed through the chief cities of the Hans,[15] and treated with the burgomasters. There were sent messengers and agents, in the behalf of the common society of the Hans merchants, unto the town of Dordrecht, to confer with the ambassadors of England, about the redressing of injuries.

Forasmuch as divers articles propounded, as well on the behalf of England, as of Prussia, and of the cities of the Hans, were so obscure, that in regard of their obscurity, there could no resolute answer be made unto them: it was appointed and concluded, that all obscure articles ought before the end of Easter next ensuing, and within one whole year after, to be declared before the Chancellor of England, for the time being. Which being done accordingly, complete justice shall be administered on both parts.

IX

Mr Robert Thorne in the year 1527 in Seville, to Doctor Ley, Lord Ambassador for King Henry the Eighth, to Charles the Emperor, being an information on the parts of the world, discovered by him and the King of Portugal: and also of the way to the Moluccas by the North.

Of the new trade of Spicery of the Emperor, there is no doubt but that the islands are fertile of cloves, nutmegs, mace, and cinnamon: and that the said islands, with other there about abound with gold, rubies, diamonds, jacynths, and other stones and pearls, as all other lands, that are under or near the equinoctial. In this navigation of the spicery, it was discovered that these islands set nothing by gold, but set more by a knife and a nail of iron. And I doubt not but to them should be as precious our corn and seeds if they might have them, as to us their spices: and likewise the pieces of glass that we have counterfeited are as precious to them, as to us their stones: which by experience is seen daily by them that have trade thither.

In a fleet of three ships and a caravel that went from this city armed by the merchants of it, which departed in April last, I and my partner have one thousand four hundred ducats that we employed in the said fleet, principally for that two Englishmen, friends of mine, which are somewhat learned in cosmography, should go in the same ships, to bring me certain relation of the situation of the country, and to be expert in the navigation of those seas.

In the year 1484 the King of Portugal minded to arm certain caravels to discover this spicery. So he armed a fleet, and in the year 1497 were discovered the Islands of Calicut, from whence is brought all the spice he has. After this in the year 1492 the King of Spain willing to discover lands towards the Occident, armed certain caravels, and then discovered this India Occidental, especially two islands of the said India, naming the one la Dominica, and the other Cuba, and brought certain gold from thence. Of the which, when the King of Portugal had knowledge, he sent to the King of Spain, and consented to the King of Spain, that touching this discovering they should divide the world between them two.

The King of Portugal had already discovered certain islands that lie against Cape Verde, and also a certain part of the mainland, towards the South, and called it the land of Brazil.

It appeareth plainly that the newfound land that we discovered, is all a mainland with the Indies Occidental, from whence the Emperor hath all the gold and pearls: and so continueth of coast, more than 5000 leagues of length, as by this card appeareth. For from the said new lands it proceedeth towards the Occident to the Indies, and from the Indies returneth towards the Orient, and after turneth southward up till it come to the straits of Todos Santos, which I reckon to be more than 5000 leagues.

So that to the Indies it should seem that we have some title, at least that for our discovering we might trade thither as other do. But all this is nothing near the spicery.

Now then if from the said newfound lands the sea be navigable, there is no doubt, but sailing northward and passing the Pole, descending to the equinoctial line, we shall hit these islands, and it should be a much shorter way, than either the Spaniards or the Portuguese have.

And though we went not to the said islands, for that they are the Emperor's or King of Portugal's, we should by the way and coming once to the line equinoctial, find lands no less rich of gold and spicery, as all other lands are under the said line equinoctial: and also should, if we may pass under the north, enjoy the navigation of all Tartary. Which should be no less profitable to our commodities of cloth, than these spiceries to the Emperor, and King of Portugal.

But it is a general opinion of all cosmographers, that passing the seventh clime,[16] the sea is all ice, and the cold so great that none can suffer it. And hitherto they had all the same opinion, that under the equinoctial line for much heat the land was uninhabitable.

Yet since (by experience is proved) no land so much habitable nor more temperate. And to conclude, I think the same should be found under the north, if it were experimented.

So I judge, there is no land unhabitable, nor sea innavigable.

X

Two voyages made by William Hawkins of Plymouth, father of Sir John Hawkins knight, late Treasurer of Her Majesty's Navy, in the years 1530 and 1532.

Old Mr William Hawkins of Plymouth, a man for his wisdom, valour experience, and skill in sea causes much esteemed, and beloved of King Henry the Eighth, and being one of the principal sea-captains in the west parts of England in his time, not contented with the short voyages commonly then made only to the known coasts of Europe, armed out a tall and goodly ship of his own of the burthen of 250 tons, called the *Paul* of Plymouth, wherewith he made three long and famous voyages unto the coast of Brazil, a thing in those days very rare, especially to our nation. He touched at the river of Sestos[17] upon the coast of Guinea, where he trafficked with the negroes, and took of them elephants' teeth, and other commodities which that place yieldeth: and so arriving on the coast of Brazil, he used there such discretion, and behaved himself so wisely with those savage people, that he grew into great familiarity and friendship with them. Insomuch that in his second voyage, one of the savage kings of the country of Brazil, was contented to take ship with him, and to be transported hither into England: whereunto Mr Hawkins agreed, leaving behind in the country as a pledge for his safety and return again, one Martin Cockeram of Plymouth. This Brazilian king being arrived, was brought up to London and presented to King Henry the Eighth, lying as then at Whitehall: at the sight of whom the

King and all the nobility did not a little marvel, and not without cause: for in his cheeks were holes made according to their savage manner, and therein small bones were planted, standing an inch out from the said holes, which in his own country was reputed for a great bravery. He had also another hole in his nether lip, wherein was set a precious stone about the bigness of a pea: all his apparel, behaviour, and gesture, were very strange to the beholders.

Having remained here the space almost of a whole year, Mr Hawkins according to his promise and appointment, purposed to convey him again into his country: but it fell out in the way, that by the change of air and alteration of diet, the said savage king died at sea, which was feared would turn to the loss of the life of Martin Cockeram his pledge. Nevertheless, the savages being fully persuaded of the honest dealing of our men with their prince, restored again the said pledge, without any harm to him, or any man of the company: which pledge of theirs they brought home again into England, with their ship freighted, and furnished with the commodities of the country. Which Martin Cockeram, being an officer of the town of Plymouth, was living within these few years.

XI

The voyage of Roger Bodenham with the great bark Aucher *to Candia and Chios, in the year 1550.*

In the year 1550 the 13 of November I Roger Bodenham Captain of the Bark *Aucher* departed, and that night came to Dover, and there came to an anchor, and there remained until Tuesday, meeting with the worthy knight Sir Anthony Aucher owner.

The 16 [of January 1551] we had sight of Cape Finisterre on the coast of Spain. The 30 we arrived at Cadiz, and there discharged certain merchandise. We arrived at Messina in Sicily, and there discharged much goods, and remained there until Good Friday in Lent.

The chief merchant that laded the bark was a merchant stranger called Anselm Salvago, and because the time was then very dangerous, and no going into Levant, especially to Chios without a safe conduct from the Turk, Anselm promised the owner that we should receive the same at Messina. But I was posted from thence to Candia, and there I was answered that I should send to Chios, and there I should have my safe conduct. I was forced to send one, and he had his answer that the Turk would give none.

I was bound to deliver the goods that were in the ship at Chios, or send them at mine adventure. The merchants without care of the loss of the ship would have compelled me to go, and plainly I would not go, because the Turk's galleys were come forth to go against Malta.[18] There were in Candia certain Turkish vessels which had brought wheat thither to sell, and were ready to depart for Turkey. And they departed in the morning betimes, carrying news that I would not go forth: the same night I prepared beforehand what I thought good, without making any man privy, until I saw time. Then I had no small business to cause my mariners to venture with the ship in such a manifest danger. Nevertheless I won them to go all with me, except three which I set on land.

I was ready to set forth about eight of the clock at night, being a fair moon shine night, and went out. Then my three mariners made such requests unto the rest of my men to come aboard, as I was constrained to take them in. And so with good wind we put into the archipelago, and being among the islands the wind scanted, and I was forced to anchor at an island called Mykonos, where I tarried 10 or 12 days. I weighed and set sail for the island of Chios, with which place I fell in the afternoon. The small vessels which came in my company, departed from me to win the shore, to get in the night, but upon a sudden they espied 3 foists of Turks coming upon them to spoil them. My pilot, having a son in one of those small vessels, entreated me, I caused my gunner to shoot a demi-culverin at a foist that was ready to enter one of the boats. This was so happy a shot, that it made the Turk to fall astern of the boat, and to leave him, by the which means he escaped. Then they all came to me, and requested that they might hang at my stern until daylight, by which time I came before the mole of Chios, and sent my boat on land to the merchants of that place to send for their goods out of hand, or else I would return back with all to Candia.

I was entreated to come into the harbour, and had a safe assurance for 20 days against the Turk's army, with a bond of the city in the sum of 12000 ducats. So I made haste and sold such goods as I had to Turks that came thither, and put all in order with as much speed as I could, fearing the coming of the Turk's Navy. Said they, we be not able to defend you, for the Turk where he cometh, taketh what he will, and leaveth what he list.

I determined to go forth. But the merchants regarding more their gains than the ship, hindered me very much in my purpose of going forth, and made the mariners come to me to demand their wages to be paid them out of hand, and to have a time to employ the same there. But God provided so for me, that I paid them their money that night, and then charged them, that if they would not set the ship forth, I would make them answer the same in England, with danger of their heads. Many were married in London and had somewhat to lose, those did stick to me. I had twelve gunners: the master-gunner who was a mad brained fellow, and the owner's servant had a parliament between themselves, and he upon the same came up to meet me with his sword drawn, swearing that he had promised the owner Sir Anthony Aucher, to live and die in the said ship against all that should offer any harm to the ship, and that he would fight with the whole army of the Turks, and never yield: with this fellow I had much to do, but at the last I made him confess his fault, and follow mine advice. Thus with much labour I got out of the mole of Chios, into the sea by warping forth, with the help of Genoese boats, and being out God sent me a special gale of wind to go my way.

About half an hour past two of the clock there came seven galleys into Chios to stay the ship: and the admiral of them was in a great rage because she was gone. Whereupon they put some of the best in prison, and took all the men of the three ships which I left in the port, and put them into the galleys. The next day came thither a hundred more of galleys, and there tarried for their whole company, which being together were about two hundred and fifty sail, taking their voyage for to surprise the island of Malta. The next day after I departed, I had the sight of Candia.

There be in that island of Candia many banished men, that live continually in the mountains, they came down to serve, to the number of four or five thousand, they are good archers, every one with his bow and arrows, a sword and a dagger, long long hair, and boots that reach up to their groin, and a shirt of mail, hanging the one half before, and the other half behind. They would drink wine out of all measure. Then the army being past, I laded my ship with wines and other things; and departed for Messina. And so from thence I came to London with the ship and goods in safety, God be praised. Richard Chancellor, who first discovered Russia, was with me in that voyage, and Matthew Baker, who afterward became the Queen's Majesty's chief shipwright.

XII

Ordinances for the direction of the intended voyage for Cathay, compiled and delivered by the right worshipful Sebastian Cabot Esquire, governor of the mystery and company of the Merchant Adventurers for the discovery of regions, dominions islands and places unknown, the 9th day of May, in the year of Our Lord God 1553.

First, the captain general, with the pilot major, the masters, merchants and other officers, to be so knit and accorded in unity, love, conformity, and obedience in every degree on all sides, that no dissension may rise betwixt them and the mariners of this company, for that dissension (by many experiences) hath overthrown many notable and likely enterprises.

Forasmuch as every person hath given an oath to be true, faithful and loyal subjects, and liege men to the King's most excellent

Majesty, and to serve His Grace, the realm, and this present voyage truly, and not to leave off the said voyage and enterprise until it shall be accomplished, so far forth as possibility and the life of man may serve or extend: therefore it behoveth every person, as well for conscience, as for duty's sake to remember his said charge.

Every person by virtue of his oath, to do effectually and with good will, every such act and deed as shall be to him from time to time commanded by the captain general.

All courses in navigation to be set and kept, by the advice of the captain, pilot major, masters and masters' mates, with the assents of their counsellors, the captain general shall in all councils and assemblies have a double voice.

The fleet shall keep together, and not separate themselves asunder, as much as by wind and weather may be done and permitted.

The merchants, and other skilful persons in writing, shall daily describe the navigation of every day and night, with the points, and observation of the lands, tides, elements, altitude of the sun, course of the moon and stars.

The steward and cook of every ship to render to the captain weekly (or oftener) a just and plain account of expenses of the victuals, as well flesh, fish, biscuit, meat, or bread, as also of beer, wine, oil, or vinegar, and all other kind of victualling under their charge, that no waste be made otherwise than reason and necessity shall command.

No blaspheming of God, or detestable swearing be used in any ship, nor communication of ribaldry, filthy tales, or ungodly talk to be suffered in the company of any ship, neither dicing, carding, tabling nor other devilish games to be frequented, whereby ensueth not only poverty to the players, but also strife, variance, brawling, fighting and oftentimes murder to the utter destruction of the parties, and provoking of God's most just wrath and sword of vengeance.

Morning and evening prayer, with other common services appointed by the King's Majesty, and laws of this realm to be read and said in every ship daily by the minister in the Admiral, and the merchant or some other person learned in other ships, and the Bible to be read devoutly and Christianly to God's honour.

Every officer is to be charged by inventory with the particulars of his charge, and to render a perfect account of the defraying of the same together with modest and temperate dispending of powder, shot, and use of all kind of artillery, which is not to be misused, but diligently to be preserved for the necessary defence of the fleet and voyage.

No liquor to be spilt on the ballast, nor filthiness to be left within board: the cook room, and all other places to be kept clean for the better health of the company, the pages to be brought up according to the laudable order and use of the sea, as well in learning of navigation, as in exercising of that which to them appertaineth.

The sick to be tended, comforted and holpen in the time of his infirmity, and every manner of person, without respect, to bear another's burden.

If any person shall fortune to die in the voyage, such apparel and other goods as he shall have at the time of his death, is to be kept, and an inventory made, and conserved to the use of his wife, and children, or otherwise according to his mind and will, and the day of his death to be entered in the books: to the intent it may be known what wages he shall have deserved to his death.

Not to disclose to any nation the state of our religion, but to pass it over in silence, without any declaration of it, seeming to bear with such laws and rites, as the place hath, where you shall arrive.

For as much as our people, and ships may appear unto them strange and wondrous, and theirs also to ours: it is to be considered how they may be used, learning much of their natures and dispositions, by some one such person, as you may first either allure, or take to be brought aboard your ships, and there to learn as you may, without violence or force, and no woman to be tempted, or entreated to incontinency, or dishonesty.

The person so taken, to be well entertained, used and apparelled, to be set on land, to the intent that he or she may allure other to draw nigh to show the commodities: and if the person taken may be made drunk with your beer, or wine, you shall know the secrets of his heart.

Every nation is to be considered advisedly, and not to provoke them by any disdain, laughing, contempt, or suchlike, but to use them with prudent circumspection, with all gentleness, and courtesy.

The names of the people of every island, are to be taken in writing, with the commodities of the same, their natures, qualities, and dispositions, what commodities they will most willingly depart with, and what metals they have in hills, mountains, streams, or rivers, in, or under the earth.

If people shall appear gathering of stones, gold, metal, or other like, on the sand, your pinnaces may draw nigh, marking what things they gather, using or playing upon the drum, or such other instruments, as may allure them, but keep you out of danger, and show to them no sign of hostility.

If you shall be invited into any lord's house, to dinner, or other parlance, go in such order of strength, that you may be stronger than they, and be wary of ambushes, and that your weapons be not out of your possession.

If you see them wear lions' or bears' skins, having long bows and arrows, be not afraid of that sight: for such be worn oftentimes more to fear strangers, than for any other cause.

There be people that can swim in the sea, havens, and rivers, naked, having bows and shafts, coveting to draw nigh your ships, which if they shall find not well watched, or warded, they will assault, desirous of the bodies of men, which they covet for meat: if you resist them they dive, and so will flee, and therefore diligent watch is to be kept both day and night, in some islands.

If occasion serve, that you may give advertisements of your proceedings, and likelihood of success in the voyage, passing such dangers of the seas, perils of ice, intolerable colds, and other impediments, by which sundry authors have ministered matter of suspicion that this voyage could not succeed for the extremity of the North Pole, lack of passage, and suchlike, which have caused wavering minds, and doubtful heads, to withdraw themselves from the adventure of this voyage, when you shall have tried by experience (most certain master of all worldly knowledge), then for declaration of the truth you may send two or one, as you shall think may pass in safety: whereby the company being advertised of your proceedings, may further provide that which may seem most beneficial for the public wealth of the same: in which things your wisdoms and discretions are to be used, for that you be not ignorant, how many persons, as well the King's Majesty, the lords of his honourable council, this whole company, as also your wives, children, kinsfolk, allies, friends and familiars, be replenished in their hearts with ardent desire to learn and know your conditions and welfares, and in what likelihood you be in, to obtain this notable enterprise, which is hoped no less to succeed to you, than the Orient or Occident Indias have to the high benefit of the Emperor, and Kings of Portugal, whose subjects, industries, and travels by sea, have enriched them, by those lands and islands, which were to all cosmographers unknown.

No conspiracies, factions, false tales, untrue reports, which be the very seeds and fruits of contention, discord, and confusion, by evil tongues to be suffered, but the same, and all other ungodliness to be chastened charitably with brotherly love, and always obedience to be used and practised by all persons in their degrees, not only for duty and conscience' sake towards God, under whose merciful hand navigators above all other creatures naturally be most nigh, but also for prudent and worldly policy, and public weal, so to endeavour yourselves as that you may satisfy the expectation of them, who at their great costs, charges and expenses, have so furnished you in good sort, and plenty of all necessaries, as the like was never in any realm seen, for such an exploit, which is most likely to be achieved, and brought to good effect, if every person in his vocation shall endeavour himself according to his most bounden duty: praying to the living God, to give you his grace, to accomplish your charge to his glory, whose merciful hand shall prosper your voyage, and preserve you from all dangers.

In witness whereof I Sebastian Cabot, Governor aforesaid, to these present ordinances have subscribed my name.

XIII

The new navigation and discovery of the kingdom of Moscovy, by the northeast, in the year 1553: enterprised by Sir Hugh Willoughby knight, and performed by Richard Chancellor pilot major of the voyage.

At what time our merchants perceived the commodities and wares of England to be in small request with the countries and people about us, and near to us, and that those merchandises were now neglected, and the price thereof abated, certain grave citizens of London, and men careful for the good of their country, began to think with themselves, how this mischief might be remedied.

Seeing that the wealth of the Spaniards and Portuguese, by the discovery and search of new trades and countries was marvellously increased, supposing the same to be a course and mean for them also to obtain the like, they thereupon resolved upon a new and strange navigation. After much speech and conference together, it was at last concluded that three ships should be prepared and furnished out, for the search and discovery of the northern part of the world, to open a way and passage to our men for travel to new and unknown kingdoms. Every man willing to be of the society, should disburse the portion of twenty and five pounds apiece: so that in short time by this means the sum of six thousand pounds being gathered, the three ships were bought, newly built and trimmed. Whereas they afore determined to have the eastern part of the world sailed unto, and yet that the sea towards the same was not open, whereas yet it was doubtful whether there were any passage yea or no, they resolved to victual the ships for eighteen months, six months victual to sail to the place, so much more to remain there if the extremity of the winter hindered their return, and so much more also for the time of their coming home.

One Sir Hugh Willoughby a most valiant gentleman, and well born, very earnestly requested to have that care and charge committed unto him. At last they concluded and appointed him the admiral. And for the government of the other ships, one Richard Chancellor, a man of great estimation for many good parts of wit in him, was elected. By the twentieth day of May, the captains and mariners should take shipping, and depart from Radcliffe upon the ebb, if it pleased God. They having saluted their acquaintance, one his wife, another his children, another his kinsfolks, and having weighed anchor, they departed with the turning of the water, and sailing easily, came first to Greenwich. The greater ships are towed down with boats, and oars, and the mariners being all apparelled in watchet or sky coloured cloth, made way with diligence. And being come near to Greenwich (where the Court then lay), presently upon the news, the courtiers came running out, and the common people flocked together, standing very thick upon the shore: the Privy Council, they looked out at the windows of the Court, and the rest ran up to the tops of the towers: the ships hereupon discharged their ordnance. One stood in the poop of the ship, and by his gesture bids farewell to his friends in the best manner he could. Another walks upon the hatches, another climbs the shrouds, another stands upon the mainyard, and another in the top of the ship. To be short, it was a very triumph. They departed and came to Harwich: at the last, with a good wind they hoisted up sail, and committed themselves to the sea, giving their last adieu to their native country, which they knew not whether they should ever return to see again or not.

Sir Hugh Willoughby, a man of good foresight called together the chiefest men of the other ships. They conclude and agree, that if any great tempest should arise at any time, and happen to disperse and scatter them, every ship should endeavour his best to go to Wardhouse,[19] a haven or castle of some name in the kingdom of Norway, and that they that arrived there first in safety, should stay and expect the coming of the rest.

The very same day in the afternoon, about four of the clock, so great a tempest suddenly arose, and the seas were so outrageous, that the ships could not keep their intended course, but some were perforce driven one way, and some another way, to their great peril and hazard. The said admiral bearing all his sails, was carried away with so great force and swiftness, that not long after he was quite out of sight, and the third ship also was dispersed and lost us. The ship boat of the admiral was overwhelmed in the sight and view of the mariners of the *Bonaventure*.

Now Richard Chancellor with his ship and company being thus left alone, shapeth his course for Wardhouse in Norway, there to expect and abide the arrival of the rest of the ships. Having stayed there the space of seven days, and looked in vain for their coming, he determined at length to proceed alone in the purposed voyage. As he was preparing himself to depart, he fell in company and speech with certain Scottishmen, who began earnestly to dissuade him from the further prosecution of the discovery, by amplifying the dangers which he was to fall into. But he persuading himself that a man of valour could not commit a more dishonourable part than for fear of danger to avoid and shun great attempts, was nothing at all changed or discouraged, remaining steadfast and immutable in his first resolution: determining either to bring that to pass which was intended, or else to die the death.

As for them which were with Master Chancellor in his ship, they were resolute. When they saw their desire and hope of the arrival of the rest of the ships to be every day more and more frustrated, they provided to sea again, and Master Chancellor held on his course towards that unknown part of the world, and sailed so far, that he came at last to the place where he found no night at all, but a continual light and brightness of the sun shining clearly upon the huge and mighty sea. At length it pleased God to bring them into a certain great bay, whereinto they entered. Looking every way about them, they espied afar off a certain fisher boat, which Master Chancellor, accompanied with a few of his men, went towards. They being amazed by the strange greatness of his ship, began presently to avoid and to flee: but he still following them at last overtook them, they (being in great fear, as men half-dead) prostrated themselves before him, offering to kiss his feet: but he (according to his great and singular courtesy,) looked pleasantly upon them, comforting them by signs and gestures, refusing those duties and reverences of theirs, and taking them up in all loving sort from the ground. And it is strange to consider how much favour afterwards in that place, this humanity of his did purchase to himself. For they spread a report abroad of the arrival of a strange nation, of a singular gentleness and courtesy: whereupon the common people came together offering to these new-come guests victuals freely, and not refusing to traffic with them, except they had been bound by a certain religious use and custom, not to buy any foreign commodities, without the knowledge and consent of the king.

This country was called Russia, or Moscovy, and Ivan Vasilivich (which was at that time their King's name) ruled and governed far and wide in those places. And the barbarous Russians asked likewise of our men whence they were: whereunto answer was made, that they were Englishmen sent into those coasts, from the most excellent King Edward the Sixth, having from him in commandment certain things to deliver to their king, and seeking nothing else but his amity and friendship, and traffic with his people, whereby they doubted not, but that great commodity and profit would grow to the subjects of both kingdoms.

They secretly sent a message to the Emperor, to certify him of the arrival of a strange nation, and withall to know his pleasure concerning them. He granted liberty to his subjects to bargain and traffic with them: and further promised, that if it would please them to come to him, he himself would bear the whole charges of post horses. So Master Chancellor began his journey, which was very long and most troublesome, wherein he had the use of certain sleds, which in that country are very common, the people almost not knowing any other manner of carriage, the cause whereof is the exceeding hardness of the ground congealed in the winter time by the force of the cold, which in those places is very extreme and horrible. After much ado and great pains taken in this long and weary journey (for they had travelled very near fifteen hundred miles), Master Chancellor came at last to Moscow, the chief city of the kingdom, and the seat of the king.

The whole country is plain and champaign, and a few hills in it: and towards the north it hath very large and spacious woods, wherein is great store of fir trees, a wood very necessary, and fit for the building of houses: there are also wild beasts bred in those woods, as bears and black wolves. When the winter doth once begin there it doth still more and more increase by a perpetuity of cold: neither doth that cold slack, until the force of the sun-beams doth dissolve the cold, and make glad the earth. Our mariners which we left in the ship in the meantime to keep it, in their going up only from their cabins to the hatches, had their breath oftentimes so suddenly taken away, that they eftsoons fell down as men very nearly dead, so great is the sharpness of that cold climate.

The empire and government of the king is very large, and his wealth at this time exceeding great. The city of Moscow is the chiefest of all the rest. Our men say, that in bigness it is as great as the City of London, with the suburbs thereof. There are many and great buildings in it, but for beauty and fairness nothing comparable to ours. Their streets and ways are not paved with stone as ours are: the walls of their houses are of wood: the roofs for the most part are covered with shingle boards.

After they had remained about twelve days in the city, there was then a messenger sent unto them, to bring them to the King's house: within the gates of the court, there sat a very honourable company of courtiers, to the number of one hundred, all apparelled in cloth of gold, down to their ankles: being conducted into the chamber of presence, our men began to wonder at the majesty of the Emperor: his seat was aloft, in a very royal throne, having on his head a diadem, or crown of gold, apparelled with a robe all of goldsmith's work, and in his hand he held a sceptre garnished, and beset with precious stones. On one side of him stood his chief Secretary, on the other side, the great Commander, both of them arrayed also in cloth of gold: and then there sat the council of one hundred and fifty in number, all in like sort arrayed.

Master Chancellor being therewithall nothing dismayed, saluted, and did his duty to the Emperor, after the manner of

England, and withall, delivered unto him the letters of our king, Edward the Sixth.

Next unto Moscow, the City of Novgorod is reputed the chiefest of Russia: the chiefest and greatest market town of all Moscovy. This town excels all the rest in the commodities of flax and hemp: it yields also hides, honey, and wax. The Flemings there sometimes had a house of merchandise, but by reason that they used ill dealing there, they lost their privileges. Those Flemings hearing of the arrival of our men in those parts, wrote their letters to the Emperor against them, accusing them for pirates and rovers, wishing him to detain and imprison them. Which things when they were known of our men, they conceived fear, that they should never have returned home. But the Emperor believing rather the King's letters, which our men brought, than the lying and false suggestions of the Flemings, used no ill entreaty towards them.

The north parts of Russia yield very rare and precious skins: and amongst the rest, those principally, which we call sables, worn about the necks of our noblewomen and ladies: it hath also martins' skins, white, black, and red fox skins, skins of hares, and ermines, and others, as beavers, minxes and minivers. They use to boil the water of the sea, whereof they make very great store of salt.

They maintain the opinions of the Greek Church. In their private houses they have images for their household saints; he that comes into his neighbour's house doth first salute his saints, although he see them not. For the articles of our faith, and the Ten Commandments, no man, or at the least very few of them do either know them or can say them: their opinion is that such secret and holy things as they are should not rashly and imprudently be communicated with the common people. Concerning the Latin, Greek and Hebrew tongues, they are altogether ignorant.

There is a certain part of Moscovy bordering upon the countries of the Tartars, wherein those Moscovites that dwell are very great idolaters. The common houses of the country are everywhere built of beams of fir tree. The form and fashion of their houses in all places is four square, with strait and narrow windows, whereby with a transparent casement made or covered with skin like to parchment they receive the light. They have stoves wherein in the morning they make a fire, and the same fire doth either moderately warm, or make very hot the whole house.

These are the things most excellent Queen, which your subjects newly returned from Russia have brought home concerning the state of that country.

XIV

The voyage to Guinea in the year 1554. The Captain whereof was Mr John Lok.

In the year of Our Lord 1554 the eleventh day of October, we departed the river of Thames with three goodly ships, the one called the *Trinity*, the other called the *Bartholomew*, the third was the *John Evangelist*, the first day of November at nine of the clock at night departing from the coast of England.

The 17 day in the morning we had sight of the Isle of Madeira, a long low land with a saddle through the midst of it, standing in two and thirty degrees: and in the west part, many springs of water running down from the mountain, and many white fields like unto cornfields, and some white houses.

The 19 day at twelve of the clock we had sight of the Canaries. Tenerife is a high land, with a great high peak like a sugar loaf, and upon the said peak is snow throughout all the whole year. And by reason of that peak it may be known above all the other islands.

Seven or eight leagues off from the river del Oro to Cape de las Barbas, there use many Spaniards and Portuguese to trade for fishing, during the month of November: and all that coast is very low lands. The fourth of December we began to set our course southeast. We fell with Cape Mensurado to the southeast. This cape may be easily known, by reason that the rising of it is like a porpoise head. Also toward the southeast there are three trees, whereof the easternmost tree is the highest, and the middlemost is like a high stack, and the southernmost like unto a gibbet. All the coast along is white sand.

On the fourth day of September, under nine degrees, we lost sight of the north star. We came to anchor three or four leagues west and by south of the Cape de Tres Puntas. Then our pinnace came aboard with all our men, the pinnace also took in more wares. They told me they would go to a place where the *Primrose* had received much gold at the first voyage, but I fearing a brigantine that was then upon the coast, did weigh and follow them. The town is called Shama,[20] where we did traffic for gold, to the northeast of Cape de Tres Puntas.

They brought from thence at the last voyage four hundred pound weight and odd of gold, of two and twenty carats and one grain in fineness: also six and thirty butts of grains, and about two hundred and fifty elephants' teeth of all quantities. Some of them were as big as a man's thigh above the knee, and weighed about four score and ten pound weight apiece. These great teeth or tusks grow in the upper jaw downwards, and not in the nether jaw upwards, wherein the painters and arras workers are deceived. At this last voyage was brought from Guinea the head of an elephant of huge bigness. This head divers have seen in the house of the worthy merchant Sir Andrew Judde, where also I saw it, considering by the work, the cunning and wisdom of the workmaster: without such consideration, the sight of such strange and wonderful things may rather seem curiosities than profitable contemplations.

The elephant (which some call an oliphant) is the biggest of all fourfooted beasts, his forelegs are longer than his hinder, he hath ankles in the lower part of his hinder legs, and five toes on his feet undivided, his snout or trunk is so long, and in such form, that it is to him in the stead of a hand: for he neither eateth nor drinketh but by bringing his trunk to his mouth, therewith he helpeth up his master or keeper, therewith he overthroweth trees. Of all beasts they are most gentle and tractable, and are of quick sense and sharpness of wit. They love rivers, and will often go into them up to the snout, wherewith they blow and snuff and play in the water. They have continual war against dragons, which desire their blood because it is very cold: and therefore the dragon lieth in wait as the elephant passeth by.

Touching the manners and nature of the people, their princes and noblemen use to pounce and raze their skins with pretty knots in divers forms, as it were branched damask, thinking that to be a decent ornament. And albeit they go in manner all naked, yet are many of them and especially their women, laden with collars, bracelets, hoops and chains, either of gold, copper or ivory. I myself have one of their bracelets of ivory, weighing two pound and six ounces of troy weight, made of one whole piece of the biggest part of the tooth, turned and somewhat carved, with a hole in the midst. Some of their women wear on their bare arms certain foresleeves made of the plates of beaten gold. On their fingers also they wear rings, made of golden wires, with a knot or wreath, like unto that which children make in a ring of a rush.

They are very wary people in their bargaining, and will not lose one spark of gold of any value. They use weights and measures, and are very circumspect in occupying the same. They that shall have to do with them, must use them gently: for they will not traffic or bring in any wares if they be evil used.

At their coming home the keels of their ships were marvellously overgrown with certain shells of such bigness that a man might put his thumb in the mouths of them. In these there groweth a certain slimy substance which at the length slipping out of the shell and falling in the sea, becometh those fowls which we call barnacles.[21] Their ships were also in many places eaten with the worms.

There died of our men at this last voyage about twenty and four, whereof many died at their return into the clime of the cold regions, as between the isles of Azores, and England. They brought with them certain black slaves, whereof some were tall and strong men. The cold and moist air doth somewhat offend them. Yet doubtless men that are born in hot regions may better abide cold, than men that are born in cold regions may abide heat.

XV

The letter of Mr George Killingworth the company's first Agent in Moscovy, touching their second voyage. Anno 1555.

It may please your worship to understand, that at the making hereof we all be in good health, thanks be to God, save only William our cook as we came from Kholmogory[22] fell into the river out of the boat, and was drowned. And the 11th day of September we came to Vologda, and there we laid all our wares up, and sold very little: but one merchant would have given us twelve roubles for a broad cloth, and he said he would have had them all, and four altines for a pound of sugar, but we did refuse it because he was the first, and the merchants were not come hither: but I fear it will not be much better. Yet notwithstanding we did for the best.

We came to Moscow the 4th day of October, and were lodged that night in a simple house: but the next day we were sent for to the Emperor his secretary, and he bade us welcome with a cheerful countenance and cheerful words. We showed him that we had a letter from our Queen's Grace to the Emperor his Grace, and then he desired to see them all; then we were appointed to a better house: and the seventh day the secretary sent for us again. The ninth day we were sent to make us ready to speak with the Emperor on the morrow. Before we came to his presence we went through a great chamber, where stood many small tuns, pails, bowls and pots of silver, I mean, like washing bowls, all parcel gilt: and within that another chamber, wherein sat (I think) near a hundred in cloth of gold, and then into the chamber where His Grace sat, and there I think were more than in the other chamber also in cloth of gold, and we did our duty. His Grace did ask how our Queen's Grace did, calling her cousin, saying that he was glad that we were come in health into this realm, and we went one by one unto him, and took him by the hand, and then His Grace bid us go in health, and come to dinner again, and we dined in his presence, and were set with our faces towards His Grace, and none in the chamber sat with their backs towards him, being I think near a hundred at dinner then, and all served with gold, as platters, chargers, pots, cups, and all not slender but very massy, and yet a great number of platters of gold, standing still on the cupboard, not moved: and divers times in the dinner time His Grace sent us meat and drink from his own table. They said His Grace's pleasure was, that his best merchants of Moscow should be spoken to, to meet and talk with us. They would know our prices of our wares: and we answered, that for our prices they must see the wares before we could make any price thereof, for the like in goodness had not been brought into the realm. Then the chancellor said, methinks you shall do best to have your house at Kholmogory, which is but one hundred miles from the right discharge of the ships. All our merchants shall bring all our merchandise to Kholmogory to you, and so shall our merchants neither go empty nor come empty: for if they lack loading homeward, there is salt, which is good ware here, that they may come laden again. So we were very glad to hear that, and did agree to his saying: for we shall nevertheless, if we list, have a house at Vologda, and at Moscow, yea, and at Novgorod, or where we will in Russia.

And thus may we continue three or four years, and in this space we shall know the country and the merchants, and which way to save ourselves best, and where to plant our houses, and where to seek for wares. I have bought five hundred weight of yarn, which stands me in eight pence farthing the Russian pound one with another. I do intend to go to Novgorod, and to Pleskau,[23] whence all the great number of the best town flax cometh, and such wares as are there I trust to buy part. Prepare fully for one ship to be ready in the beginning of April to depart off the coast of England. I pray you be not offended with these my rude letters for lack of time: I will find the means to convey you a letter with speed: which is and shall be as pleaseth God: who ever preserve your worship, and send us good sales. Written in haste.

By yours to command,

GEORGE KILLINGWORTH, *Draper.*

XVI

The navigation and discovery towards the river Ob, made by Stephen Burrough, master of the pinnace the Searchthrift, *in the year 1556.*

The 23rd of April, Saturday, being St Mark's Day, we departed from Blackwall. The right worshipful Sebastian Cabot came aboard our pinnace at Gravesend, with divers gentlemen and gentlewomen, the good old gentleman wishing them to pray for the good fortune, and prosperous success of the *Searchthrift*, our pinnace.

Friday the 15th of May we were within 7 leagues of the shore, on the coast of Norway: Saturday at an east sun we came to St Dunstan's Island, which island I so named.

June. We weighed in Corpus Christi Bay, at a northeast and by east sun: the bay is almost half a league deep: the headland which is Corpus Christi point, lieth one league from the head of the bay, where we had a great tide, like a race over the flood.

Thursday at six of the clock in the morning, there came aboard of us one of the Russian lodias, rowing with twenty oars. The master of the boat presented me with a great loaf of bread, four dried pikes, and a peck of fine oatmeal, and I gave unto the master of the boat a comb and a small glass: and he declared unto me, that he was bound to Pechora.

We weighed our anchors in the River Kola, and went into the sea seven or eight leagues, where we met with the wind far northerly, that of force it constrained us to go again back into the river, where came aboard of us sundry of their boats which declared unto me that they were also bound to the northwards, a-fishing for morse and salmon, and gave me liberally of their white and wheaten bread. There was one of them whose name was Gabriel, who showed me very much friendship, and he declared unto me, that all they were bound to Pechora, a-fishing for salmon and morses. He showed me by demonstrations, that with a fair wind we had seven or eight days' sailing to the River Pechora, so that I was glad of their company. This Gabriel promised to give me warning of shoals, as he did indeed.

Wednesday being Midsummer day, we sent our skiff a-land to sound the creek, where they found it almost dry at a low water. Although the harbour were evil, yet the stormy similitude of northerly winds tempted us to set our sails, and we let slip a cable and an anchor, and bare with the harbour. When we came upon the bark in the entrance of the creek, the wind did shrink so suddenly upon us, that we were not able to lead it in, and before we could have flatted the ship before the wind, we should have been on ground on the lee shore, so that we were constrained to let fall an anchor under our sails, and rode in a very breach, thinking to have warped in.[24] Gabriel came out with his skiff, and so did sundry others also, showing their good will to help us, but all to no purpose, for they were likely to have been drowned for their labour. We rushed in upon the other small anchor that Gabriel sent aboard, and laid that anchor to seawards: and then between these two anchors we traversed the ship's head to seawards, and set our foresail and mainsail, and when the bark had way, we cut the hawser, and so got the sea to our friend.

The next high water, Gabriel and his company departed from thence.

I sent our boat on shore to fetch fresh water and wood, and at their coming on shore this Cyril welcomed our men most gently, and also banqueted them: and in the mean time caused some of his men to fill our baricoes with water, and to help our men to bear wood into their boat: and he then put on his best silk coat and his collar of pearls, and came aboard again. I bade him welcome, and gave him a dish of figs.

July. At a northwest sun we came to an anchor within half a league of the shore, where we had good plenty of fish, both haddocks and cods, riding in 10 fathom water.

Sunday our men cut wood on shore, and brought it aboard, and we ballasted our ship with stones.

This morning Gabriel saw a smoke on the way, who rowed

unto it with his skiff, which smoke was two leagues from the place where we rowed: and at a northwest sun he came aboard again, and brought with him a Samoyed,[25] which was but a young man: his apparel was then strange unto us, and he presented me with three young wild geese.

Tuesday at a northwest sun we thought that we had seen land: which afterwards proved to be a monstrous heap of ice. Within a little more than half an hour after, we first saw this ice, we were enclosed within it before we were aware of it, which was a fearful sight to see: for, for the space of six hours, it was as much as we could do to keep our ship aloof from one heap of ice, and bear room from another, with as much wind as we might bear a course.

On St James his day, at a southwest sun, there was a monstrous whale aboard of us, so near to our side that we might have thrust a sword or any other weapon in him, which we durst not do for fear he should have overthrown our ship: and then I called my company together, and all of us shouted, and with the cry that we made he departed from us: there was as much above water of his back as the breadth of our pinnace, and at his falling down, he made such a terrible noise in the water, that a man would greatly have marvelled, but God be thanked, we were quietly delivered of him. And a little after we spied certain islands, with which we bare, and found a good harbour. We came to an anchor, and named the island St James his Island, where we found fresh water.

We saw a sail coming about the point whereunder we thought to have anchored. Then I sent a skiff aboard of him, and at their coming aboard, they took acquaintance of them, and their chief man said we were past the way, which should bring us to the Ob. This land, said he, is called Novaya Zemlya, that is to say, the New Land. I gave him a steel glass, two pewter spoons, and a pair of velvet sheathed knives: he gave me 17 wild geese. This man's name was Loshak.

There were some of their company on shore, which did chase a white bear over the high cliffs into the water.

August. I met again with Loshak, and went on shore with him, and he brought me to a heap of the Samoyeds' idols, which were in number above 300, the worst and the most unartificial work that ever I saw: the eyes and mouths of sundry of them were bloody, they had the shape of men, women and children, very grossly wrought. Before certain of their idols blocks were made as high as their mouths, being all bloody, I thought that to be the table whereon they offered their sacrifices: I saw also the instruments, whereupon they had roasted flesh, and as far as I could perceive, they made their fire directly under the spit.

Loshak being there present told me that these Samoyeds were not so hurtful as they of Ob are. They have no houses, but only tents made of deer's skins, which they underprop with stakes and poles: their boats are made of deer's skins, and when they come on shore they carry their boats with them upon their backs: for their carriages they have no other beasts to serve them, but deer only. As for bread and corn they have none, except the Russians bring it to them: their knowledge is very base, for they know no letter.

Wednesday we saw a terrible heap of ice approach near to us, and therefore we thought it good with all speed to depart from thence, and so I returned to the westward again.

At night there grew so terrible a storm, that we saw not the like, although we had endured many storms since we came out of England. It was wonderful that our bark was able to brook such monstrous and terrible seas, without the great help of God, who never faileth them at need, that put their sure trust in Him.

Saturday was calm: the latitude this day at noon was 70 degrees and a tierce, we sounded here, and had nine and forty fathoms and ooze, which ooze signified that we draw toward Novaya Zemlya. And thus being out of all hope to discover any more to the eastward this year, we thought it best to return, and that for three causes. The first, the continual northeast and northerly winds. Second, because of great and terrible abundance of ice: I adventured already somewhat too far in it, but I thank God for my safe deliverance from it. Third, because the nights waxed dark, and the winter began to draw on with his storms: and therefore I resolved to take the first best wind that God should send.

September. The eleventh day we arrived at Kholmogory, and

there we wintered, expecting the approach of the next summer to proceed farther in our intended discovery for the Ob.

XVII

A letter of the Company of the Merchants Adventurers to Russia unto George Killingworth, Richard Gray, and Henry Lane their Agents there, to be delivered in Kholmogory or elsewhere: sent in the John Evangelist.

The *Philip and Marie* arrived here ten days past: she wintered in Norway. The *Confidence* is lost there. And as for the *Bona Esperanza*, as yet we have no news of her. We fear it is wrong with her. By your bills of lading received in your general letters we perceive what wares and laden in them both. You shall understand we have freighted for the parts of Russia four good ships to be laden there by you and your order: that is to say, the *Primrose* of the load of 240 tons, master under God John Buckland: The *John Evangelist* of 170 tons, master under God Laurence Roundal: the *Anne* of London of the load of 160 tons, master under God David Philly, and the *Trinity* of London of the load of 140 tons, master under God John Robins, as by their charter parties may appear.

You shall receive, God willing, out of the said good ships, these kinds of wares following, all marked with the general mark of the Company as followeth: 25 fardels containing 207 sorting cloths, one fine violet in grain, and one scarlet, and 40 cottons for wrappers, beginning with number 1 and ending with number 52. 500 pieces of Hampshire kerseys, 9 barrels of pewter. You are to receive our said goods, and to procure the sales to our most advantage either for ready money, time or barter: having consideration that you do make good debts, and also foreseeing that you barter to profit, and for such wares as be here most vendible, as wax, tallow, train oil, hemp and flax. Of furs we desire no great plenty, because they be dead wares.

We will you set to work with all expedition in making of cables and ropes of all sort, from the smallest rope to twelve inches. Let all diligence be used, that at the return of these ships we may see samples of all ropes and cables if it be possible, that we may have good store against the next year, seeing that you have plenty of hemp there, and at a reasonable price, we trust we shall be able to bring as good stuff from thence, and better cheap than out of Denmark. We hear that there is great plenty of steel in Russia and Tartary, whereof we would you sent us part for an example. And likewise we be informed that there is great plenty of copper: we would be certified whether it be in plates or in round flat cakes, and send us some for an example. And likewise every kind of leather, wherof we be informed there is great store bought yearly by the Easterlings[26] and Dutch for Germany. Also we do understand that about the river of Pechora is great quantity of yew, which we be desirous to have knowledge of, because it is a special commodity for our realm. Therefore we have sent you a young man, whose name is Leonard Brian, that hath some knowledge in the wood: if there be none found that will serve for our purpose, then you may set the said Leonard Brian to any other business that you shall find most fittest for him, until the return of our ships the next year. For he is hired by the year only for that purpose. We doubt not that he shall do you good service there. For he hath good knowledge of wares of that country: for his bringing up hath been most in Denmark, and hath good understanding in making of ropes and cables.

Also we have sent you one Anthony Jenkinson gentleman, a man well travelled, whom we mind to use in further travelling, according to a commission delivered him. He must have forty pounds a year for four years, to be paid him by the half year, or as he will demand it of you, so let him have it from Easter last. Also the prices of wares here at this present are, bale flax twenty pound the pack and better, tow flax twenty-eight pound the hundred, train oil at nine pound the ton, wax at four pound the hundred, tallow at sixteen shillings the hundred, cables and ropes very dear: as yet there are no ships come out of Denmark.

Moreover, you had need to send new accounts, for them that

came in the *Edward* be marred and torn, so that we can make no reckoning by them: and likewise to write us a perfect note of all the goods which you received the last voyage out of the *Edward*, and herein not to fail.

XVIII

The voyage of Master Anthony Jenkinson, made from the city of Moscow in Russia, to the city of Bokhara in Bactria, in the year 1558.

The 23rd day of April, in the year 1558, I departed from Moscow by water, having with me two of your servants, namely, Richard Johnson, and Robert Johnson, and a Tartar, with divers parcels of wares, as by the inventory appeareth.

The eighth day we came unto a fair town called Murom, from Kasimov twenty leagues, where we took the sun, and found the latitude 56 degrees: and proceeding forward the eleventh day, we came unto another fair town and castle called Nijni Novgorod,[27] situated at the falling of the river Oka into the worthy river of Volga. From Ryazan to this Nijni Novgorod, on both sides the said river of Oka, is raised the greatest store of wax and honey in all the land of Russia. We tarried at the foresaid Nijni Novgorod until the nineteenth day, for the coming of a captain which was sent by the Emperor to rule at Astrakhan, who being arrived, and having the number of five hundred great boats under his conduct, some laden with victuals, soldiers, and munition: and other some with merchandise, departed all together.

Kazan is a fair town after the Russian or Tartar fashion, with a strong castle, situated upon a high hill, and was walled round about with timber and earth, but now the Emperor of Russia hath given order to pluck down the old walls, and to build them again of freestone. It hath been a city of great wealth and riches, and being in the hands of the Tartars it was a kingdom of itself, and did more vex the Russians in their wars, than any other nation: but nine years past, this Emperor of Russia conquered it, and took the king captive, who being but young is now baptized, and brought up in his court with two other princes, which were also kings of the said Kazan. Thus proceeding we passed by a goodly river called Kama, unto Astrakhan and so following the north and northeast side of the Caspian Sea, to a land of the Tartars called Turkmen, whose inhabitants are of the law of Mahomet, and were all destroyed in the year 1558, through civil wars among them, accompanied with famine, pestilence, and such plagues, in such sort that in the said year there were consumed of the people, in one sort and another, above one hundred thousand.

They were divided into divers companies called hordes, and every horde had a ruler, whom they obeyed as their king and was called a murse. Town or house they had none, but lived in the open fields, every murse or king having his hordes or people about him, with their wives, children and cattle, who having consumed the pasture in one place, removed unto another: and when they remove they have houses like tents set upon wagons or carts, which are drawn from place to place with camels, and therein their wives, children, and all their riches, which is very little, is carried about, and every man hath at the least four or five wives besides concubines. Use of money they have none, but do barter their cattle for apparel and other necessaries. They delight in no art nor science, except the wars, wherein they are expert, but for the most part they be pasturing people. They eat much flesh, and especially the horse, and they drink mare's milk, wherewith they be oftentimes drunk: they are seditious and inclined to theft and murder. Corn they sow not, neither do eat any bread, mocking the Christians for the same, saying we live by eating the top of a weed. But now to proceed forward to my journey.

The fourteenth day of July passing by an old castle, which was old Astrakhan, and leaving it upon our right hand, we arrived at new Astrakhan, which this Emperor of Russia conquered six years past. The town of Astrakhan is situated in an island upon a hill side, having a castle within the same, walled about with earth and timber, neither fair nor strong. The island is most destitute and barren of wood and pasture, and the ground will bear no corn:

the air is there most infected, by reason (as I suppose) of much fish, and especially sturgeon, by which only the inhabitants live, having great scarcity of flesh and bread. They hang up their fish in their streets and houses to dry for their provision, which causeth such abundance of flies to increase there, as the like was never seen in any land, to their great plague.

There was a great famine and plague among the people, and especially among the Tartars who came thither in great numbers to render themselves to the Russians, to seek succour at their hands, their country being destroyed: they were but ill entertained or relieved, for there died a great number of them for hunger, which lay all the island through in heaps dead, and like to beasts unburied, very pitiful to behold: many of them were also sold by the Russians, and the rest were banished from the island. At that time it had been an easy thing to have converted that wicked nation to the Christian faith, if the Russians themselves had been good Christians: but how should they show compassion unto other nations, when they are not merciful unto their own? At my being there I could have bought many goodly Tartars' children, of their own fathers and mothers, a boy or a wench for a loaf of bread worth six pence in England, but we had more need of victuals at that time than of any such merchandise. This Astrakhan is the furthest hold that this Emperor of Russia hath conquered of the Tartars towards the Caspian Sea. There is a certain trade of merchandise there used, but as yet so small and beggarly, that it is not worth the making mention.

The nineteenth day the wind being west, and we winding east southeast, we sailed ten leagues, and passed by a great river, which hath his spring in the land of Siberia. Here is no trade of merchandise used, for that the people have no use of money, and are all men of war, and pasturers of cattle, and given much to theft and murder. Thus being at an anchor against this river, and all our men being on land, saving I who lay sore sick, and five Tartars, whereof one was reputed a holy man, because he came from Mecca, there came unto us a boat with thirty men well armed and appointed, who boarded us, and began to enter into our bark, and our holy Tartar called Azy, perceiving that, asked them what they would have, and withall made a prayer: with that these rovers stayed, declaring that they were gentlemen, banished from their country, and out of living, and came to see if there were any Russians or other Christians in our bark? To whom this Azy most stoutly answered, that there were none, avowing the same by great oaths of their law (which lightly they will not break), whom the rovers believed, and upon his words departed. And so through the fidelity of that Tartar, I with all my company and goods were saved.

The third day of September 1558 we discharged our bark, and I with my company were gently entertained of the prince and of his people. But before our departure from thence, we found them to be very bad and brutish people, for they ceased not daily to molest us, either by fighting, stealing or begging, raising the price of horse and camels, and victuals, double that it was wont there to be, and forced us to buy the water that we did drink. For every camel's lading, being but 400 weight of ours, we agreed to give three hides of Russia, and four wooden dishes.

Thus being ready, the fourteenth of September we departed from that place, being a caravan of a thousand camels. And having travelled five days' journey, we came to another prince's dominion, and upon the way there came unto us certain Tartars on horseback, being well armed, and servants unto the said prince called Timor Sultan, governor of the said country. These aforesaid Tartars stayed our caravan in the name of their prince, and opened our wares, and took such things as they thought best for their said prince without money. I rode unto the same prince, and presented myself before him, requesting his favour, and passport to travel through his country, and not to be robbed nor spoiled of his people: which request he granted me, and entertained me very gently, commanding me to be well feasted with flesh and mare's milk: for bread they use none, nor other drink except water: money he had none to give me for such things as he took of me, which might be of value in Russian money, fifteen roubles, but he gave me his letter, and a horse worth seven roubles. And so I departed from him being glad that I was gone: for he was reported to be a very tyrant.

I departed and overtook our caravan and proceeded on our journey, and travelled 20 days in the wilderness from the sea side without seeing town or habitation, carrying provision of victuals with us for the same time, and were driven by necessity to eat one of my camels and a horse for our part, as the other did the like: and during the said 20 days we found no water, but such as we drew out of old deep wells, being very brackish and salt, and yet sometimes passed two or three days without the same. And the 5th day of October ensuing, we came unto a gulf of the Caspian Sea again, where we found the water very fresh and sweet.

We having refreshed ourselves departed thence the 4th day of October, and the seventh day arrived at a castle called Sellizure. This castle of Sellizure is situated upon an high hill, where the king called the Khan lyeth, whose palace is built of earth very basely, and not strong: the people are but poor, and have little trade of merchandise among them. The south part of this castle is low land, but very fruitful, where grow many good fruits. There grows a fruit called a carbuse[28] of the bigness of a great cucumber, yellow and sweet as sugar: also a certain corn called iegur, whose stalk is much like a sugar cane, and as high, and the grain like rice, which groweth at the top of the cane like a cluster of grapes; the water that serveth all the country is drawn by ditches out of the river Oxus, unto the great destruction of the said river, for which cause it falleth not into the Caspian Sea as it hath done in times past, and in short time all that land is like to be destroyed, and to become a wilderness for want of water, when the river Oxus shall fail. The sixteenth we arrived at a city called Urgenj,[29] where we paid custom as well for our own heads, as for our camels and horses. And having there sojourned one month, the king of that country called Ali Sultan, returned from a town called Khorasan, within the borders of Persia, which he lately had conquered from the Persians, with whom he and the rest of the kings of Tartary have continual wars. Before this king also I was commanded to come, to whom I presented the Emperor's letters of Russia, and he entertained me well, and demanded of me divers questions, and at my departure gave me his letters of safe conduct.

This city or town of Urgenj standeth in a plain ground, with walls of the earth, by estimation 4 miles about it. The buildings within it are also of earth, but ruined and out of good order: it hath one long street that is covered above, which is the place of their market. It hath been won and lost four times within seven years by civil wars, by means whereof there are but few merchants in it, and they very poor, and in all that town I could not sell above four kerseys.

From the Caspian Sea unto the castle of Sellizure aforesaid, and all the countries about the said sea, the people live without town or habitation in the wild fields, removing from one place to another in great companies with their cattle, whereof they have great store, as camels, horses, and sheep both tame and wild. Their sheep are of great stature with great buttocks, weighing 60 or 80 pound in weight. There are many wild horses which the Tartars do many times kill with their hawks. The hawks are lured to seize upon the beasts' necks or heads, which with chafing of themselves and sore beating of the hawks are tired: then the hunter following his game doth slay the horse with his arrow or sword. In all this land there groweth no grass, but a certain brush or heath, whereon the cattle feeding become very fat.

The Tartars never ride without their bow, arrows, and sword, although it be on hawking, or at any other pleasure, and they are good archers both on horse back, and on foot also. These people have not the use of gold, silver, or any other coin, but when they lack apparel or other necessaries, they barter their cattle for the same. Bread they have none, for they neither till nor sow: they be great devourers of flesh, which they cut in small pieces, and eat it by handfuls most greedily, and especially the horseflesh. Their chiefest drink is mare's milk soured, as I have said before. They eat their meat upon the ground, sitting with their legs double under them, and so also when they pray. Art or science have they none, but live most idly, sitting round in great companies in the fields, devising, and talking most vainly.

Thus proceeding in our journey, at night being at rest, and our watch set, there came unto us four horsemen, which we took as spies, from whom we took their weapons and bound them, and having well examined them, they confessed that they had seen the track of many horsemen, and no footing of camels, and gave us to understand, that there were rovers and thieves abroad. Whereupon we sent a post to the Sultan of Kayte, who immediately came himself with 300 men, and met these four suspected men which we sent unto him, and examined them so straitly, and threatened them in such sort, that they confessed, there was a banished prince with 40 men 3 days' journey forward, who lay in wait to destroy us, if he could, and that they themselves were of his company.

The sultan therefore understanding that the thieves were not many, appointed us 80 men well armed with a captain to go with us, and conduct us in our way. And the sultan himself returned back again, taking the four thieves with him. These soldiers travelled with us two days, consuming much of our victuals. They set out before our caravan, and having ranged the wilderness for the space of four hours, they met us, coming towards us as fast as their horses could run, and declared that they had found the tracks of horses not far from us, perceiving well that we should meet with enemies, and therefore willed us to appoint ourselves for them, and asked us what we would give them to conduct us further, or else they would return. To whom we offered as we thought good, but they refused our offer, and would have more, and so we not agreeing they departed from us, and went back to their sultan, who (as we conjectured) was privy to the conspiracy. Within 3 hours after that the soldiers departed from us, which was the 15th day of December, in the morning, we escried far off divers horsemen which made towards us, and we (perceiving them to be rovers) gathered ourselves together, being 40 of us well appointed, and able to fight, and we made our prayers together every one after his law, professing to live and die one with another, and so prepared ourselves. When the thieves were nigh unto us, we perceived them to be in number 37 men well armed, and appointed with bows, arrows and swords, and the captain a prince banished from his country. They willed us to yield our selves, or else to be slain, but we defied them, wherewith they shot at us all at once, and we at them very hotly, and so continued our fight from morning until two hours within night, divers men, horses and camels being wounded and slain on both parts: and had it not been for 4 hand guns which I and my company had and used, we had been overcome and destroyed: for the thieves were better armed, and were also better archers than we; but after we had slain divers of their men and horses with our guns, they durst not approach so nigh, which caused them to come to a truce with us until the next morning, which we accepted, and encamped ourselves upon a hill, and made the fashion of a castle, walling it about with packs of wares, and laid our horses and camels within the same to save them from the shot of arrows: and the thieves also encamped within an arrow shot of us, but they were betwixt us and the water, which was to our great discomfort, because neither we nor our camels had drunk in two days before.

When half the night was spent, the prince of the thieves sent a messenger half way unto us, requiring to talk with our captain. The message was pronounced aloud in this order, our prince demandeth that you deliver into his hands as many unbelievers (meaning us the Christians) as are among you with their goods, and in so doing, he will suffer you to depart with your goods in quietness. To the which our caravan pasha answered, that he had no Christians in his company, nor other strangers, but two Turks, which were of their law, and although he had, we would rather die than deliver them, and that we were not afraid of his threatenings, and that should he know when day appeared.

When the night was spent, in the morning we prepared ourselves to battle again: which the thieves perceiving, required to fall to agreement and asked much of us: and to be brief, the most part of our company being loth to go to battle again, and having little to lose, and safeconduct to pass, we were compelled to agree, and to give the thieves 20 ninths (that is to say 20 times 9 several things), and a camel to carry away the same, which being received, the thieves departed into the wilderness to their old habitation, and we went on our way forward.

Upon the 23rd day of December we arrived at the city of Bokhara in the land of Bactria. This Bokhara is situated in the lowest part of all the land, walled about with a high wall of earth, with divers gates into the same: it is divided into 3 partitions, whereof two parts are the king's, and the third part is for merchants and markets, and every science hath their dwelling and market by themselves. The city is very great, and the houses for the most part of earth, but there are also many houses, temples and monuments of stone sumptuously builded, and gilt, and specially bath stoves so artificially built, that the like thereof is not in the world. There is a little river running through the middle of the said city, but the water thereof is most unwholesome. And yet it is there forbidden to drink any other thing than water, and mare's milk, and whosoever is found to break that law is whipped and beaten most cruelly through the open markets, and there are officers appointed for the same, who have authority to go into any man's house, to search if he have either aquavitae, wine, or brage, and finding the same, do break the vessels, spoil the drink, and punish the master of the house most cruelly. If they perceive but by the breath of a man that he hath drunk, without further examination he shall not escape their hands.

This country of Bokhara was sometime subject to the Persians, and do now speak the Persian tongue, but yet now it is a kingdom of itself, and hath most cruel wars continually with the Persians, about their religion, although they be all Mahometists. One occasion of their wars is, for that the Persians will not cut the hair off their upper lips, as the Bokharans and all other Tartars do, which they account great sin.

The king of Bokhara hath no great power or riches, his revenues are but small, and he is most maintained by the city: for he taketh the tenth penny of all things that are there sold, as well by the craftsmen as by the merchants, to the great impoverishment of the people, whom he keepeth in great subjection, and when he lacketh money, he sendeth his officers to the shops of the said merchants to take their wares to pay his debts, and will have credit of force, as the like he did to pay me certain money that he owed me for 19 pieces of kersey. Their money is silver and copper, for gold there is none current: they have but one piece of silver, and that is worth 12 pence English, and the copper money are called pooles, and 120 of them goeth the value of the said 12 pence, and is more common payment than the silver, which the king causeth to rise and fall to his most advantage every other month.

I was commanded to come before the king, to whom I presented the Emperor of Russia his letters, who entertained us most gently, and caused us to eat in his presence, and devised with me familiarly as well of the power of the Emperor, and the Great Turk, as also of our countries, laws and religion, and caused us to shoot in hand guns before him, and did himself practise the use thereof. But after all this great entertainment before my departure he showed himself a very Tartar: for he went to the wars owing me money, and saw me not paid before his departure. And although indeed he gave order for the same, yet was I very ill satisfied, and forced to rebate part, and to take wares as payment for the rest contrary to my expectation: but of a beggar better payment I could not have, and glad I was so to be paid and dispatched.

But yet I must needs praise and commend this barbarous king, who immediately after my arrival at Bokhara, having understood our trouble with the thieves, sent 100 men well armed, and gave them great charge not to return before they had either slain or taken the said thieves. Who according to their commission ranged the wilderness in such sort, that they met with the said company of thieves, and slew part, and part fled, and four they took and brought unto the king, and two of them were sore wounded in our skirmish with our guns: and after that the king had sent for me to come to see them, he caused them all 4 to be hanged at his palace gate, because they were gentlemen, to the example of others. And of such goods as were gotten again, I had part restored me, and this good justice I found at his hands.

There is yearly great resort of merchants to this city of Bokhara, which travel in great caravans from the countries thereabouts adjoining, as India, Persia, Balkh, Russia, with divers others, and in times past from Cathay, when there was passage: but these merchants are so beggarly and poor, and bring so little quantity of wares, lying two or three years to sell the same, that there is no hope of any good trade there to be had worthy the following.

The chief commodities that are brought thither out of these foresaid countries, are these following.

The Indians do bring fine whites, which the Tartars do all roll about their heads, and all other kinds of whites, which serve for apparel made of cotton wool and crasko, but gold, silver, precious stones, and spices they bring none. I inquired and perceived that all such trade passeth to the ocean sea, and the vents where all such things are gotten are in the subjection of the Portuguese.

To speak of the said country of Cathay, and of such news as I have heard thereof, I have thought it best to reserve it to our meeting. I having made my solace at Bokhara in the winter time, and having learned by much inquisition, the trade thereof, as also of all the other countries thereto adjoining, and the time of the year being come, for all caravans to depart, I thought it good and meet, to take my journey some way, and determined to have gone from thence into Persia, and to have seen the trade of that country. Great wars did newly begin between the Sophy, and the kings of Tartary, whereby the ways were destroyed: and there was a caravan destroyed with rovers and thieves, which came out of India and Persia, by safe conduct: and about ten days' journey from Bokhara, they were robbed, and a great part slain. Also the Metropolitan of Bokhara, who is greater than the king, took the Emperor's letters of Russia from me, without which I should have been taken slave in every place: for which causes, and divers others, I was constrained to come back again to the Mare Caspium, the same way I went: so that the eighth of March 1559, we departed out of the said city of Bokhara, being a caravan of 600 camels: and if we had not departed when we did, I and my company had been in danger to have lost life and goods. For ten days after our departure, the king of Samarkand came with an army, and besieged the said city of Bokhara.

The 23rd of April, we arrived at the Mare Caspium again, where we found our bark which we came in, but neither anchor, cable, nor sail: nevertheless we brought hemp with us, and spun a cable ourselves, with the rest of our tackling, and made us a sail of cloth or cotton wool, and rigged our bark as well as we could, but boat or anchor we had none. In the meantime being devising to make an anchor of wood of a cart wheel, there arrived a bark, which came from Astrakhan, with Tartars and Russians, which had two anchors, with whom I agreed for the one: and thus being in readiness, we set sail and departed, I, and the two Johnsons being master and mariners ourselves, having in our bark the said six ambassadors, and 25 Russians, which had been slaves a long time in Tartary, nor ever had before my coming, liberty, or means to get home, and these slaves served to row when need was. Thus sailing sometimes along the coast, and sometimes out of sight of land, the 13th day of May, having a contrary wind, we came to an anchor, being three leagues from the shore, and there rose a sore storm, which continued 44 hours, and our cable being of our own spinning, broke, and lost our anchor, and being off a lee shore, and having no boat to help us, we hoisted our sail, looking for present death: but as God provided for us, we ran into a creek full of ooze, and so saved ourselves with our bark, and lived in great discomfort for a time. For although we should have escaped with our lives the danger of the sea, yet if our bark had perished, we knew we should have been, either destroyed, or taken slaves by the people of that country, who live wildly in the field, like beasts, without house or habitation. Thus when the storm was ceased, we went out of the creek again: and went directly to the place where we rode, with our bark again, and found our anchor which we lost: whereat the Tartars much marvelled, how we did it. While we were in the creek, we made an anchor of wood of cart wheels, which we had in our bark, which we threw away, when we had found our iron anchor again. Within two days after, there arose another great storm, at the northeast, and we lay a try, being driven far into the sea, and had much ado to keep our bark from sinking, the billows were so great: but at the last, having fair weather, we took the sun, and knowing how the land lay from us, we fell with the river Yaik,[30] according to our desire, whereof the Tartars were very glad, fearing that we should have been driven to the coast of Persia, whose people were unto them great enemies.

Note, that during the time of our navigation, we set up the red cross of St George in our flags, for honour of the Christians, which I suppose was never seen in the Caspian Sea before. We passed in this voyage divers fortunes: notwithstanding the 28th of May we arrived in safety at Astrakhan.

As touching the trade of Shemakha in Media[31] and Tabriz, with other towns in Persia, I have inquired, and do well understand, that it is even like to the trades of Tartary, that is little utterance, and small profit: and I have been advertised that the chief trade of Persia is into Syria, and so transported into the Levant Sea. The tenth day of June, we departed from Astrakhan towards Moscow, having an hundred gunners in our company at the Emperor's charges, for the safe conduct of the Tartar Ambassadors and me. The second of September, we arrived at the city of Moscow, and the fourth day I came before the Emperor's majesty, kissed his hand, and presented him a white cow's tail of Cathay, and a drum of Tartary. And here I cease for this time, entreating you to bear with this my large discourse, which by reason of the variety of matter, I could make no shorter, and I beseech God to prosper all your attempts.

XIX

Certain notes gathered by Richard Johnson (which was at Bokhara with Mr Anthony Jenkinson) of the reports of Russians and other strangers, of the ways of Russia to Cathay.

First from Astrakhan by sea to Serachick[32] is 15 days. From Serachick to Bokhara 30 days journey is without habitation of houses: therefore travellers lodge in their own tents, carrying with them to eat, their several provisions: and for dryness there be many wells of fair water. From Bokhara to Tashkent easy travelling with goods, is 14 days by land. From Tashkent to Occient 7 days. From Occient[33] to Cascar[34] 20 days. Cascar to Sowchick[35] 30 days journey, which Sowchick is the first border of Cathay. This land of Cathay they praise to be civil and unspeakably rich, in Cathay, the most part thereof stretching to the sun rising, are people white and of fair complexion.

XX

A letter of Master Anthony Jenkinson upon his return from Bokhara to the worshipful Master Henry Lane Agent for the Moscovy company resident in Vologda, written in Moscow the 18th of September, 1559.

Worshipful Sir, it may please you to be advertised that the fourth of this present I arrived with Richard Johnson and Robert Johnson all in health, thanks be to God. We have been as far as Bokhara, and had proceeded farther on our voyage towards the land of Cathay, had it not been for the incessant and continual wars, which are in all these brutal and wild countries, that it is at this present impossible to pass, neither went there any caravan of people from Bokhara that way these three years. And although our journey hath been so miserable, dangerous, and chargeable with losses, charges and expenses, as my pen is not able to express the same: yet shall we be able to satisfy the worshipful company's minds, as touching the discovery of the Caspian Sea, with the trade of merchandise to be had in such lands and countries as be thereabout adjacent, and have brought of the wares and commodities of those countries able to answer the principal with profit: wishing that there were utterance for as great a quantity of kerseys and other wares as there is profit to be had in the sales of a small quantity, for then it would be a trade worthy to be followed.

I commend you to the tuition of God, who send you health with heart's desire.

Your assured to command,
ANTHONY JENKINSON.

XXI

The testimony of Gerardus Mercator in his last large map of Europe, touching the notable discoveries of the English, made of Moscovy by the northeast.

The most famous navigation of the Englishmen by the northeast sea hath offered unto me a great occasion, and certain direction for the reformation of the map of Europe: which discovery hath the northern parts of Finland, Lapland, and Moscovy, laid out according to the just elevation and the quarters of the world. And further, the true observation of the latitude of the city of Moscow, made by the foresaid Englishmen, hath yielded me an infallible rule, for the correcting of the situation of the inland countries: which notable helps being ministered unto me, I thought it my duty to exhibit to the world this map, more exact and perfect than hitherto it hath been published.

XXII

The journey of Mr Anthony Jenkinson, from the famous city of London into the land of Persia, sent and employed therein by the right worshipful Society of the Merchants Adventurers, for the discovery of Lands, Islands &c. 1561.

First embarking myself in a good ship of yours, named the *Swallow*, at Gravesend, having a fair and good wind, our anchor then weighed, and committing all to the protection of our God, on the fourteenth day of July, the year aforesaid I arrived in the bay of St Nicholas in Russia: I departed from thence, and on the eighth day of August then following, I came to Vologda, which is distant from Kholmogory, seven hundred miles, where I remained four days, attending the arrival of one of your boats, wherein was laden a chest of jewels with the present, by your worships appointed for the Emperor's Majesty: I therewith departed towards the city of Moscow, where I immediately caused my coming to be signified unto the Secretary of the Imperial Majesty, with the Queen's Highness's letters addressed unto the same His Majesty, who informed the Emperor thereof. But his Highness having great affairs, and being at that present ready to be married unto a Lady of Cherkesy,[36] of the Mahometical law, commanded that during the space of three days that the same solemn feast was celebrating, the gates of the city should be shut, and that no person, stranger or native should come out of their houses during the said triumph, the cause thereof unto this day not being known.

I was commanded to come with the said letters before His Majesty, and so delivered the same into his own hands (with such presents as by you were appointed) and the same day I dined in His Grace's presence, with great entertainment. Shortly after, I desired to know whether I should be licensed to pass through His Highness's dominions into the land of Persia, according to the Queen's Majesty's request: hereunto it was answered, that I should not pass thither, for that His Majesty meant to send an army of men that way into the land of Cherkesy, whereby my journey should be both dangerous and troublesome.

There remaining a good part of the year, having in that time sold the most part of your kerseys and other wares appointed for Persia, when the time of the year required to return for England, I desired passport, and post horses for money, which was granted: but having received my passport, ready to depart, there came unto our house there Osep Napea, who persuaded me that I should not depart that day, saying that the Emperor was not truly informed.

After many allegations and objections of things, and perceiving that I would depart, I was willed to remain until the Emperor's Majesty were spoken with again touching my passage: wherewith I was content, and within three days after sending for me, he declared that the Emperor's pleasure was, that I should not only pass through his dominions into Persia, but also have His Grace's letters of commendations to foreign princes, with certain of his affairs committed to my charge, too long here to rehearse.

The 15th of March, the year aforesaid, I dined again in His Majesty's presence in company of an ambassador of Persia and others, and receiving a cup of drink at His Majesty's hands, I took my leave of His Highness. I departed from the city of Moscow the 27th day of April 1562, down by the great river of Volga, in company of the said ambassador of Persia, with whom I had great friendship and conference all the way down the same river unto Astrakhan where we arrived all in health the 10th day of June. I repaired unto the captain there, unto whom I was commended from the Emperor's Majesty, with great charge that he not only should aid and succour me with all things needful during my abode there, but also to safeconduct me with 50 gunners well appointed in two stroogs or brigantines into the Caspian Sea, until I had passed certain dangerous places which pirates and rovers do accustom to haunt.

I and my company took our voyage from the said Astrakhan, and at a west by north sun we fell with the land called Challica Ostriva,[37] being four round islands together. From thence sailing the next day, we had sight of a land called Tuke, in the country of Tumen, where pirates and rovers do use: for fear of whom we hauled off into the sea due east forty miles, and fell upon shallows out of sight of land, and there were like to have perished, escaping most hardly: the wind being contrary, and a stiff gale, we were not able to seize it: but were forced to come to an anchor in three or four fathoms of water.

And so riding at two anchors a head, having no other provision, we lost one of them, the storm and sea being grown very sore, and thereby our bark was so full of leaks, that with continual pumping we had much ado to keep her above water, although we threw much of our goods overboard, with loss of our boat, and ourselves thereby in great danger like to have perished either in the sea or else upon the lee shore, where we should have fallen into the hands of those wicked infidels, who attended our shipwreck: and surely it was very unlike that we should have escaped both the extremities, but only by the power and mercy of God, for the storm continued seven days.

The third day of August, having a fair wind, at a southeast sun we arrived at a city called Derbent in the king of Hircania's [38] dominion, where coming to land, and saluting the captain there with a present, he made to me and my company a dinner, and there taking fresh water I departed. This city of Derbent is an ancient town having an old castle therein, builded all of freestone much after our building, the walls very high and thick, and was first erected by King Alexander the Great, when he warred against the Persians.

From thence sailing southeast and southsoutheast about 80 miles, the sixth day of August, the year aforesaid, we arrived at our landing place called Shabran, where my bark discharged: the goods laid on the shore, there being in my tent keeping great watch for fear of rovers, whereof there is great plenty, the governor of the said country coming unto me, entertained me very gently, unto whom giving a present, he appointed for my safeguard forty armed men to watch and ward me. The 12th day of the same month news did come from the king with order that I should repair unto him with all speed: and for expedition, as well camels to the number of five and forty to carry my goods, as also horses for me and my company were in readiness. On the 18th of the same month I came to a city called Shemakha, in the country called Hircania and there the king hath a fair place, where my lodging being appointed, the goods were discharged: the next day I was sent for to come to the king, who gently entertained me, and having kissed his hands, he bade me to dinner, and commanded me to sit down not far from him. This king did sit in a very rich pavilion, wrought with silk and gold, placed very pleasantly upon a hill side, of sixteen fathom long, and six fathom broad, having before him a goodly fountain of fair water: whereof he and his nobility did drink, he being a prince of mean stature, and of a fierce countenance, richly apparelled with long garments of silk, and cloth of gold, embroidered with pearls and stones: upon his head was a tolipane with a sharp end standing upwards half a yard long, of rich cloth of gold, wrapped about with a piece of Indian silk of twenty yards long, wrought with gold, and on the left side of his tolipane stood a plume of feathers, set in a trunk of gold richly enamelled, and set with precious stones: his earrings had pendants of gold a handful long, with two great rubies of great value, set in the ends thereof: all the ground within his pavilion was covered with rich carpets, and under himself was spread a square carpet wrought with silver and gold, and thereupon was laid two suitable cushions. Thus the king with his nobility sitting in his pavilion with his legs across, and perceiving that it was painful for me so to sit, his highness caused a stool to be brought in, and did will me to sit thereupon, after my fashion.

Dinner time then approaching, divers cloths were spread upon the ground, and sundry dishes served, and set in a rank with divers kinds of meats, to the number of 140 dishes, as I numbered them, which being taken away with the table cloths, and others spread, a basket of fruits of sundry kinds, with other banqueting meats to the number of 150 dishes, were brought in: so that two services occupied 290 dishes. The king said unto me, Quoshe quelde, that is to say, Welcome: then he proposed unto me sundry questions, both touching religion, and also the state of our countries, and further questioned whether the Emperor of Almaine, the Emperor of Russia, or the Great Turk, were of most power, to whom I answered as I thought most meet. Then he demanded whether I intended to go any further, and the cause of my coming: unto that I answered, that I was sent with letters from the Queen's most excellent Majesty of England unto the Great Sophy, to entreat friendship and free passage, and for his safeconduct to be granted unto English merchants to trade into his seigneuries, with the like also to be granted to his subjects, when they should come into our countries, to the honour and wealth of both realms, and commodity of both their subjects. The four and twentieth day of August he sent for me again: unto whom I repaired in the morning, and the king not being risen out of his bed (for his manner is, that watching in the night, and then banqueting with his women, being an hundred and forty in number, he sleepeth most in the day) did give one commandment that I should ride on hawking with many gentlemen of his court, and that they should show me so much game and pastime as might be: which was done, and many cranes killed. We returned from hawking about three of the clock at the afternoon: the king then risen, and ready to dinner, I was invited thereunto, and approaching near to the entering in of his tent, and being in his sight, two gentlemen encountered me with two garments of that country's fashion, side, down to the ground, the one of silk, and the other of silk and gold, sent unto me from the king, and after that they caused me to put off my upper garment, being a gown of black velvet furred with sables, they put the said two garments upon my back, and so conducted me unto the king, before whom doing reverence, and kissing his hand, he commanded me to sit not far from him, and so I dined in his presence, he at that time being very merry, and demanding of me many questions, and amongst others, how I liked the manner of their hawking. Dinner so ended, I required his highness's safeconduct for to depart towards the Sophy, who dismissing me with great favour, and appointing his ambassador to safeconduct me, he gave me at my departure a fair horse with all furniture, and custom free from thence with all my goods. This country of Hircania, now called Shirvan in times past was of great renown, having many cities, towns and castles in it: but now it is otherwise, the king is subject to the said Sophy who conquered them not many years past, and caused not only all the nobility and gentlemen of that country to be put to death, but also over and besides, razed the walls of the cities, towns, and castles of the said realm, to the intent that there should be no rebellion.

Another city called Arrash[39] bordering upon the Georgians, the chiefest and most opulent in the trade of merchandise, and thereabouts is nourished the most abundant growth of raw silk, and thither the Turks, Syrians, and other strangers do resort and traffic. There be also divers good and necessary commodities to be provided and had in this said realm: viz. galls, rough and smooth, cotton wool, alum, and raw silk of the natural growth of that country: besides, near all kinds of spices and drugs, and some other commodities, which are brought thither from out of East India, but in the lesser quantity, for that they be not assured to have vent or utterance of the same: but the chiefest commodities be there, raw silks of all sorts, whereof there is great plenty. Not far from the said city there was an old castle called Gullistone. And not far from the said castle was a nunnery of sumptous building, wherein was buried a king's daughter, named Ameleck Channa, who slew herself with a knife, for that her father would have forced her (she professing chastity) to have married with a king of Tartary: upon which occasion the maidens of that country do resort thither once every year to lament her death.

The 6th of October I with my company departed and having journeyed threescore miles, came to a town called Yavate, wherein the king hath a fair house, with orchards and gardens well replenished with fruits of all sorts. Now passing ten days' journey, coming into no town or house, the sixteenth day of October we arrived at a city called Ardabil, in the latitude of eight and thirty degrees, an ancient city in the province of Azerbaijan, wherein the princes of Persia are commonly buried: and there Alexander the Great did keep his court when he invaded the Persians. Four days' journey to the westward is the city Tabriz in old time called Tauris, the greatest city in Persia, but not of such trade of merchandise as it hath been, by mean of the great invasion of the Turk. The 21st day we departed travelling for the most part over mountains all in the night season, and resting in the day, being destitute of wood, and therefore were forced to use for fuel the dung of horses and camels, which we bought dear of the pasturing people. The second day of November we arrived at Qazvīn,[40] where the Sophy keepeth his court, and were appointed to a lodging not far from the king's palace, and within two days after the Sophy commanded a prince to send for me to his house, who asked me in the name of the said Sophy how I did, and whether I were in health, and after did welcome me, and invited me to dinner, whereat I had great entertainment, and so from thence I returned to my lodging. At this time, the Great Turk's ambassador arrived four days before my coming, who was sent thither to conclude a perpetual peace betwixt the same Great Turk and the

V.D.H–5

Sophy, and brought with him a present in gold, and fair horses with rich furnitures, and other gifts, esteemed to be worth forty thousand pounds. And thereupon a peace was concluded with joyful feasts, triumphs and solemnities, corroborated with strong oaths.

Discoursing at my first arrival with the king of Shirvan of sundry matters, the king demanding whether that we of England had friendship with the Turks or not: I answered, that we never had friendship with them, and that therefore they would not suffer us to pass through their country into the Sophy his dominions, and that there is a nation named the Venetians, not far distant from us, which are in great league with the said Turks, who trade into his dominions with our commodities, chiefly to barter the same for raw silks, which come from thence: and that if it would please the said Sophy and other princes of that country to suffer our merchants to trade into those dominions, and to give us passport and safeconduct for the same, I doubted not but that it should grow to such a trade to the profit of them as never before had been the like, and that they should be both furnished with our commodities, and also have utterance of theirs.

This king understanding the matter liked it marvellously, saying, that he would write unto the Sophy concerning the same: as he did in very deed, assuring me that the Sophy would grant my request, and that at my return unto him he would give me letters of safeconduct, and privileges. The Turk's Ambassador was not then come into the land, but it chanced otherwise. For the Turk's Ambassador being arrived and the peace concluded, the Turkish merchants there declared to the same Ambassador, that my coming thither (naming me by the name of Frank) would in great part destroy their trade, and that it should be good for him to persuade the Sophy not to favour me as His Highness meant to observe the league and friendship with the Great Turk his master.

The 20th day of November aforesaid, I was sent for to come before the Sophy, otherwise called Shah Thomas, and about three of the clock at afternoon I came to the court, and in lighting from my horse at the court gate, before my feet touched the ground, a pair of the Sophy's own shoes, such as he himself weareth when he ariseth in the night to pray were put upon my feet, for without the same shoes I might not be suffered to tread upon his holy ground, being a Christian, and called amongst them *Gaiour*, that is unbeliever, and unclean. At the court gate the things that I brought to present His Majesty with, were divided by sundry parcels to sundry servitors of the court, to carry before me, for none of my company or servants might be suffered to enter into the court with me, my interpreter only excepted. Thus coming before His Majesty with such reverence as I thought meet to be used, I delivered the Queen's Majesty's letters with my present, which he accepting, demanded me of what country of Franks I was, and what affairs I had there to do? Unto whom I answered that I was of the famous City of London within the noble realm of England, and that I was sent thither from the most excellent and gracious sovereign Lady Elizabeth Queen of the said realm for to treat of friendship, and free passage of our merchants and people, to repair and traffic within his dominions, for to bring in our commodities, and to carry away theirs to the honour of both princes, the mutual commodity of both realms, and wealth of the subjects. He then demanded me in what language the letters were written, I answered, in the Latin, Italian and Hebrew: well said he, we have none within our realm that understand those tongues. Whereupon I answered that such a famous and worthy prince (as he was) wanted not people of all nations within his large dominions to interpret the same. Then he questioned with me of the state of our countries, and of the power of the Emperor of Almaine, King Philip, and the Great Turk, and which of them was of most power: whom I answering to his contentation, not dispraising the Great Turk, their late concluded friendship considered. Then he reasoned with me much of religion, demanding whether I were an unbeliever, or a Muslim, that is of Mahomet's law. Unto whom I answered, that I was neither an unbeliever nor Mahometan, but a Christian. What is that, said he unto the King of the Georgians' son, who being a Christian was fled unto the said Sophy, and he answered that a Christian was he that believeth in Jesus Christ, affirming him to be the Son of God, and the greatest Prophet. Doest thou believe so, said the Sophy unto me? Yea that I do, said I: Oh thou unbeliever, said he, we have no need to have friendship with the unbelievers, and so willed me to depart. I being glad thereof did reverence and went my way, being accompanied with many of his gentlemen and others, and after me followed a man with a bassinet of sand, sifting all the way that I had gone within the said palace, even from the said Sophy's sight unto the court gate. Thus I repaired unto my lodging.

Thus I continued for a time, daily resorting unto me divers gentlemen sent by the Sophy to confer with me, especially touching the affairs of the Emperor of Russia, and to know by what way I intended to return into my country, either by the way that I came, or by the way of Ormuz, and so with the Portuguese ships. Unto whom I answered, that I durst not return by the way of Ormuz, the Portuguese and we not being friends, fully perceiving their meaning: for I was advertised that the said Sophy meant to have wars with the Portuguese, and would have charged me that I had been come for a spy to pass through his dominions unto the said Portuguese, thinking them and us to be all one people, and calling all by the name of Franks, but by the providence of God this was prevented.

After this the said Sophy conferred with his nobility and council concerning me, who persuaded that he should not entertain me well, neither dismiss me with letters or gifts, considering that I was a Frank, and of that nation that was enemy to the Great Turk his brother. But the king of Hircania's son, understanding this deliberation, sent a man in post unto his father, for to declare and impart the purpose unto him, who as a gracious prince, did write to the Sophy that it should not stand with His Majesty's honour to do me any harm or displeasure, but rather to give me good entertainment, seeing I was come into his land of my free will, and not by constraint, and that if he used me evil, there would be few strangers resort into his country, which would be greatly unto his hindrance, which after that the said Sophy had well and thoroughly pondered and digested, changed his determined purpose, and the twentieth of March 1562 he sent to me a rich garment of cloth of gold, and so dismissed me without any harm.

During the time that I sojourned at the said city of Qazvīn, divers merchants out of India came thither unto me, with whom I conferred for a trade of spices: whereunto they answered that they would bring of all sorts, if they were sure of vent, whereof I did promise to assure them. But before I proceed any further to speak of my return, I intend somewhat to treat of the country of Persia, of the Great Sophy, and of his country, laws and religion.

This land of Persia is great and ample, divided into many kingdoms and provinces. Every province hath his several king, or sultan, all in obedience to the Great Sophy. The country for the most part toward the sea side is plain and full of pasture, but into the land, high, full of mountains, and sharp. To the south it bordereth upon Arabia and the East Ocean. To the north upon the Caspian Sea and the lands of Tartary. To the east upon the provinces of India, and to the west upon the confines of Chaldea, Syria, and other the Turk's lands. This Sophy that now reigneth is nothing valiant, although his power be great, and his people martial: and through his pusillanimity the Turk hath much invaded his countries. This prince is of the age of fifty years, and of a reasonable stature, having five children. His eldest son he keepeth captive in prison, for that he feareth him for his valiantness and activity. These persons are comely and of good complexion, proud and of good courage, esteeming themselves to be best of all nations, both for their religion and holiness, which is most erroneous, and also for all other their fashions. They be martial, delighting in fair horses and good harness, soon angry, crafty and hard people.

My bark being ready at the Caspian Sea, having a fair wind, and committing ourselves unto God the 30th day of May 1563, we arrived at Astrakhan, having passed no less dangers upon the sea in our return, than we sustained in our going forth, and remaining at the said Astrakhan, until the tenth day of June, one hundred gunners being there admitted unto me for my safeguard up the River Volga, the fifteenth of July I arrived at the city of Kazan. I was conducted from place to place unto the city of

Moscow, where I arrived the twentieth day of August 1563 in safety, thanks be to God, with all such goods, merchandises, and jewels, all which goods I was commanded to bring into the Emperor's treasury before it was opened, which I did, and delivered those parcels of wares which were for His Majesty's account, videlicet, precious stones, and wrought silks of sundry colours and sorts, much to his highness's contentation, and the residue belonging to you, viz. crasko, and raw silks, with other merchandises (as by account appeareth), were brought unto your house, whereof part there remained, and the rest was laden in your ships lately returned.

Coming to Kholmogory and so down to the sea side, I found your ships laden and ready to depart, where I embarked myself in your good ship called the *Swallow*, the 9th of July, one thousand five hundred sixty four, and having passed the seas with great and extreme dangers of loss of ship, goods and life, the 28th day of September last (God be praised) we arrived here at London in safety.

XXIII

The voyage into Persia made by Tho. Alcock, who was slain there, and by George Wrenne, and Ric. Cheinie servants to the worshipful company of the Muscovy merchants in 1563 written by Richard Cheinie.

In the year 1563 I was appointed by Mr Antho. Jenkinson, and Mr Thomas Glover your Agent in Russia, to go for Persia in your worship's affairs, one Thomas Alcock having the charge of the voyage committed to him. The 10 of May anno 1563 we departed from a town called Yaroslavl[41] upon our voyage towards Persia. The 24 of July we arrived at Astrakhan : and the 21 of August we arrived at Shemakha, whereas the king Obdolocan lay in the field. The third day after our arrival at Shemakha we were called before the king: we gave him a present, and he entertained us very well.

We were commanded to come before the king, who sat in his tent upon the ground with his legs across, and all his dukes round about his tent, the ground being covered with carpets: we were commanded to sit down, the King appointing every man his place to sit. And the king commanded the Emperor of Russia's merchants to rise up, and to give us the upper hand. The 20 of October Thomas Alcock departed from Shemakha towards Qazvīn, leaving me at Shemakha to recover such debts as the dukes of Shemakha owed for wares which they took of him. In the time I lay there I could recover but little. And at Thomas Alcock's coming from Qazvīn, I hearing of his arrival there, departed from Shemakha, finding him there in safety with all such goods as he had with him. During his abode there for seven days he made suit to the king for such money as the dukes owed him. But the king was displeased. Thomas Alcock seeing the king would show us no favour, willed me to depart to Shemakha with all such goods as he had brought with him from Qazvīn, I leaving him at the court.

The third day after mine arrival at Shemakha, I had news that Thomas Alcock was slain coming on his way towards me. When this fell out, your worships had no other servant there but me among those heathen people. Who having such a sum of goods lying under my hands, and seeing how the Russians sent their goods with as much haste as they might to the sea side, and having but four men to send our wares to the sea side, I used such diligence, that within two days after Thomas Alcock was slain, I sent all your worships' goods with a mariner, William August, and a Swede, for that they might the safer arrive at the sea side, being safely laid in. All which goods afterwards arrived in Russia in good condition, Master Glover having the receipt. I remained after I had sent the goods into Russia six weeks in Shemakha, for the recovery of such debts as were owing, and at last with much trouble recovered to the sum of fifteen hundred roubles or there about, which Mr Glover received of me at my coming to Moscow.

I have sown the seed, and other men have gathered the harvest:

I have travelled both by land and by water full many a time with a sorrowful heart, as well for the safeguard of their goods as yours, how to frame all things to the best, and they have reaped the fruits of my travail. But ever my prayer was to God, to deliver me out of those miseries which I suffered for your service among those heathen people. Therefore knowing my duty which I have done, as a true servant ought to do I beseech your worships (although I have but small recompense for my service,) yet let me have no wrong, and God will prosper you the better.

Also, to inform your worships of your Persian voyage what I judge: it is a voyage to be followed. The king of Gilan,[42] whereas yet you have had no traffic, liveth all by merchandise: and it is near Qazvīn, and not past six weeks travel from Ormuz, whither all the spices be brought: and here (I mean at Gilan), a trade may be established: but your worships must send such men as are no riotous livers, nor drunkards. For if such men go, it will be to your dishonour and great hindrance, as appeared by experience the year 1565 whenas Richard Johnson went to Persia, whose journey had been better stayed than set forward. His vicious living there hath made us to be counted worse than the Russians.

Again, if such men travail in your affairs in such a voyage, you shall never know what gain is to be gotten. For how can such men employ themselves to seek the trade, that are inclined to such vices? Or how can God prosper them in your affairs? But when a trade is established by wise and discreet men, then will it be for your worships to traffic there, and not before: for a voyage or market made evil at the first, is the occasion that your worships shall never understand what gain is to be gotten thereby hereafter.

XXIV

The voyage made by Mr John Hawkins to the coast of Guinea and the Indies of Nova Hispania, 1564.

Master John Hawkins with the *Jesus* of Lubeck, a ship of 700 and the *Salomon* a ship of 140 the *Tiger* a bark of 50 and the *Swallow* of 30 tons, being all well furnished with men to the number of one hundred threescore and ten, as also with ordnance and victual requisite for such a voyage, departed out of Plymouth the 18 day of October, in the year of our Lord 1564 with a prosperous wind.

The fourth of November they had sight of the island of Madeira, and the sixth day of Tenerife, which they thought to have been the Canary. To speak somewhat of these islands being called in old time *insulae fortunatae*, by the means of the flourishing thereof, the fruitfulness of them doth surely exceed far all other: for they make wine better than any in Spain: for sugar, sweets, raisins of the sun, and many other fruits, abundance: for resin and raw silk, there is great store: they have many camels also, which being young, are eaten of the people for victuals, and being old, they are used for carriage of necessaries: whose property is as he is taught to kneel at the taking of his load, and unloading again: his nature is to engender backwards contrary to other beasts: of shape very deformed, with a little belly, long misshapen legs, and feet very broad. This beast liveth hardly, and is contented with straw and stubble, but of force strong, being well able to carry five hundredweight.

The 25 he came to Cabo Blanco,[43] which is upon the coast of Africa, and a place where the Portuguese do ride, that fish there in the month of November especially, and is a very good place of fishing, for mullet, and dogfish. The people of that part of Africa are tawny, having long hair without any apparel, saving before their privy members. Their weapons in wars are bows and arrows.

The 29 we came to Cape Verde. These people are all black, and are called negroes, without any apparel, saving before their privities: of stature goodly men.

The two and twentieth [of December] the captain went into the river, called Callowsa,[44] with the two barks, and the *John*'s pinnace, and the *Salomon*'s boat, leaving at anchor in the river's mouth the two ships, the river being twenty leagues in, where the Portuguese rode, and dispatched his business, and so returned with two caravels, laden with negroes.

The captain was advertised by the Portuguese of a town of the negroes, where was not only great quantity of gold, but also that there were not above forty men, and an hundred women and children in the town, so that he might get an hundred slaves: he determined to stay before the town three or four hours, to see what he could do: and thereupon prepared his men in armour and weapon together, to the number of forty men well appointed, having to their guides certain Portuguese: we landing boat after boat, and divers of our men scattering themselves, contrary to the captain's will, by one or two in a company, for the hope that they had to find gold in their houses, ransacking the same, in the meantime the negroes came upon them, and hurt many being thus scattered, whereas if five or six had been together, they had been able, as their companions did, to give the overthrow to 40 of them. While this was doing the captain who with a dozen men, went through the town, returned, finding 200 negroes at the water's side, shooting at them in the boats, and cutting them in pieces which were drowned in the water. Thus we returned back somewhat discomforted, although the captain in a singular wise manner carried himself, with countenance very cheerful outwardly: having gotten by our going ten negroes, and lost seven of our best men, and we had 27 of our men hurt.

We departed with all our ships from Sierra Leone, towards the West Indies, and for the space of eighteen days, we were becalmed, having now and then contrary winds, which happened to us very ill, being but reasonably watered, for so great a company of negroes, and ourselves, which pinched us all, and that which was worst, put us in such fear that many never thought to have reached to the Indies, without great death of negroes and of themselves: but the Almighty God, who never suffereth his elect to perish, sent us the sixteenth of February, the ordinary breeze, which is the northwest wind, which never left us, till we came to an island of the cannibals, called Dominica, where we arrived the ninth of March, upon a Saturday. The cannibals of that island, and also others adjacent are the most desperate warriors that are in the Indies, by the Spaniards' report, who are never able to conquer them, and they are molested by them not a little, when they are driven to water there in any of those islands: of very late, not two months past, in the said island, a caravel being driven to water, was in the night set upon by the inhabitants, who cut their cable in the hawser, whereby they were driven ashore, and so taken by them, and eaten.

We came to a place in the main called Cumana,[45] whither the captain going in his pinnace, spake with certain Spaniards, of whom he demanded traffic, but they made him answer, they were but soldiers newly come thither, and were not able to buy one negro: whereupon he asked for a watering place, and they pointed him a place two leagues off, called Santa Fe, where we found marvellous good watering. Near about this place, inhabited certain Indians, who the next day after we came thither, came down to us, presenting mill and cakes of bread, which they had made of a kind of corn called maize, in bigness of a pease, the ear whereof is much like to a teasel, but a span in length, having thereon a number of grains. Also they brought down to us hens, potatoes and pines, which we bought for beads, pewter whistles, glasses, knives, and other trifles.

These potatoes be the most delicate roots that may be eaten, and do far exceed our parsnips or carrots. These Indians being of colour tawny like an olive, having every one of them both men and women hair all black, and no other colour, neither men nor women suffering any hair to grow in any part of their body, but daily pull it off as it groweth. They go all naked, the men covering no part of their body but their yard, upon the which they wear a gourd or piece of cane, made fast with a thread about their loins, leaving the other parts of their members uncovered, whereof they take no shame. The women also are uncovered, saving with a cloth which they wear a hand-breadth, wherewith they cover their privities both before and behind. These men carry every man his bow and arrows, whereof some arrows are poisoned for wars, which they keep in a cane together: the experience whereof we saw not once or twice, but daily, for they are so good archers that the Spaniards for fear thereof arm themselves and their horses with quilted canvas of two inches thick, and leave no place of their body open to their enemies, saving their eyes which they may not hide, and yet oftentimes are they hit in that so small a scantling: their poison is of such a force, that a man being stricken therewith dieth within four and twenty hours.

The Indian women delight not when they are young in bearing of children, because it maketh them have hanging breasts which they account to be great deforming in them, and upon that occasion while they be young, they destroy their seed, saying that it is fittest for old women. Moreover, when they are delivered of child, they go straight to wash themselves, without making any further ceremony for it, not lying in bed as our women do. The beds which they have are made of cotton, and wrought artificially of divers colours, which they carry about with them when they travel, and making the same fast to two trees, lie therein they and their women.

We passed between the mainland, and the island called Tortuga,[46] and sailed along the coast, the captain saw many Caribs on shore, and some also in their canoes, which made tokens unto him of friendship, and showed him gold, meaning thereby that they would traffic for wares. Whereupon he stayed to see the manners of them, and so for two or three trifles they gave such things as they had about them, and departed: but the Caribs were very importunate to have them come on shore, which if it had not been for want of wares to traffic with them, he would not have denied them, because the Indians which he saw before were very gentle people. These were no such kind of people as we took them to be, but more devilish and are eaters and devourers of any man they can catch. A Captain General sent by the King sailing along in his pinnace, as our captain did to descry the coast, was by the Caribs called ashore, and gold showed as though they desired traffic, with the which the Spaniard, suspecting no deceit at all, went ashore amongst them: who was no sooner ashore, but with four or five more was taken, the rest of his company being invaded by them, saved themselves by flight, but they that were taken, paid their ransom with their lives, and were presently eaten. And this is their practice to toll with their gold the ignorant to their snares. Their policy in fight with the Spaniards is marvellous: for they choose for their refuge the mountains and the woods where the Spaniards with their horses cannot follow them, and if they fortune to be met in the plain where one horseman may overrun 100 of them, they have devised to pitch stakes of wood in the ground, and also small iron pikes to mischief their horses. They have more abundance of gold than all the Spaniards have, and live upon the mountains where the mines are in such number, that the Spaniards have much ado to get any of them.

We kept our course along the coast, and came to a town called Burburata, where his ships came to an anchor, and he himself went ashore to speak with the Spaniards, to whom he declared himself to be an Englishman, and came thither to trade with them by the way of merchandise, and therefore required licence for the same. They were contented he should bring his ships into harbour, and there they would deliver him any victuals he would require. The next day divers of them came to cheapen, but could not agree of price, because they thought the price too high. Whereupon the captain perceiving they went about to bring down the price, did send for the principals of the town, and made a show he would depart, declaring himself to be very sorry that he had so much troubled them, seeing now his pretence was to depart, whereat they asked him what cause moved him thereunto, seeing by their working he was in possibility to have his licence.

To the which he replied that it was not only a licence that he sought, but profit, which he perceived was not there to be had, and therefore would seek further, and withall showed him his writings what he paid for his negroes, declaring also the great charge he was at in his shipping, and men's wages, and therefore to countervail his charges, he must sell his negroes for a greater price than they offered. So they put him in comfort to sell better there. And if it fell out that he had no licence that he should not lose his labour in tarrying, for they would buy without licence. Whereupon the captain promised them to stay, so that he might make sale of his lean negroes, which they granted unto. And the next day did sell some of them.

The captain perceiving that they would neither come near his price he looked for by a great deal, nor yet would abate the King's custom of that they offered, so that either he must be a great loser by his wares, or else compel the officers to abate the same King's custom which was too unreasonable: therefore, the sixteenth of April he prepared one hundred men well armed with bows, arrows, arquebuses and pikes, with which he marched to the townwards, and being perceived by the Governor, he straight with all expedition sent messengers to know his request, desiring him to march no further forward until he had answer again, which incontinent he should have. So our captain declaring how unreasonable a thing the King's custom was, requested to have the same abated, and to pay seven and a half per centum, which is the ordinary custom for wares through his dominions there, and unto this if they would not grant, he would displease them.

Having ended our traffic here we departed the night before the which the Caribs, whereof I have made mention before, being to the number of 200 came in their canoes to Burboroata, intending by night to have burned the town, and taken the Spaniards, who being more vigilant because of our being there, perceiving them coming, raised the town and took one, but the rest escaped away. But this one had for his travail a stake thrust through his fundament, and so out at his neck.

We came to an island called Curaçao. In this place we had traffic for hides, and found great refreshing both of beef, mutton and lambs, whereof there was plenty. The increase of cattle in this island is marvellous, which from a dozen of each sort brought thither by the governor, in 25 years he had a hundred thousand at the least, and of other cattle was able to kill without spoil of the increase 1500 yearly, which he killeth for the skins, and of the flesh saveth only the tongues, the rest he leaveth to the fowl to devour.

The captain sailing by the shore in the pinnace, came to the Rancheria, a place where the Spaniards used to fish for pearls, and there spoke with a Spaniard, who told him how far off he was from Rio de la Hacha: where having talked with the King's Treasurer of the Indies resident there, he declared his quiet traffic in Burboroata and desired to have the like there also: but the Treasurer made answer that they were forbidden by the Viceroy, who having intelligence of our being on the coast, did send express commission to resist us, with all the force they could. Our captain replied, that he was in an armada of the Queen's Majesty's of England, and sent about other her affairs, but enforced by contrary winds to come into those parts, where he hoped to find such friendship as he should do in Spain, in that there was amity betwixt their princes. But seeing that they would contrary to all reason go about to withstand his traffic, having the force that he hath, he therefore willed them to determine either to give him a licence to trade, or else to stand to their own harms: so upon this it was determined he should have licence to trade, but they would give him such a price as was the one half less than he had sold for before. Whereupon the captain wrote to them a letter, that they dealt too rigorously with him, to go about to cut his throat in the price of his commodities, which were so reasonably rated. But seeing they had sent him this to his supper, he would in the morning bring them as good a breakfast. In the morning, he shot off a whole culverin to summon the town, and preparing one hundred men in armour, went ashore, having in his great boat two falcons of brass, which being perceived by the townsmen, they incontinent in battle array with their drum and ensign displayed, marched from the town to the sands, of footmen to the number of a hundred and fifty, making great brags with their cries. Our captain perceiving them so bragful, commanded the two falcons to be discharged at them, which put them in no small fear. At every shot they fell flat to the ground, and as we approached near unto them, they broke their array, and dispersed. The horsemen also being about thirty, made as brave a show as might be, coursing up and down with their horses, their brave white leather targets in one hand, and their javelins in the other. But when we landed, they gave ground, for little they thought we would have landed so boldly: and therefore as the captain was putting his men in array, they sent a messenger on horseback with a flag of truce: but the captain not well contented with this messenger, marched forwards. Upon this we made our traffic quietly with them. In the meantime we watered a good breadth off from the shore, whereby the strength of the fresh water running into the sea, the salt water was made fresh. In this river we saw many crocodiles of sundry bignesses, but some as big as a boat, with 4 feet, a long broad mouth, and a long tail, whose skin is so hard, that a sword will not pierce it. His nature is to live out of the water as a frog doth, but he is a great devourer, and spareth neither fish, which is his common food, nor beasts, nor men.

The fifth of July we had sight of certain islands of sand, called the Tortugas, where the captain went in with his pinnace, and found such a number of birds, that in half an hour he laded her with them; and if they had been ten boats more, they might have done the like. These islands bear the name of Tortoises, because of the number of them, which there do breed, whose nature is to live both in the water and upon land also, but breed only upon the shore, in making a great pit wherein they lay eggs, to the number of three or four hundred, and covering them with sand, they are hatched by the heat of the sun; and by this means cometh the great increase. Of these we took very great ones, which have both back and belly all of bone, of the thickness of an inch: the flesh whereof we proved, eating much like veal; and finding a number of eggs in them, tasted also of them, but they did eat very sweetly.

They kept on their way along the coast of Florida, and the fifteenth day came to an anchor, and so where the Frenchmen abode, ranging all the coast along, seeking for fresh water, anchoring every night, because we would overshoot no place of fresh water, and in the day time the captain in the ship's pinnace sailed along the shore, went into every creek, speaking with divers of the Floridians. They found sorrel to grow as abundantly as grass, and where their houses were, great store of maize and mill, and grapes of great bigness, but of taste much like our English grapes. Also deer great plenty, which came upon the sands before them. Their houses are not many together, for in one house an hundred of them do lodge; they being made much like a great barn, having no place divided, but one small room for their king and queen. For the making of their fire, not only they but also the negroes do the same, which is made only by two sticks rubbing them one against another. In their apparel the men only use deer skins, wherewith some only cover their privy members, othersome use the same as garments to cover them before and behind; which skins are painted some yellow and red, some black and russet, and every man according to his own fancy. They do not omit to paint their bodies also with curious knots, or antic work, as every man in his own fancy deviseth.

The women also for their apparel use painted skins, but most of them gowns of moss, somewhat longer than our moss, which they sew together artificially, and make the same surplice wise, wearing their hair down to their shoulders, like the Indians. In this river, the captain entering with his pinnace, found a French ship of four score ton, and two pinnaces, and speaking with the keepers, they told him of a fort two leagues up, which they had built, in which their captain Monsieur Laudonnière was, with certain soldiers, into the which our captain went; where he was by the general very gently entertained, who declared unto him the extremity they were driven to for want of victuals, having brought very little with them; they being two hundred men at their first coming, had in short space eaten all the maize they could buy of the inhabitants about them, and therefore were driven to serve a king of the Floridians against other his enemies, for victuals: they were fain to gather acorns, which being stamped small, and often washed, to take away the bitterness of them, they did use for bread, eating withall roots, whereof they found many good and wholesome. This hardness not contenting some of them, who would not take the pains so much as to fish in the river before their doors, but would have all things put in their mouths, they did rebel against the captain, and so to the number of fourscore of them, departed with a bark and a pinnace, and so went to the islands of Hispaniola and Jamaica a roving, where they spoiled and pilled the Spaniards; and having taken two caravels laden with wine and cassava, which is a bread made of roots, and much other victuals and treasure, had not the grace to depart therewith, but were of such haughty stomachs, that they thought their force to be such that no man durst meddle with them, and so kept harbour in Jamaica, going daily ashore at their pleasure.

A ship and a galliass being made out of Santo Domingo came thither into the harbour, and took twenty of them, whereof the most part were hanged, and the rest carried into Spain, and some (to the number of five and twenty) escaped in the pinnace, and came to Florida; where at their landing they were put in prison, and incontinent four of the chiefest being condemned, at the request of the soldiers, were hanged upon a gibbet. This lack of threescore men was a great discourage and weakening to the rest, for they were the best soldiers that they had: for they had now made the inhabitants weary of them by their daily craving of maize, having no wares left to content them withall, and thereforce were enforced to rob them, and to take away their victuals perforce, which was the occasion that the Floridians (not well contented therewith) did take certain of their company in the woods, and slew them; whereby there grew great wars betwixt them and the Frenchmen. The Frenchmen having not above forty soldiers left unhurt, had not above ten days' victual, left before we came. In which perplexity our captain spared them out of his ship twenty barrels of meal, and four pipes of beans, with divers other necessaries which he might conveniently spare: and to help them the better homewards, whither they were bound before our coming, at their request we spared them one of our barks of fifty ton. Notwithstanding the great want that the Frenchmen had, the ground doth yield victuals sufficient; but they being soldiers, desired to live by the sweat of other men's brows. The ground yieldeth naturally grapes in great store, for in the time that the Frenchmen were there, they made 20 hogsheads of wine. Also it yieldeth roots passing good, deer marvellous store, with divers other beasts, and fowl, serviceable to the use of man. These be things wherewith a man may live, having corn or maize wherewith to make bread: for maize maketh good savoury bread, and cakes as fine as flour. And this maize was the greatest lack they had, because they had no labourers to sow the same, and therefore to them that should inhabit the land it were requisite to have labourers to till and sow the ground. The Indians with the head of maize roasted, will travel a whole day; in this order I saw threescore of them feed, who were laden with wares, and came fifty leagues off. The Floridians when they travel, have a kind of herb dried, who with a cane and an earthen cup in the end, with fire, and the dried herbs put together, do suck through the cane the smoke thereof, which smoke satisfieth their hunger, and therewith they live four or five days without meat or drink. At the Frenchmen's first coming thither they received for a hatchet two pound weight of gold, because they know not the estimation thereof: but the soldiers being greedy of the same, did take it from them, giving them nothing for it. How they came by this gold and silver the Frenchmen know not as yet. It seemeth they had estimation of their gold and silver, for it is wrought flat and graven, which they wear about their necks; othersome made round like a pancake, with a hole in the midst, to bolster up their breasts withall, because they think it a deformity to have great breasts. As for mines either of gold or silver, the Frenchmen can hear of none. The Frenchmen obtained pearls of them of great bigness. The Spaniards used to keep daily fishing some two or three hundred Indians: and their order is to go in canoes, or rather great pinnaces, with thirty men, whereof the one half be divers, the rest do open the same for the pearls. The oysters which have the smallest sort of pearls are found in seven or eight fathom water, but the greatest in eleven or twelve fathom.

Of beasts in this country besides deer, foxes, hares, polecats, coneys, ounces, and leopards, I am not able certainly to say: but it is thought that there are lions and tigers as well as unicorns. Of fish also they have in the river, pike, roach, salmon, trout, and divers other small fishes, and of great fish, some of the length of a man and longer, being of bigness accordingly, having a snout much like a sword a yard long. Of the sea fowl above all other not common in England, I noted the pelican, which is feigned to be the lovingest bird that is: which rather than her young should want, will spare her heart blood out of her belly: having a body like a hern, with a long neck, and a thick long beak, from the nether jaw whereof down to the breast passeth a skin of such a bigness, as is able to receive a fish as big as one's thigh, and this her big throat and long bill doth make her seem so ugly.

From thence we departed the 28 of July upon our voyage homewards: and took leave of the Frenchmen that there still remained, who with diligence determined to make as great speed after, as they could. Thus by means of contrary winds oftentimes, we prolonged our voyage in such manner that victuals scanted with us, so that we were in divers times in despair had not God of his goodness better provided for us. We were provoked to call upon him by fervent prayer, which moved him to hear us, so that we had a prosperous wind, which did set us so far as to be upon the bank of Newfoundland, on Saint Bartholomew's Eve, and took a great number of fresh codfish, which greatly relieved us. With a good large wind the twentieth of September we came to Padstow in Cornwall, God be thanked, in safety, with the loss of twenty persons in all the voyage, and with great profit to the venturers of the said voyage, as also to the whole realm, in bringing home both gold, silver, pearls and other jewels great store. His name therefore be praised for evermore. Amen.

XXV

Letter of Mr Arthur Edwards, written the 26 of April 1566 in Shemakha in Media, to the right worshipful Sir Thomas Lodge Knight and Alderman: and in his absence to Mr Thomas Nichols, Secretary to the right worshipful Company trading into Russia, Persia, and other the North and East parts.

Worshipful Sir, my bounden duty remembered, with hearty prayer unto God for the preservation of you and yours in perfect health with increase of worship. My last letter I sent you was from Astrakhan the 26th of July 1565. From whence Richard Johnson, myself, and Alexander Kitchin, departed as the 30th of the same. And by means of contrary winds, it was the 23rd of August before we came to our desired port named Nizovaya.[47] There, after we had gotten your goods on land, with much labour and strength of men, as also windlasses devised and made, we hauled your bark over a bar of beach or pebble stones into a small river, sending your ship's apparel with other things to an house hired in a village thereby. And as soon as we might get camels, being the fifth of September we departed thence, and came to this town of Shemakha, unto Abdollocan the king of this country. He received our presents with giving us thanks for our good wills, demanding if Mr Jenkinson were in good health, and whether he would return into these parts again. He willed us also himself to sit down before him the distance of a quoit's cast from his tent, where he sat with divers of his council and nobility, sending us from his table such meat as was before him. At the taking of our leave, he willed us to put our whole minds and requests in writing, that he might further understand our desires. But while we were about to do so, God took this good king our friend out of this present life the 2nd of October past. The want of him hath been the cause that as yet we cannot receive certain debts. Howbeit, we doubt not but we shall recover all such sums of money as are owing us for this voyage. Great troubles have chanced in these parts. Of those which were of the old king's council or bare any rule about him in these quarters, some are in prison, some are pinched by the purse, and others sent for unto the Shah. These troubles have partly been the let that wares were not sold as they might, to more profit. Your agent Richard Johnson bought four horses, minding to have sent to Qasvín Alexander Kitchin, whom God took to his mercy the 23rd of October last: and before him departed Richard Davis one of your mariners. We are now destitute of others to supply their rooms. Four mariners were few enough to sail your bark, whereof at this present we have but one, whose name is William Smith, an honest young man, and one that doth good service here. For want and lack of mariners that should know their labours, we all were like to be cast away in a storm. For all the broad side of our bark lay in water, and we had much ado to recover it, but God of his mercy delivered us. Mariners here may do you good service all the winter otherways: and merchants here will be gladder to ship their goods in us giving good freight. One merchant at this present is content to pay 20 roubles for twenty camels lading freight to Astrakhan. Such barks as must pass these seas, may not draw above five foot of water, because that in many places are very shallow waters. We mind hereafter to make the Russian boats more strong, and they shall serve our turns very well. Some in times past took great pains in the getting of the Shah's letters or privilege: now, I trust (with God's help) they may be obtained: this privilege once gotten and obtained, we shall live in quietness and rest, and shall shortly grow into a great trade for silks both raw and wrought, with all kind of spices and drugs, and other commodities here, as to Mr Anthony Jenkinson is well known. The truth of the slaughter of Thomas Alcock your servant, is not certainly known. Some think it was by the means of a noble man, with whom your said servant was earnest in demanding of your debts: upon whose words he was so offended, that he procured his death. But others do think verily, that in riding from the court without company, false knaves lay in wait, thinking he had much about him, and so slew him. Honest merchants are glad of our being here, and seek to grow in acquaintance with us, being glad to further us in that they may, and have spoken in our favours to the chiefest of this country.

For all kinds of wares bought or sold, you shall thoroughly be advertised by your agent Richard Johnson, whose reckonings or accounts at no hands I might see or be privy unto. Your kerseys were good and well sorted, they are and will be sold from 150 shahis[48] to 160 the piece. Two hundred pieces were sold under, that needed not: one 100 pieces at 146 and 147 the piece but more would have been given, if circumspection had been used. Here is at this time bought for England 11 packs of raw silk, 25 and 26 batmans being in every pack: the batman being 7 pound, which may be six pound and a half of English weight, being bought here from 66 to 70 shahis the batman. It is fine and good, little coarse at this time was to be had. I would God the Company could find the means to have vent to make sales for the one half that we may buy here. The Company may have for 30 or 40 thousand pounds yearly. From Astrakhan in 7 or 8 days, we may sail with our bark to a place named Gilan: alum is there good cheap, being brought from thence hither to Shemakha, and sold here for two bists their batman, which may be five pence in our money: and so I have bought to be sent home 223 batmans for example. And at Gilan there is raw silk enough for the Company's stock. From Qasvín to Ormuz, is about 30 days travelling with camels. I have written the prices of wares in my letter to the governor both for spices and some drugs which I do know. Also you shall understand here is plenty of yew for bow staves. I caused three horse loads to be bought for us to know the truth: but they were cut out of season this month of April, the sap being in them. Your agent will send some home for example. I have written my mind to Mr Glover your agent, what Russian wares I think best be bought for this country, and to send someone hither that hath the Russian tongue, for we have need. And the company shall do well hereafter in taking of servants to be sent hither, to see that they be such as have discretion, and be something broken in the world, and seen in the trade of merchandise, and one as can speak the

Portuguese tongue, for then we may buy a slave that can speak this language and the Portuguese tongue also, which shall then interpret unto us in all your secret doings, not making the Russians privy: for they are sorry that we do trade into these parts, for we are better beloved than they are: because they are given to be drunkards, they are much hated of these people. It is to be wished that none should serve your worships in these parts that be given to that kind of vice. Your London reds are not to be sent hither, for they will not give above 18 shahis their arshine. Here be reds of more orient colour, being Venice dye. The people are given much to wear cloth: the common people specially wear kerseys,[49] and the merchants of more wealth wear broadcloth. You shall do well to send five or six broadcloths, some blacks, puces, or other sad colours, that may be afforded at 20 shahis the arshine, and not above. It is here reported that King Philip hath given the Turks a great overthrow at Malta, and taken 70 or 80 of his chief captains.

Thus wishing I had more time to write, I pray you to bear with this my scribbled letter, and after you have read it, that Mr Nichols may have a sight thereof.

By your servant to command,
ARTHUR EDWARDS.

XXVI

The voyage into Persia, made by Mr Arthur Edwards, Agent, John Sparke, Laurence Chapman, Christopher Faucet, and Richard Pingle, in the year 1568.

May it please you to understand that your agent Mr Arthur Edwards and we departed from Yaroslavl in July 1568 and the 14 of August arrived at our port with your ship, the *Grace of God*, and the goods in her in good safety, God be thanked for it, finding there neither the people so ready to aid us for the unlading of the goods, nor yet so obedient to the Shah's privilege, as the worshipful company have been informed. Our goods brought upon land, we were compelled to open and sell as they would set the price, or otherwise it would have been worse for us. We were speedily aided with camels by the Sultan, to carry our goods to Shemakha, to which place we attained the first of September, finding it so thoroughly furnished with all manner of commodities by such as came before us, that no man would ask to buy any one piece of kersey of us, such as came out of Russia afterwards had brought their goods to that and other places, and spoiled those sales we might have made.

[Your agent] bent himself for Qazvín, taking with him the greatest sum of the goods, and two of the worshipful's servants, John Sparke and myself, to help and procure the better sale for the same: and leaving at Shemakha Christopher Faucet and Richard Pingle with three hundred and fifty pieces of kerseys in their hands, which so far forth as I can understand, lie for the greatest part unsold. Upon our way, at a certain town, we chanced to barter nine pieces of kerseys for fourscore and four batmans of cinnamon, selling the kerseys at one hundred and fifty shahis the piece.

Being not far from Tiflis, the principal place in this country for uttering of cloth I persuaded your Agent to send thither to prove what might be done, and receiving from him four and fifty pieces of kerseys, I proceeded on that voyage myself, finding in that place great store of broadcloth and kerseys brought thither, some part by the Turks who be resident there, some by the Armenians, who fetch them at Aleppo, and some by the townsmen, who travel unto Venice and there buy them, so that no man offered me one penny more than a hundred and forty shahis for a kersey: having charge from your agent not to stay there above the space of seven days, but to repair to Qazvín with all speed, and furthermore, having regard to keep up the price of the worshipfuls' commodities, I found means to barter them away for spices, neither in goodness nor yet in price to my content: nevertheless, considering the cold sales which were there, as well for your kerseys, as also the hot news, that Ormuz way was shut up by occasion that the Indians do war against them, which is true in deed: I thought it necessary to buy them, the prices and weight whereof appeareth by my accompt.

It chanced me in that place to meet with the governor's merchant of Grozny, not a little desirous to bargain with me for a hundred pieces of kerseys for his master, and offering me so good bonds for the payment of the money: and offering me besides his own letter in the behalf of his master, that no custom should be demanded for the same, and hearing good report made of him by the Armenians also, and that he was a Christian, I was much more the willing to bargain with him, and sold him a hundred pieces for a hundred and threescore shahis apiece. I sent my tolmach from me back to Shemakha, with such goods as I bought at Tiflis. At whose arrival there, as I do perceive, the captain would not accomplish his bargain to take them, but saith, he hath no need of them: such is the constancy of all men in this country. If the ware be bought, and they do mislike it afterwards, they will bring it again, and compel you to deliver the money, regarding the Shah's letters, which manifesteth the contrary, as a straw in the wind.

The way once open to Ormuz, from whence cometh no such store of spices as the worshipful doth look for, here will be put away in Tiflis, some for money and other some for barter, to the number of three hundred or four hundred pieces of kerseys, being in colours and goodness to the examples here sent you, the rest of the kerseys to make them up a thousand, and broadcloths to the sum of a hundred, be as many as will be put away yearly in this country, so far as yet I can perceive.

To break the trade between the Venetians and the whole company of the Armenians it is not possible, unless the worshipful will find some means to receive of them yearly to the numbers of a hundred mules' lading, and deliver them for the same one third part money, the rest cloth and kerseys fitted in colours meet for this country.

Six days journey from Tiflis, grow abundance of galls,[50] which are brought up yearly by the Venetians, and be sold there for two bists the Tiflis batman, which as your Agent here saith, maketh six pound English weight. It is supposed much good will be done by buying of them: which might at this present have partly been proved, if so be that some could do but half that which hath been written.

Touching drugs, I find many as well at Tiflis, as also in Qazvīn, but the goodness nothing like to such as be brought into England out of other places: and the price is so high that small gain will be had in buying of them. At my coming to Qazvīn I found news given out (as your Agent saith) that the Shah would buy all such commodities as he had, and give him silk and spices for the same: but by report the Shah never took cloth into his treasury all the days of his life, and will not now begin: his whole trade is in raw silk, which he selleth always for money to the Armenians and Turks.

Babylon is from hence fifteen days' journey, whereas by true report be great store of dates, the commodity fit for England, and the place so near unto us might easily have been known, if he, whose deeds and sayings differ much, had been willing to the same. Kashan also is but seven days' journey from hence, and a place by report where most store of spices be at all times to be had, over and above any place in this country.

To travel in this country is not only miserable and uncomfortable for lack of towns and villages to harbour in when night cometh, and to refresh men with wholesome victuals in time of need, but also such scarcity of water, that sometimes in three days' journey together, is not to be found any drop fit for man or beast to drink, besides the great danger we stand in for robbing by these infidels, who do account it remission of sins to wash their hands in the blood of one of us. Better it is therefore in my opinion to continue a beggar in England during life, than to remain a rich merchant seven years in this country, as some shall well find at their coming hither.

By the commandment of the Agent also I went to Gilan, as well to see what harbour was there for your ship, as also to understand what commodity is there best sold, and for what quantity. I found the way from hence so dangerous and troublesome, that with my pen I am not able to note it unto you: no man travelleth from hence thither, but such poor people as need constraineth to buy rice for their relief to live upon, and he escapeth very hardly that cometh there with the same.

The town of Lahilan,[51] which was the chiefest place in all that land, have I seen, which be now overrun by the Shah and his power, and be so spoiled, and the people so robbed, that not one of them is able to buy one kersey. The best commodity there to be bought, is raw silk, also there is to be had what store of alum you will.

In these parts be many Turkish merchants resident, which give an outward show, as though they were glad of our coming hither, but secretly they be our mortal enemies, searching by all means to hinder our sales. They wish us to go to Hallap[52] with the rest of our commodities unsold, where they say we shall have good entertainment in spite of the great number of Venetians which be there resident, and the custom but two in the hundred, and our kerseys to be sold presently, had we never so many, for twelve ducats,[53] which maketh of this money 165 shahis: but by such as know the place, market and custom, it is reported to us credibly to the contrary, and that such kerseys as ours be, are not sold for above 8 ducats there: the custom thirty in the hundred and more, that no place in the world is so well furnished with good cloth and kerseys, and of so brave colour as that place is, supposing it to be craftily purposed of them, to bring us into trouble, which God defend us from.

The price of spices be these, at this present enhanced by reason the way is shut to Ormuz, which when God shall send open, I purpose (God willing) to see, and at my return to advertise the worshipful what benefit is there to be had. Pepper 25 shahis the Tiflis batman: cloves 50 shahis, long pepper 25 shahis, maces large 50 shahis, ginger 24 shahis, ready money all, or else look not upon them. And the best sort of raw silk is sold for 60 shahis the Tiflis batman. Thus I end for this time, beseeching God to preserve you in continual health.

By your obedient servant,
LAWRENCE CHAPMAN.

XXVII

The embassage of the right worshipful Thomas Randolfe, Esquire, to the Emperor of Russia, in the year 1568

The 22 day of June, in the year of our Lord 1568, I went aboard the *Harry*, lying in the road at Harwich with my company, being to the number of forty persons or thereabout: of which one half were gentlemen, desirous to see the world.

We sailed for the space of twelve days with a prosperous wind, without tempest or outrage of sea: having compassed the North Cape we directed our course flat southeast, and so sailing between two bays, the two and thirtieth day after our departure from Harwich, we cast anchor at Saint Nicholas road.[54] In all the time of our voyage, more than the great number of whales engendering together, and the spermaceti, which we might see swimming upon the sea, there was no great thing to be wondered at. At St Nicholas we landed where there standeth an abbey of monks (to the number of twenty) built all of wood: the apparel of the monks is superstitious, in black hoods, as ours have been. Their church is fair, but full of painted images, tapers, and candles. Their own houses are low, and small rooms. They lie apart, they eat together, and are much given to drunkenness, unlearned, write they can, preach they do never, ceremonious in their church, and long in their prayers.

At my first arrival I was presented from their prior with two great rye loaves, fish both salt and fresh of divers sorts, both sea fish and fresh water, one sheep alive, black, with a white face, and with so many solemn words inviting me to see their house, they took their leave.

The river that runneth there into the sea is called Dvina, very large, but shallow. This river taketh his beginning about 700 miles within the country, and upon this river standeth Kholmogory, and many pretty villages, well situated for pasture, arable land, wood and water. The river pleasant between high hills of either side inwardly inhabited, and in a manner a wilderness of high fir trees.

At Kholmogory we tarried three weeks, not being suffered to depart before the Emperor had word of our coming, who sent to meet us a gentleman of his house, to see us furnished of victuals, and all things needful, upon his own charge.

The allowance of meat and drink was for every day two roubles, besides the charge of boats by water, and four score post horses by land, with above 100 carts to carry my wines, and other carriage.

Kholmogory is a great town builded all of wood, not walled, but scattered house from house. The people are rude in manners, and in apparel homely, saving upon their festival, and marriage days. The people of this town are much given to drunkenness.

In this town the Englishmen have lands of their own, given them by the Emperor, and fair houses, with offices for their commodity.

I was five whole weeks upon the river of Dvina till I came to Vologda,[55] being drawn with men against the stream, for other passage there is none. In this town the Emperor hath built a castle environed with a wall of stone, and brick, the walls fair and high, round about. Here (as in all other their towns) are many churches, some built of brick, the rest of wood, many monks and nuns in it: a town also of great traffic, and many rich merchants there dwelling.

From hence we passed by land towards Moscow in post. The country is very fair, plain and pleasant, well inhabited, corn, pasture, meadows enough, rivers, and woods, fair and goodly.

At Yaroslavl we passed the river of Volga, more than a mile over. This river descendeth into Mare Caspium, portable through of very great vessels with flat bottoms, which far pass any that our country useth. To sail by this river into Mare Caspium the English company have caused a bark to be built of 27 tons, which there was never seen before.

To Moscow we came about the end of September, received by

no man, not so much as our own countrymen suffered to meet us, which bred suspicion in me.

We were brought to a house built of purpose by the Emperor for ambassadors, fair and large. Two gentlemen were appointed to attend upon me, the one to see us furnished of victuals, the other to see that we should not go out of the house, nor suffer any man to come unto us. He that looked to our persons so straitly handled us, that we had no small cause to doubt that some evil had been intended unto us. No supplication, suit or request could take place for our liberty, nor yet to come to his presence.

Having passed 17 weeks in this sort, the Emperor sendeth word that we should be ready against Tuesday the 20 of February, at eight a clock in the morning.

The hour being come that I should go to the court, the two gentlemen came unto me apparelled more princely than before I had ever seen them. They press us to depart, and mounted up their own horses, and the ambassador upon such a one as he had borrowed, his men marching on foot, to their great grief.

The ambassador (being myself) was conveyed into an office accompanied with the two gentlemen: I tarried two long hours before I was sent for to the Emperor. I was conveyed by my gentlemen up a pair of stairs through a large room, where sat by my estimation 300 persons, all in rich attire, taken out of the Emperor's wardrobe for that day, upon three ranks of benches, set round about the place, rather to present a majesty, than that they were either of quality or honour.

At the first entry into the chamber I with my cap gave them the reverence, such as I judged their stately sitting, grave countenances, and sumptuous apparel required, and seeing that it was not answered again of any of them I covered my head, and so passing to a chamber where the Emperor was, there received me from the door from my two gentlemen, two of the Emperor's counsellors, and shewed me to the Emperor, and brought me to the middle of the chamber, where I was willed to stand still, and to say that which I had to say. I by my interpreter opened my message from the Queen my mistress, from whom I came, at whose name the Emperor stood up, and demanded divers questions of her health and state: he gave me his hand in token of my welcome, and caused me to sit down, and further asked me divers questions.

I delivered Her Majesty's present, which was a notable great cup of silver curiously wrought, with verses graven in it. He licensed me and my whole company to depart, who were all in his presence, and were saluted by him with a nod of the head, and said unto me: I dine not this day openly for great affairs I have, but I will send thee my dinner, and give leave to thee and thine to go at liberty, and augment our allowance to thee, in token of our love and favour to our sister the Queen of England.

I with reverence took my leave, being conveyed by two other of greater calling than those that brought me to the Emperor's sight, and so returned to my lodging.

Within one hour after in comes a duke richly apparelled, accompanied with fifty persons, each of them carrying a silver dish with meat, and covered with silver. The duke first delivered twenty loaves of bread of the Emperor's own eating, having tasted the same, and delivered every dish into my hands, and tasted of every kind of drink that he brought. This being done, the duke and his company sat down with me, and took part of the Emperor's meat, and went not away from me unrewarded.

Within few nights after the Emperor had will to speak secretly with me, and sent for me in the night: the place was far off and the night cold, and I having changed my apparel into such as the Russians do wear, found great incommodity thereby.

Having talked with him above three hours, towards the morning I was dismissed, and so came home to my lodging, where I remained above six weeks. About the beginning of April, the Emperor sent for me again and being come, I dealt effectually with him in the behalf of our English merchants, and found him so graciously inclined towards them, that I obtained at his hands my whole demand for larger privileges in general, together with all the rest my particular requests. And then he commended to my conduct into England, a nobleman of his, called Andrew Savin, as his ambassador, for the better confirmation of his privileges granted, and other negotiations with Her Majesty. And thus being dispatched with full contentment, the said ambassador and myself departed, and embarked at St. Nicholas about the end of July, and arrived safely at London in the month of September following.

XXVIII

Letters in verse, written by Master George Turberville out of Moscovy, 1568, to certain friends of his in London.

I left my native soil, full like a reckless man,
And unacquainted of the coast, among the Russians ran:
A people passing rude, to vices vile inclined,
Folk fit to be of Bacchus trained, so quaffing is their kind.
Drink is their whole desire, the pot is all their pride,
The soberest head doth once a day stand needful of a guide...
The house that hath no god, or painted saint within,
Is not to be resorted to, that roof is full of sin.
Besides their private gods, in open places stand
Their crosses unto which they crouch, and bless themselves with hand.
Devoutly down they duck, with forehead to the ground,
Was never more deceit in rags, and greasy garments found...
Their gait is very brave, their countenance wise and sad,
And yet they follow fleshly lusts, their trade of living bad.
It is no shame at all accounted to defile
Another's bed, they make no care their follies to conceal...
It is a sandy soil, no very fruitful vain,
More waste and woody grounds there are, than closes fit for grain.
Yet grain there growing is, which they untimely take,
And cut or ere the corn be ripe, they mow it on a stack
And laying sheaf by sheaf, their harvest so they dry,
They make the greater haste, for fear the frost the corn destroy.
For in the winter time, so glare is the ground,
As neither grass, nor other grain, in pastures may be found.

In comes the cattle then, the sheep, the colt, the cow,
Fast by his bed the mujik then a lodging doth allow,
Whom he with fodder feeds, and holds as dear as life:
And thus they wear the winter with the mujik and his wife.
Seven months the winter dures, the glare it is so great,
As it is May before he turn his ground to sow his wheat.
The bodies eke that die unburied lie they then,
Laid up in coffins made of fir, as well the poorest men,
As those of greater state: the cause is lightly found,
For that in winter time, they cannot come to break the ground...
No pewter to be had, no dishes but of wood,
No use of trenchers, cups cut out of birch are very good.
They use but wooden spoons, which hanging in a case
Each mujik at his girdle ties, and thinks it no disgrace...
In every room a stove, to serve the winter turn,
Of wood they have sufficient store, as much as they can burn...
Where he is wont to have a bear's skin for his bed,
And must, instead of pillow, clap his saddle to his head.
In Russia other shift there is not to be had,
For where the bedding is not good, the bolsters are but bad.
I mused very much, what made them so to lie,
Sith in their country down is rife, and feathers out of cry:
Unless it be because the country is so hard,
They fear by niceness of a bed their bodies would be marred...
I write not all I know, I touch but here and there,
For if I should, my pen would pinch, and eke offend I fear.
Who so shall read this verse, conjecture of the rest,
And think by reason of our trade, that I do think the best.
But if no traffic were, then could I boldly pen
The hardness of the soil, and eke the manners of the men...
The Russian men are round of bodies, fully fac'd,
The greatest part with bellies big that overhang the waist,
Flat headed for the most, with faces nothing fair,
But brown, by reason of the stove, and closeness of the air:
It is their common use to shave or else to shear
Their heads, for none in all the land long lolling locks doth
wear...

Verse, written by George Turberville in Moscow

Their garments be not gay, nor handsome to the eye,
A cap aloft their heads they have, that standeth very high ...
Their shirts in Russia long, they work them down before,
And on the sleeves with coloured silks, two inches good and
more ...
Within his boots the Russian wears, the heels they underlay
With clouting clamps of steel, sharp pointed at the toes,
And over all a shuba furred, and thus the Russian goes ...
These are the Russian robes. The richest use to ride
From place to place, his servant runs, and follows by his side ...
For when the Russian is pursued by cruel foe,
He rides away, and suddenly betakes him to his bow,
And bends me but about in saddle as he sits,
And therewithall amidst his race his following foe he hits.
Their bows are very short, like Turkey bows outright,
Of sinews made with birchen bark, in cunning manner dight.
Small arrows, cruel heads, that fell and forked be,
Which being shot from out those bows, a cruel way will flee ...
The common game is chess, almost the simplest will
Both give a check and eke a mate, by practice comes their skill ...
Conceive the rest yourself, and deem what lives they lead,
Where lust is law, and subjects live continually in dread.
And where the best estates have none assurance good
Of lands, of lives, nor nothing falls unto the next of blood.
But all of custom doth unto the prince redown,
And all the whole revenue comes unto the King his crown.
Good faith I see thee muse at what I tell thee now,
But true it is, no choice, but all at prince's pleasure bow ...
The strangeness of the place is such for sundry things I see,
As if I would I cannot write each private point to thee.
The cold is rare, the people rude, the prince so full of pride,
The realm so stored with monks and nuns, and priests on every
side:
The manners are so Turkey like, the men so full of guile,
The women wanton, temples stuffed with idols that defile
The seats that sacred ought to be, the customs are so quaint,
As if I would describe the whole, I fear my pen would faint ...

If thou be wise, as wise thou art, and wilt be ruled by me,
Live still at home, and covet not those barbarous coasts to see.
No good befalls a man that seeks, and finds no better place,
No civil customs to be learned, where God bestows no grace.

XXIX

A discourse written by one Miles Philips Englishman, put on shore in the West Indies by Mr John Hawkins 1568.

Upon Monday the second of October 1567 the weather being reasonable fair, our general Mr John Hawkins, having commanded all his captains and masters to be in readiness to make sail with him, he himself being embarked in the *Jesus*, hoisted sail, and departed from Plymouth upon his intended voyage for the parts of Africa, and America, being accompanied with five other sail of ships, as namely the *Minion*, the *William and John*, the *Judith*, in whom was captain Mr Francis Drake afterwards knight, the *Angel*, and the *Swallow*. Coming to the island of Gomera being one of the islands of the Canaries, where according to an order before appointed, we met with all our ships which were before dispersed, we then took fresh water and upon the eighteenth day of the same month we came to an anchor upon the coast of Africa, at Cape Verde in twelve fathom water; and here our general landed certain of our men, to the number of 160 or thereabout, seeking to take some negroes. And they going up into the country for the space of six miles, were encountered with a great number of the negroes: who with their envenomed arrows did hurt a great number of our men, so that they were enforced to retire to the ships, in which conflict they recovered but a few negroes, and of our men there died seven or eight in very strange manner, with their mouths shut, so that we were forced to put sticks into their mouths to keep them open. Upon the coast of Guinea, we obtained 150 negroes. There was a negro sent as an ambassador to our general, from a king of the negroes, which was oppressed with other kings his bordering neighbours, desiring our general to grant him succour and aid against those his enemies, which our general granted unto, and went himself in person a land, with the number of two hundred of our men, and the king which had requested our aid, did join his force with ours, so that thereby our general assaulted, and set fire upon a town in which there was at least the number of eight or ten thousand negroes, and they perceiving that they were not able to make any resistance sought by flight to save themselves, in which their flight there were taken prisoners to the number of eight or nine hundred, which our general ought to have had for his share: howbeit the negro king falsifying his word and promise, secretly in the night conveyed himself away with as many prisoners as he had in his custody: but our general notwithstanding finding himself to have now very near the number of 500 negroes thought it best to depart with them, and such merchandise as he had from the coast of Africa, towards the West Indies, and therefore commanded with all diligence to take in fresh water and fuel, and so with speed to prepare to depart. In a storm we lost one of our ships, the *William and John*, of which ship and of her people, we heard no tidings.

Upon the third day of February 1568 we departed from the coast of Africa, having the weather somewhat tempestuous, which made our passage the more hard: and sailing so for the space of 52 days, upon the 27 of March 1568 we came in sight of an island called Dominica, in the West Indies. From thence we departed for Cartagena, we could not obtain any traffic there, and so our general thought it best to depart from thence the rather for the avoiding of certain dangerous storms called the hurricanes, so the 24 of July 1568 we departed from thence directing our course north: sailing toward Florida upon the 12 of August an extreme tempest arose, which dured for the space of 8 days, in which our ships were most dangerously tossed and beaten hither, and thither, so that we were in continual fear to be drowned, and in the end we were constrained to flee for succour to the port of St Juan de Ulloa, or Vera Cruz, which is the port that serveth for the city of Mexico: the sixteenth of September 1568 we entered St Juan de Ulloa. The Spaniards there supposing us to have been the King of Spain's fleet, the chief officers of the country thereabouts came presently aboard, perceiving themselves to have made an unwise adventure, they were in great fear: howbeit our general did use them all very courteously. In the said port there were twelve ships which by report had in them in treasure to the value of two hundred thousand pound all which being in our general his power, he did freely set at liberty, as also the passengers, not taking from any of them all the value of one groat: only he stayed two men of credit and account, the one named Don Lorenzo de Alva, and the other Don Pedro de Rivera, presently our general sent to the Viceroy to Mexico which was three score leagues off, certifying him of our arrival there by force of weather, desiring that he would, considering our necessities and wants, furnish us with victuals, and quietly suffer us to repair our ships. And furthermore that at the arrival of the Spanish fleet which was there daily expected and looked for, to the end that there might no quarrel arise between them, and our general and his company for the breach of amity, he humbly requested of His Excellency, that there might in this behalf some special order be taken.

The next morning we descried 13 sail of great ships: it was the King of Spain's fleet then looked for. Our general was in a great perplexity of mind, considering with himself that if he should keep out that fleet from entering into the port, a thing which he was very well able to do with the help of God, then should that fleet be in danger of present shipwreck and loss of all their substance, which amounted unto the value of one million and eight hundred thousand pounds. Again he saw that if he suffered them to enter, they would practise by all manner of means to betray him. Therefore did he choose the least evil, which was to suffer them to enter under assurance, and so to stand upon his guard, and to defend himself and his from their treasons, the messenger being returned from Don Martin de Henriques, the new Viceroy, who came in the same fleet, did certify our general, that for the better maintenance of amity between the King of Spain and our

Sovereign, all our requests should be both favourably granted, and faithfully performed: signifying further that he heard and understood of the honest and friendly dealing of our general, toward the King of Spain's subjects in all places where he had been: our requests were articled, and set down in writing. Viz.

1. We might have victuals for our money, and licence to sell as much wares, as might suffice to furnish our wants.
2. We might be suffered peaceably to repair our ships.
3. The island might be in our possession during the time of our abode there, in which island our general had already planted and placed certain ordnance which were eleven pieces of brass, he required that the same might so continue, and that no Spaniard should come to land in the said island, having or wearing any kind of weapon about him.
4. For the better and more sure performance and maintenance of peace, there might twelve gentlemen of credit be delivered of either part as hostages.

These conditions agreed upon in writing by the Viceroy and signed with his hand, 10 hostages upon either part were received. And further it was concluded that the two generals should meet, and give faith each to other for the performance of the promises. All which being done, the same was proclaimed by the sound of a trumpet, and commandment was given that none of either part should violate or break the peace upon pain of death: the fleet entered the port, the ships saluting one another as the manner of the sea doth require: the morrow after being Friday we laboured on all sides in placing the English ships by themselves, and the Spanish ships by themselves, the captains and inferior persons of either part, offering, and showing great courtesy one to another. The Viceroy and governor thereabout had secretly at land assembled to the number of 1000 chosen men, and well appointed, meaning the next Thursday being the 24 of September at dinner time to assault us. This port was then at our being there, a little island of stones, not past three foot above water in the highest place, and not past a bow shot over any way at the most, and it standeth from the mainland, two bowshoots or more: there is not in all this coast any other place for ships safely to arrive at: also the north winds in this coast are of great violence and force, and unless the ships be safely moored in, with their anchors fastened in this island, there is no remedy, but present destruction and shipwreck. All this our general wisely foreseeing, did provide that he would have the said island in his custody, or else the Spaniards might at their pleasure, have but cut our cables, and so with the first northwind that blew we had our passport, for our ships had gone ashore.

The time approaching that their treason must be put in practice, the same Thursday morning, some appearance began to show itself, as shifting of weapons from ship to ship, and bending their ordnance against our men upon the land: which apparent shows caused our general to send one to the Viceroy, to enquire of him what was meant thereby, which presently sent that the ordnance aforesaid, should be removed, returning answer to our general in the faith of a Viceroy, that he would be our defence, and safety from all villainous treachery: this was upon Thursday in the morning. Our general not being therewith satisfied, seeing that they had secretly conveyed a great number of men aboard a great hulk or ship of theirs of six hundred ton, which ship rode hard by the *Minion*, he sent again to the Viceroy Robert Barret the master of the *Jesus*, a man that could speak the Spanish tongue very well, and required that those men might be unshipped again, which were in that great hulk. The Viceroy then perceiving that their treason was thoroughly espied, stayed our master, and sounded the trumpet, and gave order that his people should upon all sides charge upon our men, on shore, and elsewhere, which struck such amaze and sudden fear among us, that many gave place, and sought to recover our ships for the safety of themselves. The Spaniards which secretly were hid in ambush at land were quickly conveyed over to the island in their long boats, and so coming to the island, they slew all our men that they could meet with, without mercy. The *Minion* which had somewhat before prepared herself to avoid the danger, hauled away and abode the first brunt of the 300 men that were in the great hulk: then they sought to fall aboard the *Jesus*, where was a cruel fight, and many of our men slain: but yet our men defended themselves, and kept them out: so the *Jesus* also got loose, and joining with the *Minion*, the fight waxed hot upon all sides: but they having won and got our ordnance on shore, did greatly annoy us. In this fight there were two great ships of the Spaniards sunk, and one burnt, so that with their ships they were not able to harm us, but from the shore they beat us cruelly with our own ordnance, in such sort that the *Jesus* was very sore spoiled: and suddenly the Spaniards having fired two great ships of their own, they came directly against us, which bred among our men a marvellous fear. Howbeit the *Minion* which had made her sails ready, shifted for herself without consent of the general, so that very hardly could our general be received into the *Minion*: the most of our men that were in the *Jesus* shifted for themselves, and followed the *Minion* in the boat, and those which that small boat was not able to receive, were most cruelly slain by the Spaniards. Of our ships none escaped saving the *Minion* and the *Judith*: and all such of our men as were not in them were enforced to abide the tyrannous cruelty of the Spaniards. For it is a certain truth, that whereas they had taken certain of our men ashore, they took and hung them up by the arms upon high posts until the blood burst out of their fingers' ends: of which men so used, there is one Copstow, and certain others yet alive, who by the merciful providence of the Almighty, were long since arrived here at home in England, carrying still about with them (and shall to their graves) the marks and tokens of those inhumane and more than barbarous cruel dealings.

Most of his men were by the Spaniards slain and drowned, and all his ships sunk and burned, saving the *Minion*, and the *Judith*, which was a small bark of fifty ton, wherein was then captain Francis Drake: the same night the said bark lost us, we being in great necessity, and enforced to remove with the *Minion* two bowshoot from the Spanish fleet, where we anchored all that night: and the next morning we weighed anchor a storm took us with a north wind, in which we were greatly distressed, having but two cables and two anchors left. The morrow after, the storm being ceased and the weather fair, we weighed, and set sail, being many men in number, and but small store of victuals to suffice us for any long time: by means whereof we were in despair and fear that we should perish through famine, hunger constrained us to eat hides, cats and dogs, mice, rats, parrots and monkeys: to be short our hunger was so great, that we thought it savoury and sweet whatsoever we could get to eat.

On the eight of October we came to land again, in the bottom of the Bay of Mexico, where we hoped to have found some inhabitants, that we might have had some relief of victuals, and a place where to repair our ship, which was so greatly bruised, that we were scarce able with our weary arms to keep forth the water: being thus oppressed with famine on the one side and danger of drowning on the other, not knowing where to find relief, we began to be in wonderful despair, there were a great many that did desire our general to set them on land, making their choice rather to submit themselves to the mercy of the savages than longer to hazard themselves at sea, where if they perished not by drowning, yet hunger would enforce them in the end to eat one another: to which request our general did very willingly agree, considering with himself that it was necessary for him to lessen his number: being resolved to set half his people on shore, it was a world to see how suddenly men's minds were altered: for they which a little before desired to be set on land, were now of another mind, and requested rather to stay: our general was forced for the more contentation of all men's minds, and to take away all occasions of offence, to take this order: first he made choice of such persons as were needful to stay, and that being done, of those which were willing to go he appointed such as he thought might be best spared, and presently appointed that by the boat they should be set on shore, our general promising us that the next year he would either come himself, or else send to fetch us home.

In the evening it being Monday the eight of October, 1568, when we were all come to shore, we found fresh water, whereof some of our men drunk so much, that they had almost cast themselves away, for we could scarce get life of them for the space of two or three hours after: some were so cruelly swollen, what with the drinking in of the salt water, and what with the eating of the fruit which we found on land, having a stone in it much like an almond that they were all in very ill case, so that we were in a manner all of us both feeble, faint and weak.

The next morning, we thought it best to travel along by the sea coast, to seek out some place of habitation: and so departing from an hill where we had rested all night, not having any dry thread about us (for all the night it rained cruelly:) as we went from the hill, and were come into the plain, we were greatly troubled to pass for the grass and weeds that grew there higher than any man. On the left hand we had the sea, and upon the right hand great woods, so that of necessity we must needs pass on our way westward, through those marshes; and going thus, suddenly we were assaulted by the Indians, a warlike kind of people.

These people are called Chichimici, they wear their hair long, even down to their knees, they do also colour their faces green, yellow, red and blue, which maketh them to seem very ugly and terrible to behold. These people keep wars against the Spaniards, of whom they have been oftentimes very cruelly handled: for with the Spaniards there is no mercy. They perceiving us at our first coming on land, supposed us to have been their enemies, the Spaniards, and having by their forerunners described what number we were, and how feeble and weak without armour or weapon, they suddenly raised a terrible and huge cry, and so came running fiercely upon us, shooting off their arrows as thick as hail, unto whose mercy we were constrained to yield, not having amongst us any kind of armour, nor yet weapon, saving one caliver, and two old rusty swords: when they perceived, that we sought not any other than favour and mercy at their hands, and that we were not their enemies the Spaniards, they had compassion on us, and came and caused us all to sit down: they came to all such as had any coloured clothes amongst us, and those they did strip stark naked, and took their clothes away with them, but those that were apparelled in black they did not meddle withall, and so went their ways, and left us without doing us any further hurt, only in the first brunt they killed eight of our men. Shortly after they had left us stripped we thought it best to divide ourselves into two companies, and so being separated, half of us went under the leading of one Anthony Godard, who is yet a man alive, and dwelleth at this instant in the town of Plymouth, whom before we chose to be captain over us all, and those which went under his leading, of which number I Miles Philips was one, travelled westward that way which the Indians with their hands had before pointed us to go. The other half went under the leading of one John Hooper, took their way and travelled northward: we began to reckon amongst ourselves, how many we were that were set on shore, and we found the number to be a hundred and fourteen, whereof two were drowned in the sea, and eight were slain at the first encounter, so that there remained an hundred and four, of which 25 went westward with us, and 52 to the north with Hooper and Ingram: and as Ingram since hath often told me, they were not past three of their company slain, and there were but six and twenty of them that came again to us, so that of the company that went northward, there is yet lacking, and not certainly heard of, the number of three and twenty men. We travelled on still westward, sometimes through such thick woods, that we were enforced with cudgels to break away the brambles and bushes from tearing our naked bodies: other sometimes we should travel through the plains, in such high grass that we could scarce see one another, and as we passed in some places, we should have of our men slain, and fall down suddenly, being struck by the Indians, which stood behind trees and bushes, in secret places, and so killed our men as they went by, for we went scatteringly in seeking of fruits to relieve ourselves. We were also often greatly annoyed with a kind of fly, which the Spaniards called mosquitoes. They are scarce so big as a gnat; they will suck one's blood marvellously, and if you kill them while they are sucking, they are so venomous that the place will swell extremely: but if you let them suck their fill, and to go away of themselves, then they do you no other hurt, but leave behind them a red spot somewhat bigger than a flea-biting. As we travelled thus for the space of ten or twelve days, our captain did oftentimes cause certain to go up into the tops of high trees, to see if they could descry any town or place of inhabitants, but they could not perceive any, at length they descried a great river that fell from the northwest into the main sea, and presently after, we heard an arquebus shot off, which did greatly encourage us, for thereby we knew that we were near to some Christians, and did therefore hope shortly to find some succour and comfort, and within the space of one hour after, as we travelled, we heard a cock crow, which was also no small joy unto us, and so we came to the north side of the river of Panuco[56]: of this river we drank very greedily, for we had not met with any water in six days before, and as we were here by the river resting ourselves, we perceived many Spaniards upon the other side of the river, riding up and down on horseback: the Spaniards made out about the number of twenty horsemen, and embarking themselves in the canoes, they led their horses by the reins swimming over after them, and being come over to that side of the river where we were, they saddled their horses, and came very fiercely running at us. Our captain Anthony Godard did persuade us to submit unto them, for being naked, as we at this time were, and without weapon, we could not make resistance, they perceived us to be Christians, and did call for more canoes, and carried us over by four and four in a boat, they understanding by our captain how long we had been without meat, imparted between two and two a loaf of bread made of that country wheat, which the Spaniards call maize, of the bigness of our halfpenny loaves. This bread was very sweet and pleasant unto us, for we had not eaten any in a long time before: and what is it that hunger doth not make to have a savoury and delicate taste?

When we were all come to the town, the governor there showed himself very severe unto us, and threatened to hang us all: and then he demanded what money we had, which in truth was very little: we accounted that amongst us all we had the number of five hundred pesos.

When he had taken all that we had, he caused us to be put into a little house much like a hogsty, where we were almost smothered. Many of our men which had been hurt by the Indians at our first coming on land, whose wounds were very sore and grievous, desired to have the help of their surgeons to cure their wounds. The governor answered, that we should have none other surgeon but the hangman, which should sufficiently heal us of all our griefs: and thus reviling us, and calling us English dogs, and Lutheran heretics, we remained the space of three days in this miserable state, not knowing what should become of us.

Upon the fourth day looking every hour when we should suffer death, there came a great number of Indians and Spaniards weaponed to fetch us out of the house, and amongst them we espied one that brought a great many of new halters, at the sight whereof we made no other account but that we should presently have suffered death, and so crying and calling to God for mercy and forgiveness of our sins, we prepared ourselves, making us ready to die: when we were come out of the house, with those halters they bound our arms behind us, and so coupling us two and two together, they commanded us to march on through the town, and so along the country from place to place towards the city of Mexico, which is distant the space of ninety leagues, having only but two Spaniards to conduct us, they being accompanied with a great number of Indians warding on either side with bows and arrows, lest we should escape. Upon the second day at night we came unto a town which the Indians call Nohele, and the Spaniards Santa Maria: in which town there is a house of white friars, which did very courteously use us, and gave us hot meat, as mutton and broth, and garments also to cover ourselves withall, made of white baize: we fed very greedily of the meat. Our greedy feeding caused us to fall sick of hot burning agues. And here at this place one Thomas Baker one of our men died of a hurt: for he had been shot before with an arrow into the throat at the first encounter.

The next morrow about ten of the clock, we departed from thence, bound two and two together, and guarded as before, and so travelled on our way towards Mexico, till we came to a town named Mestitlan, where is a house of black friars: and in this town there are about the number of three hundred Spaniards, both men, women, and children. The friars sent us meat from the the house ready dressed, and the friars, and the men and women used us very courteously, and gave us some shirts and other such things as we lacked. Here our men were very sick of their agues, and with eating of another fruit which did bind us so sore, that for the space of ten or twelve days we could not ease ourselves. The next morning we departed from thence with our two Spaniards and Indian guard, as aforesaid. Of these two Spaniards the one was an aged man, who all the way did very courteously entreat us, and would carefully go before to provide for us both meat and things necessary to the uttermost of his power: the other was a young man who all the way travelled with us, and never departed from us, who was a very cruel caitiff, and he carried a javelin in his hand, and sometimes when as our men with very feebleness and faintness were not able to go so fast as he required them, he would take his javelin in both his hands, and strike them with the same between the neck and the shoulders so violently, that he would strike them down; then would he cry, and say, *marchad, marchad Ingleses perros, Luterianos, enemigos de Dios*: which is as much to say in English, as march, march on you English dogs, Lutherans, enemies to God.

On our journey towards Mexico, within two leagues of it, there was built by the Spaniards a very fair church, called Our Lady's church, in which there is an image of Our Lady of silver and gilt, being as high, and as large as a tall woman, and before this image, there are as many lamps of silver as there be days in the year, which upon high days are all lighted. Whensoever any Spaniards pass by this church, although they be on horseback, they will alight, and come into the church, and kneel before the image: which image they call in the Spanish tongue, Nuestra Señora de Guadalupe. At this place there are certain cold baths; the water thereof is somewhat brackish in taste: but very good for any that have any sore or wound: and every year once upon our Lady Day the people use to repair thither to offer, and pray in that Church before the image, and they say that Our Lady of Guadalupe doth work a number of miracles.

Here we met with a great number of Spaniards on horseback, which came from Mexico to see us, both gentlemen, and men of occupations, and they came as people to see a wonder: there was brought us by the Spaniards from the market place great store of meat, sufficient to have satisfied five times so many as we were: some also gave us hats, and some gave us money: which place we stayed for the space of two hours, and from thence we were conveyed by water in two large canoes to an hospital. After our coming thither, many of the company that came with me from Panuco died within the space of fourteen days: soon after we were put altogether into Our Lady's hospital, in which place we were courteously used, and visited oftentimes by virtuous gentlemen and gentlewomen of the city, who brought us divers things to comfort us withall, as suckets and marmalades, and that very liberally. In which hospital we remained for the space of six months, until we were all whole and sound of body, and then we were appointed by the Viceroy to be carried unto the town of Texcoco: in which town there are certain houses of correction, like to Bridewell here in London: into which place divers Indians are sold for slaves, some for ten years, and some for twelve. It was no small grief unto us when we understood that we should be carried thither, and to be used as slaves, we had rather be put to death: howbeit there was no remedy, yet by the good providence of our merciful God, we happened there to meet one Robert Sweeting, who was the son of an Englishman born of a Spanish woman: this man could speak very good English, and by his means we were helped very much with victuals from the Indians as mutton, hens, and bread. Continuing thus straitly kept in prison there for the space of two months, at the length we agreed amongst ourselves to break forth of prison, come of it what would, for we were minded rather to suffer death than longer to live in that miserable state. And so having escaped out of prison, we knew not what way to fly for the safety of ourselves, the night was dark, and it rained terribly, and not having any guide, we went we knew not whither, and in the morning, at the appearing of the day, we perceived ourselves to be come hard to the city of Mexico, which is 24 English miles from Texcoco. The day being come we were espied by the Spaniards, and pursued and taken, and brought before the Viceroy and head justices, who threatened to hang us for breaking of the King's prison. Yet in the end they sent us into a garden belonging to the Viceroy, and coming thither, we found there our English gentlemen which were delivered as hostages when as our general was betrayed at

San Juan de Ulloa, as is aforesaid, and with them we also found Robert Barret, the master of the *Jesus*, in which place we remained labouring and doing such things as we were commanded, for the space of four months, having but two sheep a day allowed to suffice us all, being very near a hundred men, and for bread we had every man two loaves a day, of the quantity of one halfpenny loaf. At the end of which four months, they having removed our gentlemen hostages, and the master of the *Jesus*, did cause it to be proclaimed, that what gentleman Spaniard soever was willing, or would have any English man to serve him should repair to the said garden, and there take their choice: happy was he that could soonest get one of us.

The gentlemen that thus took us for their servants or slaves did new apparel us throughout, with whom we abode, doing such service as they appointed us unto, which was for the most part to attend upon them at the table, and to be as their chamberlains, and to wait upon them when they went abroad. In this sort we remained and served in the city of Mexico, and thereabouts for the space of a year and somewhat longer. Afterwards many of us were by our masters appointed to go to sundry of their mines and to be as overseers of the negroes and Indians that laboured there. In which mines many of us did profit and gain greatly: the Indians and negroes which wrought under our charge, upon our well using of them, would at times as upon Saturdays when they had left work, labour for us, and blow as much silver as should be worth unto us 3 marks. Sundry weeks we did gain so much by this means besides our wages, that many of us became very rich, for we lived and gained thus in those mines some three or four years. As concerning those gentlemen which were delivered as hostages, the said gentlemen were sent away into Spain with the fleet, where as I have heard it credibly reported, many of them died with the cruel handling of the Spaniards in the Inquisition House. Robert Barret also master of the *Jesus*, was sent away with the fleet into Spain the next year following, where afterwards he suffered persecution in the Inquisition, and at the last was condemned to be burnt.

In the year of Our Lord one thousand five hundred seventy

four, the Inquisition began to be established in the Indies, very much against the minds of many of the Spaniards themselves. The Chief Inquisitor, thought it best to call us that were Englishmen first in question, and so much the rather, for that they had perfect knowledge and intelligence that many of us were become very rich, and therefore we were a very good prey to the Inquisitors: so that now began our sorrows afresh, for we were sent for, and sought out in all places of the country, and proclamation made upon pain of losing of goods and excommunication, that no man should hide or keep secret any Englishmen or any part of their goods. By means whereof we were all soon apprehended in all places, and all our goods seized and taken for the Inquisitors' use, and so from all parts of the country we were conveyed and sent as prisoners to the city of Mexico, and there committed to prison in sundry dark dungeons, where we could not see but by candle light, and were never past two together in one place, so that we saw not one another, neither could one of us tell what was become of another. Thus we remained close imprisoned for the space of a year and a half. We were often called before the Inquisitors alone, and there severely examined our faith, and commanded to say the Paternoster, the Ave Maria, and the Creed in Latin, which God knoweth a great number of us could not say, otherwise than in the English tongue. And having Robert Sweeting who was our friend at Texcoco always present with them for an interpreter, he made report for us, that in our own country speech we could say them perfectly, although not word for word as they were in Latin. Then did they proceed to demand of us upon our oaths what we did believe of the Sacrament, and whether there did remain any bread or wine after the words of consecration, yea or no, and whether we did not believe that the host of bread which the priest did hold up over his head, and the wine that was in the chalice, was the very true and perfect body and blood of our Saviour Christ, yea or no: to which if we answered not yea, then there was no way but death. Then they would demand of us what opinions we had been taught to hold contrary to the same whiles we were in England: to which for the safety of our lives we were constrained to say, that we never did believe, nor had been taught otherwise. Then would they charge us that we did not tell them the truth, that they knew the contrary, and therefore we should make a better answer at the next time, or else we should be racked, and made to confess the truth whether we would or no. And so coming again before them the next time, we were still demanded of our belief while we were in England, and also what we thought or did know of such of our own company as they did name unto us, and at other times they would promise us, that if we would tell them truth, then we should have favour and be set at liberty, although we very well knew their fair speeches were but means to entrap us, to the hazard and loss of our lives: howbeit God so mercifully wrought for us by a secret means that we had, that we kept us still to our first answer, and would still say that we had told the truth unto them, and that for our sins and offences in England against God and Our Lady or any of his blessed saints, we were heartily sorry for the same, and did cry God mercy, and besought the Inquisitors for God's sake, considering that we came into those countries by force of weather, and against our wills, and that never in all our lives we had either spoken or done anything contrary to their laws, and therefore they would have mercy upon us. Yet all this would not serve; for still from time to time we were called upon to confess, and about the space of three months before they proceeded to their severe judgement, we were all racked, and some enforced to utter that against themselves, which afterwards cost them their lives. And thus having gotten from our own mouths matter sufficient for them to proceed in judgement against us, they caused a large scaffold to be made in the midst of the market place in Mexico right over against the head church, and 14 or 15 days before the day of their judgement, with the sound of a trumpet, they did assemble the people in all parts of the city: before whom it was then solemnly proclaimed, that whosoever would upon such a day repair to the market place, they should hear the sentence of the Holy Inquisition against the English heretics. The night before they came to the prison where we were, bringing with them certain fool's coats which they had prepared for us, being called in their language *sanbenitos*, which coats were made of yellow cotton and red crosses upon them: they were so busied in putting on their coats about us, and bringing us out into a large yard, and placing and pointing us in what order we should go to the scaffold or place of judgement upon the morrow, that they did not once suffer us to sleep all that night long. The next morning being come, there was given to every one of us for our breakfast a cup of wine, and a slice of bread fried in honey, and so about eight of the clock in the morning, we set forth of the prison, every man alone in his yellow coat, and a rope about his neck, and a great green wax candle in his hand unlighted, having a Spaniard appointed to go upon either side of every one of us: and so marching in this order and manner toward the scaffold in the market place, which was a bowshoot distant or thereabouts, we found a great assembly of people all the ways and so coming to the scaffold we went up by a pair of stairs and found seats ready prepared for us to sit down on, every man in order as he should be called to receive his judgement. Presently the Inquisitors came up another pair of stairs, and the Viceroy and all the chief justices with them. When they were set down, then came up also a great number of friars, white, black and grey, about the number of three hundred persons. Then was silence commanded, and then presently began their severe and cruel judgement.

The first man that was called was one Roger the chief armourer of the *Jesus*, and he had judgement to have three hundred stripes on horseback, and after condemned to the galleys as a slave for 10 years.

After him were called John Gray, John Browne, John Rider, John Moone, James Collier, and one Thomas Browne: these were adjudged to have 200 stripes on horseback and after to be committed to the galleys for the space of 8 years.

Then was called John Keyes, and was adjudged to have 100 stripes on horseback, and condemned to serve in the galleys for the space of 6 years.

Then were severally called the number of 53 one after another, and every man had his several judgement, some to have 200 stripes on horseback, and some 100, and condemned for slaves to the galleys, some for 6 years, some for 8 and some for 10.

And then was I Miles Philips called, and was adjudged to serve in a monastery for 5 years, without stripes, and to wear a fool's coat or *sanbenito* during all that time.

Then were called John Storey, Richard Williams, David Alexander, Robert Cooke, Paul Horsewell and Thomas Hull: the six were condemned to serve in monasteries without stripes, some for three years and some for four, and to wear the *sanbenito* during all the said time. Which being done, and it now drawing toward night, George Rively, Peter Momfrie, and Cornelius the Irishman, were called and had their judgement to be burnt to ashes, and so were presently sent away to the place of execution in the market place but a little from the scaffold, where they were quickly burnt and consumed. And as for us that had received our judgement, being 68 in number, we were carried back that night to prison again. And the next day in the morning being Good Friday, the year of Our Lord 1575, we were all brought into a court of the Inquisitors' palace, where we found a horse in readiness for every one of our men which were condemned to have stripes, and to be committed to the galleys, which were in number 60 and so they being forced to mount up on horseback naked from the middle upward, were carried to be showed as a spectacle for all the people to behold throughout the chief and principal streets of the city, and had the number of stripes to every one of them appointed, most cruelly laid upon their naked bodies with long whips by sundry men appointed to be the executioners thereof: and before our men there went a couple of criers which cried as they went: behold these English dogs, Lutherans, enemies to God. They returned to the Inquisitor's House, with their backs all gore blood, and swollen with great bumps, and were then taken from their horses, and carried again to prison, where they remained until they were sent into Spain to the galleys, there to receive the rest of their martyrdom: and I and the 6 other with me which had judgement, and were condemned amongst the rest to serve an apprenticeship in the monastery, were taken presently and sent to certain religious houses.

I Miles Philips and William Lowe were appointed to the black friars, to be an overseer of Indian workmen, who wrought there in building of a new church: amongst which Indians I learned their language or Mexican tongue very perfectly, and had great familiarity with many of them, whom I found to be a courteous and loving kind of people, ingenious, and of great understanding, and they hate and abhor the Spaniards with all their hearts, they have used such horrible cruelties against them, and do still keep them in such subjection and servitude, that they and the negroes also do daily lay in wait to practise their deliverance out of that thraldom. We served out the years that we were condemned for, with the use of our fool's coats, and we must needs confess that the friars did use us very courteously: for every one of us had his chamber with bedding and diet, and all things clean and neat: yea many of the Spaniards and friars themselves do utterly abhor and mislike of that cruel Inquisition, and would comfort us the best they could. We were then brought again before the Chief Inquisitor, and had all our fool's coats pulled off and hanged up in the head church, and every man's name and judgement written thereupon with this addition, an heretic Lutheran reconciled. And there are also all their coats hanged up, which were condemned to the galleys, with their names and judgements, and underneath his coat, heretic Lutheran reconciled. And also the coats and names of the three that were burnt, whereupon were written, an obstinate heretic Lutheran burnt. Then were we suffered to go up and down the country, and to place ourselves as we could, and yet not so free, but that we very well knew that there was good espial always attending us and all our actions, so that we dared not once speak or look awry. David Alexander and Robert Cooke returned to serve the Inquisitor, who shortly after married them both to two of his negro women: Richard Williams married a rich widow of Biscay with 4000 pesos: for mine own part I could never thoroughly settle myself to marry in that country, although many fair offers were made unto me, but I could have no liking to live in that place, where I must everywhere see and know such horrible idolatry committed, and durst not once for my life speak against it: and therefore I had always a longing and desire to this my native country. I made my choice to learn to weave grograins and taffetas, and so compounding with a silk-weaver, I bound myself for three years to serve him, and gave him a hundred and fifty pesos to teach me the science, and by this means I lived the more quiet, and free from suspicion. I was called before the Inquisitor, and demanded why I did not marry: I answered that I had bound myself at an occupation. Well said the Inquisitor, I know thou meanest to run away, and therefore I charge thee here upon pain of burning as an heretic relapsed, that thou depart not out of this city, nor come near to the port of San Juan de Ulloa, nor to any other port: to which I answered, that I would willingly obey. Yea said he, see thou doth so.

So I remained, and learned the art, at the end whereof there came news to Mexico that there were certain Englishmen landed with a great power at the port of Acapulco, upon the South Sea, and that they were coming to Mexico to take the spoil thereof, which wrought a marvellous great fear amongst them, and many of those that were rich, began to shift for themselves, their wives and children: upon which hurly burly the Viceroy caused a general muster to be made of all the Spaniards in Mexico. Then was I Miles Philips sent for before the Viceroy, and were examined if we did know an English man named Francis Drake, which was brother to Captain Hawkins: to which we answered, that Captain Hawkins had not any brother but one, which was a man of the age of threescore years or thereabouts, and was now governor of Plymouth in England. And then he demanded of us if we knew one Francis Drake, and we answered, no.

There were eight hundred men made out under the leading of several captains, whereof two hundred were sent to Acapulco, the port where it was said that Captain Drake had been. They had for captain Doctor Robles Alcalde de Corte, with whom I Miles Philips went as interpreter, having licence given by the Inquisitors. When we were come to Acapulco, we found that Captain Drake was departed from thence, more than a month before. Our captain embarked himself in a small ship of threescore ton or thereabout, with whom I went as interpreter in his own ship: we being embarked kept our course and ran southward towards Panama, and have coasted thus, and being more to the south than Guatemala, we met at last with other ships which came from

Panama, of whom we were certainly informed that he was clean gone off that coast: and so we returned back to Acapulco again. All the while I was at sea, with them, I was a glad man, for I hoped that if we met with Master Drake, we should all be taken, so that then I should have been freed out of that danger and misery wherein I lived, and should return to mine own country of England again. Little doth any man know the sorrow and grief that inwardly I felt, although outwardly I was constrained to make fair weather of it. Our captain made report to the Viceroy what he had done. Then again I was commanded by the Viceroy that I should not depart the city of Mexico, but always be at my master's house in a readiness at an hour's warning, notwithstanding within one month after certain Spaniards going 18 leagues from Mexico, to send away certain hides and cochineal, and my master having leave of the secretary for me to go with them, I took my journey with them being very well horsed and appointed. We had perfect intelligence that the fleet was ready to depart, I not being past 3 days journey from the port of San Juan de Ulloa, thought it to be the meetest time for me to make an escape, and I was the bolder, presuming upon my Spanish tongue, which I spoke as naturally as any of them all, thinking with myself, that when I came to San Juan de Ulloa, I would get to be entertained as a soldier, and so go home into Spain in the same fleet, and therefore secretly one evening late, the moon shining fair, I conveyed myself away, and riding so for the space of two nights and two days sometimes in, and sometimes out, resting very little all that time, upon the second day at night I came to the town of Vera Cruz. I was no sooner alighted, but within the space of one half hour after, I was by ill hap arrested, and brought before justices there, being taken and suspected to be a gentleman's son of Mexico, that was run away from his father. There being a great hurly burly about this matter, every man charging me that I was the son of such a man dwelling in Mexico, which I flatly denied, affirming that I knew not the man, yet would they not believe me, but urged still upon me that I was he that they sought for, and so I was conveyed away to prison. And as I was thus going to prison, to the further increase of my grief, it chanced that at that very instant there was a poor man in the press that was come to town to sell hens, who told the justices that they did me wrong, and that in truth he knew very well that I was an Englishman, and no Spaniard, and one of Captain Hawkins' men, and that he had known me wear the *sanbenito* in the black friars at Mexico, for 3 or 4 whole years together: when they perceived that I could not deny, and perceiving that I was run from Mexico, and came thither of purpose to convey myself away with the fleet, I was presently committed to prison with a sorrowful heart, often wishing myself that that man which knew me had at that time been further off: howbeit he in sincerity had compassion of my distressed estate, thinking by his speech, and knowing of me, to have set me free from that present danger which he saw me in: howbeit, contrary to his expectation, I was thereby brought into my extreme danger, and to the hazard of my life. I was no sooner brought into prison, but I had a great pair of bolts clapped on my legs, and thus I remained in that prison for the space of three weeks, where were also many other prisoners, condemned to the galleys. My prison fellow caused his friend which came often unto him to the grate of the prison, to bring him wine and victuals, to buy for him 2 knives which had files in their backs, which files were so well made that they would suffice any prisoner to file off his irons, and of those knives or files he brought one to me, and told me that he had caused it to be made for me, and let me have it at the price it cost him, which was 2 pesos, the value of 8 shillings of our money: which knife when I had it, I was a joyful man, and conveyed the same into the foot of my boot, upon the inside of my left leg, within 3 or 4 days after that I was suddenly called for, and brought before the head justice which caused those my irons with the round bolt to be stricken off and sent to a smith's in the town, where was a new pair of bolts made ready for me of another fashion, which had a broad iron bar coming between the shackles, and caused my hands to be made fast with a pair of manacles, and so was I presently laid into a wagon all alone, which was there ready to depart with sundry other wagons. All were loaded with sundry merchandise which came in the fleet out of Spain.

The wagon that I was in was foremost and as we travelled I being alone in the wagon, began to try if I could pluck my hands out of the manacles, although it were somewhat painful for me, yet my hands were so slender that I could pull them out, and put them in again, and ever as we went, when the wagon made most noise, and the men were busiest, I would be working to file off my bolts, 8 leagues from Vera Cruz, we came to a high hill, one of the wheels of the wagon broke, so that the other wagons went afore, and the wagon man that had charge of me set an Indian carpenter to work to mend the wheel. As it drew towards night I being alone had quickly filed off all my bolts, and so espying my time in the dark of the evening before they returned down the hill again, I conveyed myself into the woods, carrying my bolts and manacles with me, and a few biscuits, and two small cheeses. Being come into the woods I threw my irons into a thick bush, and then covered them with moss and then shifted for myself as I might all that night. Thus by the good providence of Almighty God, I was freed from mine irons all saving the collar that was about my neck, and so got my liberty the second time.

The next morning I perceived by the sun rising what way to take to escape their hands: I thought to keep my course as the woods and mountains lay, still direct south as near as I could, to convey myself far enough from that way that went to Mexico. Travelling thus in my boots with mine iron collar about my neck, and my bread and cheese, the very same forenoon I met with a company of Indians which were hunting of deer: to whom I spoke in the Mexican tongue, and told them how that I had of a long time been kept in prison by the cruel Spaniards, and did desire them to help me to file off mine iron collar, which they willingly did: rejoicing greatly with me, that I was thus escaped out of the Spaniards' hands. Then I desired that I might have one of them to guide me out of those desert mountains towards the south, which they also most willingly did: and so they brought me to an Indian town 8 leagues distant where I stayed three days, for that I was somewhat sickly. At which town (with the gold that I had quilted in my doublet) I bought me a horse of one of the Indians, which cost me six pesos, and so travelling south, within the space of two leagues I happened to overtake a grey friar, one that I had been familiar withall in Mexico, whom then I knew to be a zealous good man, and one that did much lament the cruelty used against us by the Inquisitors, and truly he used me very courteously: and I having confidence in him did indeed tell him, that I was minded to adventure to see if I could get out of the said country if I could find shipping, and did therefore pray him of his aid, which he faithfully did, not only in directing me which was my safest way to travel, but he also of himself kept me company for the space of three days, and ever as we came to the Indians' houses, he gathered among them in money to the value of 20 pesos, which at my departure from him he freely gave unto me. So came I to the city of Guatemala upon the South Sea, which is distant from Mexico about 250 leagues, where I stayed 6 days, for that my horse was weak. And from thence I travelled still south and by east seven days journey: here I hired two Indians to be my guides, and I bought hens, and bread to serve us so long time, and took with us things to kindle fire every night, because of wild beasts, and to dress our meat: and every night when we rested, my Indian guides would make two great fires, between the which we placed ourselves, and my horse. And in the night time we should hear the lions roar, with tigers, ounces, and other beasts, and some of them we should see in the night, which had eyes shining like fire. And travelling thus for the space of twelve days, we came at last to the port of Cavallos[57] upon the east sea. This is a good harbour for ships, and is without either castle or bulwark. I having dispatched away my guides, went down to the haven, where I saw certain ships laden chiefly with Canary wines, where I spake with one of the masters, who asked me what country man I was, and I told him that I was born in Granada, and he said that then I was his countryman. I required him that I might pass home with him in his ship, paying for my passage: and he said yea, so that I had a safe conduct, or letter testimonial to show, that he might incur no danger: for he said, it may be that you have killed some man or be indebted, and would therefore run away. To that I answered that there was not any such cause. Well, in the end we grew to a price, that for 60 pesos he would carry me into Spain: a glad man was I at this good hap, and I quickly sold my horse, and made my provision of hens and bread to serve me in my passage; and thus within 2 days after we set sail, and never stayed until we came to Havana, where we found the whole fleet of Spain, which was bound home from the Indies. And here I was hired for a soldier to serve in the admiral ship of the same fleet.

Thus we set sail, and had a very ill passage home, the weather was so contrary. We kept our course in manner north east, and never saw land till we fell with the Arenas Gordas hard by San Lucar. And there was an order taken that none should go on shore until he had licence: as for me, I was known by one in the ship, who told the master that I was an Englishman, which (as God would) it was my good hap to hear: for if I had not heard it, it had cost me my life. Notwithstanding, I would not take any knowledge of it, and seemed to be merry and pleasant, that we were all come so well in safety. Presently after, licence came that we should go on shore, and I pressed to be gone with the first: howbeit, the master came unto me, and said, Sirrah, you must go with me to Seville by water: I knew his meaning well enough, and that there he meant to offer me up as a sacrifice to the Holy House. For the ignorant zeal of a number of those superstitious Spaniards is such, that they think that they have done God good service, when they have brought a Lutheran heretic to the fire to be burnt. Well, I perceiving all this, took upon me not to suspect anything, but was still jocund and merry: howbeit, I knew it stood me upon to shift for myself. And so waiting my time when the master was in his cabin asleep, I conveyed myself secretly down by the shrouds into the shipboat, and made no stay but cut the rope wherewithal she was moored, and so by the cable hauled on shore, where I leapt on land, and let the boat go whither it would. Thus by the help of God I escaped that day, and then never stayed at San Lucar, but went all night by the way which I had seen other take towards Seville: so that the next morning I came to Seville, and sought me out a workmaster, that I might fall to weaving of taffetas; and I set myself close to my work, and durst not for my life once to stir abroad for fear of being known:

and being thus at my work, within 4 days after I heard one of my fellows say, that he heard there was great inquiry made for an Englishman that came home in the fleet: what an heretic Lutheran (quoth I) was it, I would to God I might know him, surely I would present him to the Holy House. And thus I kept still within doors at my work, and feigned myself not well at ease, and that I would labour as I might to get me new clothes. And continuing thus for the space of three months I called for my wages, and bought me all things new, different from the apparel that I did wear at sea, and yet durst not be overbold to walk abroad: after understanding that there were certain English ships at San Lucar bound for England, I took a boat and went aboard one of them and desired the master that I might have passage with him to go into England, and told him secretly that I was one of those which Captain Hawkins did set on shore in the Indies: he very courteously prayed me to have him excused, for he durst not meddle with me, and prayed me therefore to return from whence I came. Which when I perceived, with a sorrowful heart, God knoweth, I took my leave of him, not without watery cheeks. And then I went to St Mary Port[58] which is 3 leagues from San Lucar, where I put myself to be a soldier to go in the King of Spain's galleys, which were bound for Majorca, and coming thither in the end of the Christmas holidays, I found there two English ships, the one of London, and the other of the West Country, which were ready freighted and stayed but for a fair wind. To the master of the one, which was of the West Country went I, and told him that I had been 2 years in Spain to learn the language, and that I was now desirous to go home and see my friends, for that I lacked maintenance: and so having agreed with him for my passage, I took shipping. And thus through the providence of Almighty God, after 16 years absence, having sustained many and sundry great troubles and miseries, I came home to this my native country of England in the year 1582 in the month of February, in the ship called the *Landret*, and arrived at Poole.

XXX

The names of such countries as I Anthony Jenkinson have travelled unto, from the second of October 1546, at which time I made my first voyage out of England, until the year of our Lord 1572, when I returned last out of Russia.

First, I passed into Flanders, and travelled from thence through Germany, passing over the Alps I travelled into Italy, and from thence made my journey through Piedmont into France, throughout all which realm I have thoroughly journeyed.

I have also travelled through the kingdoms of Spain and Portugal, I have sailed through the Levant seas every way, and have been in all the chief islands within the same sea, as Rhodes, Malta, Sicily, Cyprus, Candia and divers others.

I have been in many parts of Greece, Morea, Achaia, and where the old city of Corinth stood.

I have travelled through a great part of Turkey, Syria, and divers other countries in Asia Minor.

I have passed over the mountains of Lebanon to Damascus, and travelled through Samaria, Galilee, Philistine or Palestine, unto Jerusalem, and so through all the Holy Land.

I have been in divers places of Africa, as Algiers, Cola, Bona, Tripolis, the gullet within the gulf of Tunis.

I have sailed far northward within the mare glaciale, where we have had continual day, and sight of the sun ten weeks together, and that navigation was in Norway, Lapland, Samogitia,[59] and other very strange places.

I have travelled through all the ample dominions of the Emperor of Russia and Moscovy, which extend from the north sea, and the confines of Norway and Lapland, even to the Mare Caspium.

I have been in divers countries near about the Caspian Sea, as

Kazan, Crimea, Nogaia,[60] with divers others of strange customs and religions.

I have sailed over the Caspian Sea, and discovered all the regions thereabout adjacent, as Cherkesy, Shirvan, with many others.

I have travelled 40 days' journey beyond the said sea, towards the Oriental India, and Cathay, through divers deserts and wildernesses, and passed through 5 kingdoms of the Tartars, and all the lands of Turcoman[61] and Zagatay,[62] and so to the great city of Bokhara in Bactria, not without great perils and dangers sundry times.

After all this, in An. 1562, I passed again over the Caspian Sea another way, and landed in Armenia, at a city called Derbent, built by Alexander the Great, and from thence travelled through Media, Parthia,[63] Hircania, into Persia to the court of the great Sophy called Shah Tamasso, unto whom I delivered letters from the Queen's Majesty, and remained in his court 8 months, and returning homeward, passed through divers other countries. Finally I made two voyages more after that out of England into Russia, the one in the year 1566, and the other in the year 1571. And thus being weary and growing old, I am content to take my rest in mine own house, chiefly comforting myself in that my service has been honourably accepted, and rewarded of Her Majesty and the rest by whom I have been employed.

XXXI

The request of an honest merchant to a friend of his, to be advised and directed in the course of killing the whale. An. 1575. These requests were thus answered.

A proportion for the setting forth of a ship of 200 tons, for the killing of the whale.

There must be 55 men who departing for Wardhouse in the month of April, must be furnished with 4 quintals and a half of bread for every man.

250 hogsheads to put the bread in.
150 hogsheads of cider.
6 quintals of oil.
8 quintals of bacon.
6 hogsheads of beef.
10 quarters of salt.
150 pound of candles.
8 quarters of beans and peas.
Saltfish and herring, a quantity convenient.
4 tuns of wines.
Half a quarter of mustard seed, and a quern.
A grindstone.
800 empty shaken hogsheads.
350 bundles of hoops.
800 pair of heads for the hogsheads.
10 estachas for harpoon irons.
3 pieces of baibens[64] for the javelins small.
2 tackles to turn the whales.
A hawser of 27 fathoms long to turn the whales.
15 great javelins.
18 small javelins.
50 harpoon irons.
6 machicos to cut the whale withall.
2 dozen of machetos to mince the whale.
2 great hooks to turn the whale.
3 pair of canhooks.
6 hooks for staves.
3 dozen of staves for the harpoon irons.
6 pulleys to turn the whale with.
10 great baskets.
10 lamps of iron to carry light.
5 kettles of 150 lbs the piece, and 6 ladles.
1000 of nails for the pinnaces.
500 of nails for the houses, and the wharf.
18 axes and hatchets to cleave wood.
12 pieces of lines, and 6 dozen of hooks.
2 beetles of rosemary.

4 dozen of oars for the pinnaces.
6 lanterns.
Item, gunpowder and matches for arquebuses as shall be needful.
Item, there must be carried from hence 5 pinnaces, five men to strike with harpoon irons, two cutters of whale, 5 coopers, and a purser or two.

XXXII

Certain arguments to prove a passage by the Northwest, learnedly written by Mr Richard Willes gentleman.

Four famous ways there be spoken of to those fruitful and wealthy islands, which we do usually call Moluccas, continually haunted for gain, and daily travelled for riches therein growing. These islands stand east from the meridian, distant almost half the length of the world, in extreme heat, under the equinoctial line, possessed of infidels and barbarians: yet great abundance of wealth there is painfully sought in respect of the voyage dearly bought, and from thence dangerously brought home. The Portuguese voyage is very well understood of all men, and the southeastern way round about Africa by the Cape of Good Hope more spoken of, better known and travelled, than it may seem needful to discourse thereof.

The second way lieth southwest, between the West Indies or South America, and the south continent, through that narrow strait where Magellan first passed these latter years, leaving thereunto his name. The way no doubt the Spaniards would commodiously take, for that it lieth near unto their dominions there, could the current and winds as easily suffer them to return: for the which impossibility of striving against the force both of wind and stream, this passage is little or nothing used, although it be very well known.

The third way by the northeast, beyond all Europe and Asia,

that worthy and renowned knight Sir Hugh Willoughby sought to his peril, enforced there to end his life for cold, congealed and frozen to death. And truly this consisteth rather in the imagination of geographers, than allowable either in reason, or approved by experience, as well it may appear by the unlikely sailing in that northern sea always clad with ice and snow, the foul mists and dark fogs in the cold climate, the little power of the sun to clear the air, the uncomfortable nights so near the Pole, five months long.

A fourth way to go unto these aforesaid happy islands Molucca Sir Humphrey Gilbert a learned and valiant knight discourseth at large. But the way is dangerous, the passage doubtful, the voyage not thoroughly known.

First, who can assure us of any passage rather by the north west, than by the north east? Do not both ways lie in equal distances from the North Pole? Is not the ocean sea beyond America farther distant from our meridian by 30 or 40 degrees west, than the extreme points of Cathay eastward, if Ortelius' general card of the world be true? In the northeast that noble knight Sir Hugh Willoughby perished for cold: and can you then promise a passenger any better hap by the northwest?

Grant the West Indies not to continue continent unto the Pole, grant there be a passage between these two lands, let the gulf lie nearer us than commonly in cards we find it set. Let the way be void of all difficulties, yet doth it not follow that we have free passage to Cathay. For example's sake: in the Mediterranean sea, we sail to Alexandria in Egypt, the barbarians bring their pearls and spices from the Moluccas up the Red Sea or Arabian gulf to Suez, scarcely three days journey from the aforesaid haven: yet have we no way by sea from Alexandria to the Moluccas, for that isthmus or little strait of land between the two seas. In like manner although the northern passage be free at 61 degrees of latitude, and the west ocean beyond America, usually called Mar del Sur, known to be open at 40 degrees elevation from the island Japan, yet three hundred leagues northerly above Japan: yet may there be land to hinder the through passage that way by sea, America there being joined together in one continent.

Furthermore it were to small purpose to make so long, so painful, so doubtful a voyage by such a new found way, if in Cathay you should neither be suffered to land for silks and silver, nor able to fetch the Molucca spices and pearl for piracy in those seas.

Finally, all this great labour would be lost, all these charges spent in vain, if in the end our travellers might not be able to return again, and bring safely home into their own native country that wealth and riches, which they in foreign regions with adventure of goods, and danger of their lives have sought for. By the northeast there is no way, the southeast passage the Portuguese do hold as the lords of those seas. At the southwest Magellan's experience hath taught us the eastern current striketh so furiously on that strait, and falleth with such force into that narrow gulf, that hardly any ship can return that way.

To answer the objection, besides Cabot and all other travellers' navigations, the only credit of Mr Frobisher may suffice, who lately through all these islands of ice and mountains of snow, passed that way, even beyond the gulf that tumbleth down from the north, and in some places though he drew one inch thick ice, as he returning in August did, yet came he safely home again.

Whence I pray you came the contrary tide, that Mr Frobisher met withall after that he had sailed no small way in that passage, if there be any isthmus or strait of land betwixt the aforesaid northwestern gulf, and Mar del Sur, to join Asia and America together?

The rude Indian canoe hauleth those seas, the Portuguese, the Saracens, and Moors travel continually up and down that reach from Japan to China, from China to Malacca, from Malacca to the Moluccas: and shall an Englishman, better appointed than any of them all (that I say no more of our navy) fear to sail in that ocean? What seas at all do want piracy? What navigation is there void of peril?

Our travellers need not to seek their return by the northeast, neither shall they be constrained, except they list, either to attempt Magellan's strait at the southwest, or to be in danger of the Portuguese for the southeast: they may return by the northwest, that same way they do go forth.

XXXIII

The worthy enterprise of John Fox an Englishman in delivering 266 Christians out of the captivity of the Turks at Alexandria, the 3 of January 1577.

Among our merchants here in England, it is a common voyage to traffic into Spain: whereunto a ship, being called *The Three Half Moons*, manned with 38 men, and well fenced with munitions, the better to encounter their enemies withal, and having wind and tide, set from Portsmouth, 1563, and bended her journey toward Seville a city in Spain, intending there to traffic with them. And falling near the straits, they perceived themselves to be beset round with eight galleys of the Turks, in such wise, that there was no way for them to fly or escape away, but that either they must yield or else be sunk. Which the owner perceiving, manfully encouraged his company, exhorting them valiantly to show their manhood, putting them likewise in mind of the old and ancient worthiness of their country men, who in the hardest extremities have always most prevailed and gone away conquerors, yea, and where it hath been almost impossible. Such (quoth he) hath been the valiantness of our country men, and such hath been the mighty power of our God.

Then stood up one Grove the master, being a comely man, with his sword and target, holding them up in defiance against his enemies. So likewise stood up the owner, the master's mate, boatswain, purser, and every man well appointed. Now likewise sounded up the drums, trumpets and flutes, which would have encouraged any man, had he never so little heart or courage in him.

Then taketh him to his charge John Fox the gunner in the disposing of his pieces in order to the best effect, and sending his bullets towards the Turks. Shortly they drew near, so that the bowmen fell to their charge in sending forth their arrows so thick amongst the galleys, and also in doubling their shot so sore upon the galleys, that there were twice so many of the Turks slain, as the number of the Christians were in all. But the Turks discharged twice as fast against the Christians, and so long, that the ship was very sore stricken and bruised under water. Which the Turks perceiving, made the more haste to come aboard the ship: which ere they could do, many a Turk bought it dearly with the loss of their lives. Yet was all in vain, and boarded they were, where they found so hot a skirmish, that it had been better they had not meddled with the feast. For the Englishmen showed themselves men in deed, in working manfully with their brown bills and halberds: where the owner, master, boatswain, and their company stood to it so lustily, that the Turks were half dismayed. But chiefly the boatswain showed himself valiant above the rest: for there was none of them that either could or durst stand in his face, till at the last there came a shot from the Turks, which broke his whistle asunder, and smote him on the breast, so that he fell down, bidding them farewell: the press and store of the Turks was so great, that they were not long able to endure, but were so overpressed, that they could not wield their weapons: by reason whereof, they must needs be taken, which none of them intended to have been, but rather to have died: except only the master's mate, who shrunk from the skirmish, like a notable coward, esteeming neither the value of his name, nor accounting of the present example of his fellows. In fine, the Turks were victors. The Christians must needs to the galleys, to serve in new offices: and they were no sooner in them, but their garments were pulled over their ears, and torn from their backs, and they set to the oars.

Nigh to the city of Alexandria, being a haven town, and under the dominion of the Turks, there is a road, being made very fencible with strong walls, whereinto the Turks do customly bring their galleys on shore every year, in the winter season, and there do trim them, and lay them up against the spring time. In which road there is a prison, wherein the captives and such prisoners as serve in the galleys, are put for all that time, until the seas be calm and passable for the galleys, every prisoner being most grievously laden with irons on their legs, to their great pain, and sore disabling of them. Into which prison were these Christians put, and fast warded all the winter season. But ere it was long, the master and the owner, by means of friends, were redeemed: the rest abiding still by the misery, saving one John Fox, who being somewhat skilful in the craft of a barber, by reason thereof made great shift in helping his fare now and then with a good meal. Insomuch, till at the last, God sent him favour in the sight of the keeper of the prison, so that he had leave to go in and out to the road, at his pleasure, paying a certain stipend unto the keeper, and wearing a lock about his leg: which liberty likewise, six more had upon like sufferance: who by reason of their long imprisonment, not being feared or suspected to start aside, or that they would work the Turks any mischief, had liberty to go in and out at the said road, in such manner, as this John Fox did, with irons on their legs, and to return again at night.

In the year of our Lord 1577 in the winter season, the galleys happily coming to their accustomed harbour, and being discharged of all their masts, sails, and other such furnitures, and all the masters and mariners of them being then nested in their own homes: there remained in the prison of the said road two hundred threescore and eight Christian prisoners, who had been taken by the Turks' force, and were of sixteen sundry nations. Among which there were three Englishmen, whereof one was named John Fox of Woodbridge in Suffolk, the other William Wickney of Portsmouth, in the county of Southampton, and the third Robert Moore of Harwich in the county of Essex. Which John Fox having been thirteen or fourteen years under their gentle entreatance, and being too too weary thereof, minding his escape, weighed with himself by what means it might be brought to pass: and continually pondering with himself thereof, took a good heart unto him, in hope that God would not be always scourging his children.

Not far from the road, and somewhat from thence, at one side of the city, there was a certain victualling house, which one Peter Unticaro had hired, paying also a certain fee unto the keeper of the road. This Peter Unticaro was a Spaniard born, and a Christian, and had been prisoner about thirty years, and never practised any means to escape, but kept himself quiet: until that now this John Fox using much thither, they broke one to another their minds, concerning the restraint of their liberty and imprisonment. So that this John Fox at length opening unto this Unticaro the device which he would fain put in practice. At seven weeks end they had sufficiently concluded how the matter should be: who making five more privy to this their device, whom they thought they might safely trust, determined in three nights after to accomplish their deliberate purpose. Whereupon the same John Fox, and Peter Unticaro, and the other six appointed to meet all together in the prison the next day, being the last day of December: where this John Fox certified the rest of the prisoners, what their intent and device was, and how and when they minded to bring their purpose to pass: who thereunto persuaded them without much ado to further their device. Which the same John Fox seeing, delivered unto them a sort of files, which he had gathered together for this purpose, by the means of Peter Unticaro, charging them that every man should be ready discharged of his irons by eight of the clock on the next day.

Next night, John Fox and his six other companions, being all come to the house of Peter Unticaro, passing the time away in mirth for fear of suspect, till the night came on, so that it was time for them to put in practice their device, sent Peter Unticaro to the master of the road, in the name of one of the masters of the city, with whom this keeper was acquainted. The keeper agreed to go with him, willing the warders not to bar the gate, saying, that he would not stay long, but would come again with all speed.

In the mean season, the other seven had provided them of such weapons, as they could get in that house: and John Fox took him to an old rusty sword blade, without either hilt or pommel, which he made to serve his turn, in bending the hand end of the sword, instead of a pommel, and the other had got such spits and glaives as they found in the house.

The keeper now being come unto the house, and perceiving no light, nor hearing any noise, straightway suspected the matter: John Fox standing behind the corner of the house, stepped forth unto him: who perceiving it said, O Fox, what have I deserved of thee, that thou shouldest seek my death? Thou villain (quoth Fox) hast been a bloodsucker of many a Christian's blood, and now thou shalt know what thou hast deserved at my hands: wherewith he lift up his sword, and stroke him so main a blow, as therewithal his head clave asunder, so that he fell stark dead to the ground. Whereupon Peter Unticaro went in, and certified the rest how the case stood with the keeper: who came presently forth, and some with their spits ran him through, and others with their glaives hewed him in sunder, cut off his head, and mangled him so, that no man should discern what he was.

Then marched they towards the road, whereinto they entered softly, where were six warders, whom one of them asked, saying, who was there? quoth Fox and his company, all friends. When they were all within, quoth Fox, my masters, here is not to every man a man, wherefore look you play your parts. Who so behaved themselves, that they had dispatched these six quickly. Then John Fox barred the gate surely, and planted a cannon against it.

Then entered they into the gaoler's lodge, where they found the keys of the fortress and prison by his bedside, and there had they all better weapons. In this chamber was a chest, wherein was a rich treasure, and all in ducats, which this Peter Unticaro, and two more opening, stuffed themselves so full as they could, between their shirts and their skin: which John Fox would not once touch and said that it was his and their liberty which he sought for, to the honour of his God, and not to make a mart of the wicked treasure of the infidels.

Now these eight being armed with such weapons as they thought well of, thinking themselves sufficient champions to encounter a stronger enemy, and coming unto the prison, Fox opened the gates and doors thereof and called forth all the prisoners, whom he set, some to ramming up the gate, some to the dressing up of a certain galley, which was the best in all the road, whereinto some carried masts, sails, oars, and other such furniture as doth belong unto a galley.

At the prison were certain warders, whom John Fox and his company slew: in the killing of whom, there were eight more

Turks, which perceived them, and got them to the top of the prison: unto whom John Fox, and his company, were fain to come to by ladders, where they found a hot skirmish. For some of them were there slain, some wounded, and some but scarred, and not hurt. As John Fox was thrice shot through his apparel, and not hurt, Peter Unticaro, and the other two, not able to wield themselves, being so pestered with the weight and uneasy carrying of the wicked and profane treasure: and also divers Christians were as well hurt about that skirmish, as Turks slain.

Amongst the Turks was one thrust through, who fell off from the top of the prison wall, and made such a lowing, that the inhabitants thereabouts understood the case, that the prisoners were paying their ransoms: wherewith they raised both Alexandria, which lay on the west side of the road, and a castle which was at the city's end, next to the road, and also another fortress which lay on the northside of the road: so that now they had no way to escape, but one, which might seem impossible to be a way for them. So was the Red Sea impossible for the Israelites to pass through. So was it impossible, that the walls of Jericho should fall down. Such impossibilities can our God make possible.

Now is the road fraught with lusty soldiers, labourers, and mariners, who are fain to stand to their tackling, in setting to every man his hand, some to the carrying in of victuals, some munitions, some oars, and some one thing, some another, but most are keeping their enemy from the wall of the road. But to be short, there was no time misspent, no man idle, nor any man's labour ill bestowed, or in vain. So that in short time, this galley was ready trimmed up. Whereinto every man leaped in all haste, hoisting up the sails lustily, yielding themselves to his mercy and grace, in whose hands are both wind and weather.

Now is this galley afloat, and out of the safety of the road: now have the two castles full power upon the galley, now is there no remedy but to sink: how can it be avoided? The cannons let fly from both sides, and the galley is even in the midst, and between them both.

There was not one of them that feared the shot, which went thundering round about their ears, nor yet were once scarred or touched, with five and forty shot, which came from the castles. Here did God hold forth his buckler, he shieldeth now this galley, and hath tried their faith to the uttermost. For they sailed away, being not once touched with the glance of a shot, and are quickly out of the Turkish cannon's reach. Then might they see them coming down by heaps to the water side, in companies like unto swarms of bees, making show to come after them with galleys, in bustling themselves to dress up the galleys, which would be a swift piece of work for them to do, for that they had neither oars, masts, sails, nor anything else ready in any galley. But yet they are carrying them into them, some into one galley, and some into another, so that, being such a confusion amongst them, without any certain guide, it were a thing impossible to overtake them: beside that, there was no man that would take charge of a galley, the weather was so rough, and there was such an amazedness amongst them.

When the Christians were safe out of the enemy's coast, John Fox called to them all, willing them to be thankful unto Almighty God for their delivery, and most humbly to fall down upon their knees, beseeching him to aid them unto their friend's land, and not to bring them into another danger, sith he had most mightily delivered them from so great a thraldom and bondage.

Thus when every man had made his petition, they fell straight way to their labour with the oars, in helping one another, when they were wearied, and with great labour striving to come to some Christian land, as near as they could guess by the stars. But the winds were so diverse, one while driving them this way, another while that way, that they were now in a new maze. Having no victuals in the galley, it might seem that one misery continually fell upon another's neck: but to be brief, the famine grew so great, that in 28 days, wherein they were on the sea, there died eight persons, to the astonishment of all the rest.

So it fell out, that upon the 29th day, after they set from Alexandria, they fell upon the Isle of Candy, and landed at Gallipoli,[65] where they were made much of by the Abbot and monks there, who caused them to stay there, while they were well refreshed and eased. They kept there the sword, wherewith John Fox had killed the keeper, esteeming it as a most precious jewel, and hung it up for a monument.

When they thought good, having leave to depart from thence, they sailed along the coast, till they arrived at Tarento, where they sold their galley, and divided it, every man having a part thereof. And then they came afoot to Naples, where they departed asunder, every man taking him to his next way home. From whence John Fox took his journey unto Rome, where he was well entertained of an Englishman, who presented his worthy deed unto the Pope, who rewarded him liberally, and gave him his letters unto the King of Spain, where he was very well entertained of him there, who for this his most worthy enterprise gave him in fee twenty pence a day. From whence, being desirous to come into his own country, he came thither at such time as he conveniently could, which was in the year of our Lord God, 1579. Who being come into England, went unto Court, and showed all his travel unto the Council: who considering of the state of this man, in that he had spent and lost a great part of his youth in thraldom and bondage, extended to him their liberality, to help to maintain him now in age, to their right honour, and to the encouragement of all true hearted Christians.

XXXIV

The famous voyage of Sir Francis Drake into the South Sea, and there hence about the whole globe of the earth, begun in the year of our Lord, 1577.

The 15 day of November, in 1577, Mr Francis Drake, with a fleet of five ships and barks, and 164 men, gentlemen and sailors, departed from Plymouth, giving out his pretended voyage for Alexandria.

Upon the coast of Barbary, the 27 day we found an island called Mogador, between which island and the main, we found a very good and safe harbour for our ships to ride in.

On this island our general erected a pinnace, whereof he brought out of England with him four already framed.

We departed from this place the last day of December, and coasting along the shore, we did descry certain Spanish fishermen, to whom we gave chase and took three of them, and proceeding further we met with three caravels and took them also.

The 17 day of January we arrived at Cabo Blanco, where we remained 4 days, and in that space our general mustered, and trained his men on land in warlike manner, to make them fit for all occasions.

We departed this harbour the 22 of January, carrying along with us one of the Portuguese caravels which was bound to the islands of Cape Verde for salt.

Upon one of those islands called Maio, we gave ourselves a little refreshing. The island is wonderfully stored with goats and wild hens, and it hath salt also without labour, the people gather it into heaps, which continually in great quantity is increased upon the sands by the flowing of the sea, and the receiving heat of the sun.

Amongst other things we found here a kind of fruit called cocos, which because it is not commonly known with us in England, I thought good to make some description of it.

The tree beareth no leaves nor branches, but at the very top the fruit groweth in clusters, hard at the top of the stem of the tree, as big every several fruit as a man's head: but having taken off the uttermost bark, which you shall find to be very full of strings or sinews, as I may term them, you shall come to a hard shell which may hold of quantity in liquor a pint commonly, or some a quart, and some less: within that shell of the thickness of half an inch good, you shall have a kind of hard substance and very white, no less good and sweet than almonds: within that again a certain clear liquor, which being drunk, you shall not only find it very delicate and sweet, but most comfortable and cordial.

Our general departed hence the 31 of this month, and sailed by the island of San Tiago, but far enough from the danger of the inhabitants, who shot and discharged at us three pieces, but they all fell short of us, and did us no harm. The mountains and high places of the island are said to be possessed by the Moors, who having been slaves to the Portuguese, made escape to the desert places of the island, where they abide with great strength.

We espied two ships under sail, to the one of which we gave chase, and in the end boarded her with a ship-boat without resistance, and she yielded unto us good store of wine.

Being departed from these islands, we drew towards the line, where we were becalmed the space of 3 weeks, but yet subject to diverse great storms, terrible lightnings and much thunder: but with this misery we had the commodity of great store of fish, as dolphins, bonitos, and flying fishes, whereof some fell into our ships, where hence they could not rise again for want of moisture, for when their wings are dry, they cannot fly.

The first land that we fell with was the coast of Brazil, which we saw the fifth of April in the height of 33 degrees towards the pole Antarctic, and being discovered at sea by the inhabitants of the country, they made upon the coast great fires for a sacrifice (as we learned) to the devils, about which they use conjurations, making heaps of sand and other ceremonies, that when any ship shall go about to stay upon their coast, not only sands may be gathered together into shoals in every place, but also that storms and tempests may arise, to the casting away of ships and men.

The place where we met, our general called the Cape of Joy, where every ship took in some water. Here we found a good temperature and sweet air, a very fair and pleasant country with an exceedingly fruitful soil, where were great store of large and mighty deer, but we came not to the sight of any people: but travelling further into the country, we perceived the footing of people in the clay-ground, showing that they were men of great stature. Being returned to our ships, we weighed anchor, and harboured ourselves between a rock and the main, where by means of the rock that broke the force of the sea, we rode very safe, and upon this rock we killed for our provision certain sea-wolves, commonly called with us seals.

From hence we went our course to 36 degrees, and entered the great river of Plate, and ran into 54 and 55 fathoms and a half of fresh water, but our general finding here no good harbour, as he thought he should, bare out again to sea the 27 of April, but we sailing along, found a fair and reasonable good bay wherein were many, and the same profitable islands, one whereof had so many seals, as would at the least have laden all our ships.

Our general being on shore in an island, the people of the country showed themselves unto him, leaping and dancing, and entered into traffic with him, but they would not receive any things at any man's hands, but the same must be cast upon the ground. They are of clean, comely, and strong bodies, swift on foot, and seem to be very active.

We watered and made new provision of victuals, as by seals, whereof we slew to the number of 200 or 300 in the space of an hour.

The next day, we harboured ourselves again in a very good harbour, called by Magellan Puerto San Julián, where we found a gibbet standing upon the main, which we supposed to be the place where Magellan did execution upon some of his disobedient and rebellious company.

In this port our general began to enquire diligently of the actions of Mr Thomas Doughty, and found them not to be such as he looked for, but tending rather to contention or mutiny, whereby (without redress) the success of the voyage might greatly have been hazarded: whereupon the company was called together and made acquainted with the particulars of the cause, which were found partly by Master Doughty's own confession, and partly by the evidence of the fact, to be true: which when our general saw, although his private affection to Mr Doughty (as he then in the presence of us all sacredly protested) was great, yet the care he had of the state of the voyage, of the expectation of Her Majesty, and of the honour of his country did more touch him (as indeed it ought), than the private respect of one man: so that the cause being thoroughly heard, and all things done in good order as near as might be to the course of our laws in England, it was concluded that Mr Doughty should receive punishment according to the quality of the offence: he seeing no remedy but patience for himself, desired before his death to receive the communion, which he did at the hands of Mr Fletcher our minister, and our general himself accompanied him in that holy action: which being done, and the place of execution made ready, he having embraced our general and taken his leave of all the company, with prayer for the Queen's Majesty and our realm, in quiet sort laid his head to the block, where he ended his life. This being done, our general made divers speeches to the whole company, persuading us to unity, obedience, love, and regard of our voyage; and for the better confirmation thereof, willed every man the next Sunday following to prepare himself to receive the communion, as Christian brethren and friends ought to do, which was done in very reverent sort, and so with good contentment every man went about his business.

The 20 day we fell with the Strait of Magellan going into the South Sea, at the cape or headland whereof we found the body of a dead man, whose flesh was clean consumed.

In this strait there be many fair harbours, with store of fresh water, but yet they lack their best commodity: for the water is there of such depth, that no man shall find ground to anchor in, except it be in some narrow river or corner, or between some rocks, so that if any extreme blasts or contrary winds do come (whereunto the place is much subject) it carrieth with it no small danger.

The land on both sides is very huge and mountainous, covered with snow. This strait is extremely cold, with frost and snow continually; the trees seem to stoop with the burden of the weather, and yet are green continually, and many good and sweet herbs do very plentifully grow and increase under them.

The 24 of August we arrived at an island in the straits, where we found great store of fowl which could not fly, of the bigness of geese,[66] whereof we killed in less than one day 3,000 and victualled ourselves thoroughly therewith.

The seventh day we were driven by a great storm from the entering into the South Sea two hundred leagues and odd in longitude, and one degree to the southward of the Strait: in which height, and so many leagues to the westward, the fifteenth day of September fell out the eclipse of the moon at the hour of six of the clock at night: But neither did the ecliptical conflict of the moon impair our state, nor her clearing again amend us a whit, but the accustomed eclipse of the sea continued in his force, we being darkened more than the moon seven fold.

From the bay (which we called the Bay of Severing of Friends) we were driven back to the southward of the straits in 57 degrees and a tierce: in which height we came to an anchor among the islands, having there fresh and very good water, with herbs of singular virtue. Not far from hence we entered another bay, where we found people both men and women in their canoes, naked, and ranging from one island to another to seek their meat, who entered traffic with us for such things as they had.

We returning hence northward again, found the 3 of October three islands, in one of which was such plenty of birds as is scant credible to report.

We ran, supposing the coast of Chile to lie as the general maps have described it, namely northwest, which we found to lie and trend to the northeast and eastwards, whereby it appeareth that this part of Chile hath not been truly hitherto discovered, or at least not truly reported for the space of 12 degrees at the least, being set down either of purpose to deceive, or of ignorant conjecture.

The 29 of November we cast anchor, and our general hoisting out our boat, went with ten of our company to shore, where we found people, whom the cruel and extreme dealings of the Spaniards have forced for their own safety and liberty to flee from the main, and to fortify themselves in this island. The people came down to us to the waterside with show of great courtesy, bringing to us potatoes, roots, and two very fat sheep, which our general received and gave them other things for them, and had promise to have water there: but the next day repairing again to the shore, and sending two men a land with barrels to fill water, the people taking them for Spaniards (to whom they use to show no favour if they take them) laid violent hands on them, and as we think, slew them.

Our general seeing this, stayed here no longer, but weighed anchor, and set sail towards the coast of Chile, and drawing towards it, we met near to the shore an Indian in a canoe, who thinking us to have been Spaniards, came to us and told us, that at a place called Santiago, there was a great Spanish ship laden from the kingdom of Peru: for which good news our general gave him divers trifles, whereof he was glad, and went along with us and brought us to the place, which is called the port of Valparaíso.

We found indeed the ship riding at anchor, having in her eight Spaniards and three negroes, who thinking us to have been Spaniards and their friends, welcomed us with a drum: as soon as we were entered, one of our company called Thomas Moon began to lay about him, and struck one of the Spaniards, and said unto him, *abajo perro*, that is in English, go down dog. One of these Spaniards seeing persons of that quality in those seas, crossed and blessed himself: but to be short, we stowed them under hatches all save one Spaniard, who suddenly and desperately leapt overboard into the sea, and swam ashore to the town of Santiago, to give them warning.

They of the town being not above nine households, presently fled away and abandoned the town. Our general manned his boat, and the Spanish ship's boat, and went to the town, we rifled it, and came to a small chapel which we entered, and found therein a silver chalice, two cruets, and one altar-cloth, the spoil whereof our general gave to Mr Fletcher his minister.

We found also in this town a warehouse stored with wine of Chile, and many boards of cedar-wood, all which wine we brought away with us, and certain of the boards to burn for fire-wood: we departed the haven, having first set all the Spaniards on land, saving one John Griego a Greek born, whom our general carried with him for his pilot to bring him into the haven of Lima.

At sea, our general rifled the ship, and found in her good store of the wine of Chile, and 25,000 pesos of very pure and fine gold of Valdivia, amounting in value to 37,000 ducats of Spanish money, we arrived next at a place called Coquimbo, where our general sent 14 of his men on land to fetch water: but they were espied by the Spaniards, who came with 300 horsemen and 200 footmen, and slew one of our men with a piece, the rest came aboard in safety, and the Spaniards departed: we went ashore again, and buried our man, and the Spaniards came down again with a flag of truce, but we set sail and would not trust them.

From hence we went to a certain port called Tarapaca,[67] where being landed, we found by the sea side a Spaniard lying asleep, who had lying by him 13 bars of silver, which weighed 4000 ducats Spanish; we took the silver, and left the man.

Not far from hence going on land for fresh water, we met with a Spaniard and an Indian boy driving 8 llamas or sheep of Peru which are as big as asses: every of which sheep had on his back 2 bags of leather, each bag containing 50 lbs. weight of fine silver: so that bringing both the sheep and their burthen to the ships, we found in all the bags 800 weight of silver.

Here hence we sailed to a place called Arica, and being entered the port, we found there three small barks which we rifled, and found in one of them 57 wedges of silver, each of them weighing about 20 pound weight, and every of these wedges were of the fashion and bigness of a brickbat. Our general contented with the spoil of the ships, left the town and put off again to sea and set sail for Lima.

To Lima we came the 13 day of February, and being entered the haven, we found there about twelve sail of ships lying fast moored at an anchor; for the masters and merchants were here most secure, having never been assaulted by enemies. Our general rifled these ships, and found in one of them a chest full of royals of plate, and good store of silks and linen cloth. In which ship he had news of another ship called the *Cacafuego* which was gone towards Paita [68], and that the same ship was laded with treasure: whereupon we stayed no longer here, but cutting all the cables of the ships in the haven, we let them drive whither they would, either to sea or to the shore, and with all speed we followed the *Cacafuego*: but she was gone from thence towards Panama, whom our general still pursued, and by the way met with a bark laden with ropes and tackle for ships, which he boarded and searched, and found in her 80 lbs. weight of gold, and a crucifix of gold with goodly great emeralds set in it which he took, and some of the cordage also for his own ship.

We departed, still following the *Cacafuego*, and our general promised our company that whosoever could first descry her, should have his chain of gold for his good news. It fortuned that John Drake going up into the top, descried her about three of the clock, and about six of the clock we came to her and boarded her, and shot at her three pieces of ordnance, and struck down her mizzen, and being entered, we found in her great riches, as jewels and precious stones, thirteen chests full of royals of plate, four-score pound weight of gold, and six and twenty ton of silver. The place where we took this prize, was called Cape de San Francisco, about 150 leagues from Panama.

We went on our course still towards the west, and not long after met with a ship laden with linen cloth and fine China-dishes of white earth, and great store of China-silks, of all which things we took as we listed.

The owner himself of this ship was in her, who was a Spanish gentleman, from whom our general took a falcon of gold, with a great emerald in the breast thereof, and the pilot of the ship he took also with him, and so cast the ship off.

This pilot brought us to the haven of Guatulco. We landed, and went presently to the town, and to the townhouse, where we found a judge sitting in judgement, being associate with three other officers, upon three negroes that had conspired the burning of the town: both which judges and prisoners we took, and brought them a shipboard, and caused the chief judge to write his letter to the town, to command all the townsmen to avoid that we might safely water there. Which being done, and they departed, we ransacked the town, and in one house we found a pot of the quantity of a bushel, full of reals of plate, which we brought to our ship.

And here one Thomas Moon one of our company, took a Spanish gentleman as he was flying out of the town, and searching him, he found a chain of gold about him, and other jewels, which he took, and so let him go.

Our general thinking himself both in respect of his private injuries received from the Spaniards, as also of their contempts and indignities offered to our country and prince in general, sufficiently satisfied and revenged: and supposing that Her Majesty at his return would rest content with this service, purposed to continue no longer upon the Spanish coasts, but began to consider and to consult the best way for his country.

He thought it not good to return by the Straits, for two special causes: the one, lest the Spaniards should there wait and attend for him in great strength, whose hands, he being left but one ship, could not possibly escape. The other cause was the dangerous situation of the mouth of the Straits in the South Sea, where continual storms blustering, as he found by experience, besides the shoals and sands upon the coast, he thought it not a good course to adventure that way: he resolved therefore to avoid these hazards, to go forward to the islands of the Moluccas, and there hence to sail the course of the Portuguese by the Cape of Buena Esperanza.[69]

Upon this resolution, he began to think of his best way to the Moluccas, and finding himself where he now was becalmed, he saw that of necessity he must sail somewhat northerly to get a wind. We therefore set sail, and sailed 600 leagues at the least for a good wind.

The 5 day of June, being in 43 degrees towards the pole Arctic, we found the air so cold, that our men being grievously pinched with the same, complained of the extremity thereof, and the further we went, the more the cold increased upon us. Whereupon we thought it best for that time to seek the land, and did so, finding it not mountainous, but low plain land, till we came within 38 degrees, it pleased God to send us into a fair and good bay,[70] with a good wind to enter the same.

In this bay we anchored, and the people of the country having their houses close by the water's side, showed themselves unto us, and sent a present to our general.

When they came unto us, they greatly wondered at the things that we brought, but our general (according to his natural and accustomed humanity) courteously entreated them, and liberally bestowed on them necessary things to cover their nakedness, whereupon they supposed us to be gods, and would not be persuaded to the contrary.

Their houses are digged round about with earth, and have clefts of wood set upon them, joining close together at the top like a spire steeple, which by reason of that closeness are very warm.

Their beds is the ground with rushes strewed on it, and lying about the house, have the fire in the midst. The men go naked, the women take bulrushes, and comb them after the manner of hemp, and thereof make their loose garments, which being knit about their middles, hang down about their hips, having also about their shoulders a skin of deer, with the hair upon it. These women are very obedient and serviceable to their husbands.

After they were departed from us, they came and visited us the second time, and brought with them feathers and bags of tobacco for presents: and when they came to the top of the hill (at the bottom whereof we had pitched our tents) they stayed themselves: where one appointed for speaker wearied himself with making a long oration, which done, they left their bows upon the hill, and came down with their presents.

In the meantime the women remaining on the hill, tormented themselves lamentably, tearing their flesh from their cheeks, whereby we perceived that they were about a sacrifice. In the meantime our general with his company went to prayer, and to reading of the Scriptures, at which exercise they were attentive, and seemed greatly to be affected with it.

The news of our being there spread through the country, the people that inhabited round about came down, and amongst them the king himself, a man of a goodly stature, and comely personage.

In the forefront was a man, who bore the sceptre or mace before the king, whereupon hanged two crowns, a lesser and a bigger, with three chains of a marvellous length: the crowns were made of knit work wrought artificially with feathers of divers colours: the chains were made of a bony substance, and few be the persons among them that are admitted to wear them. Next unto him, was the king himself, with his guard about his person, clad with coney skins, and other skins: after them followed the naked common sort of people, every one having his face painted, some with white, some with black, and other colours.

In the meantime our general gathered his men together, and

marched within his fenced place, making against their approaching a very war-like show.

In coming towards our bulwarks and tents, the sceptre-bearer began a song observing his measures in a dance, and that with a stately countenance, whom the king with his guard, and every degree of persons following, did in like manner sing and dance, saving only the women, which danced and kept silence. The general permitted them to enter within our bulwark, where they continued their song and dance a reasonable time. They made signs to our general to sit down, to whom the king, and divers others made supplications, that he would take their province into his hand, and become their king, making signs that they would resign unto him their right and title of the whole land, and become his subjects. In which, to persuade us the better, the king and the rest, with one consent, and with great reverence, joyfully singing a song, did set the crown upon his head, enriched his neck with all their chains: which thing our general thought not meet to reject, because he knew not what honour and profit it might be to our country. Wherefore in the name, and to the use of Her Majesty he took the sceptre, crown, and dignity of the said country into his hands.

Our necessary business being ended, our general with his company travelled up into the country to their villages, where we found herds of deer by 1,000 in a company, being most large, and fat of body.

Our general called this country Nova Albion, and that for two causes: the one in respect of the white cliffs, which lie towards the sea: and the other, because it might have some affinity with our country in name, which sometime was so called.

There is no part of earth here to be taken up, wherein there is not some probable show of gold or silver.

At our departure hence our general set up a monument of our being there, as also of Her Majesty's right and title to the same, namely a plate, nailed upon a fair great post, whereupon was engraved Her Majesty's name, the day and year of our arrival there, with the free giving up of the province and people into Her Majesty's hands, together with Her Highness' picture and arms, in a piece of six pence of current English money under the plate, whereunder was also written the name of our general.

It seemeth that the Spaniards hitherto had never been in this part of the country, neither did ever discover the land by many degrees, to the southwards.

After we had set sail from hence, we continued without sight of land till the 13 day of October, which day we fell with certain islands[71] 8 degrees to the northward of the line, from which came a great number of canoes, having in some of them 4 in some 6 and in some also 14 men, bringing with them cocos, and other fruits. Their canoes were hollow within, and cut with great art and cunning, being very smooth within and without, having a prow, and a stern of one sort, yielding inward circle-wise, being of a great height, and full of certain white shells for a bravery, and on each side of them lie out two pieces of timber about a yard and a half long, more or less.

This people have the nether part of their ears cut into a round circle, hanging down very low upon their cheeks, whereon they hang things of a reasonable weight. The nails of their hands are an inch long, their teeth are as black as pitch.

We continued our course by the islands of Tagulada, Zelon, and Zewarra,[72] being friends to the Portuguese, the first whereof hath growing in it great store of cinnamon.

The 14 of November we fell with the islands of Molucca, next morning early we came to anchor, at which time our general sent a messenger to the king with a velvet cloak for a present, and token of his coming to be in peace, and that he required nothing but traffic and exchange of merchandise, whereof he had good store.

The king was moved with great liking towards us, and sent to our general, that he should have what things he needed. In token whereof he sent to our general a signet, and within short time after came in his own person to our ship, to bring her into a better and safer road than she was in at present.

The king sent before 4 great and large canoes, in every one whereof were certain of his greatest, attired in white lawn of cloth of Calicut, having over their heads from the one end of the canoe to the other, a covering of thin perfumed mats, borne up with a frame made of reeds for the same use, under which every one did sit in his order according to his dignity, to keep him from the heat of the sun, divers of whom being of good age and gravity, did make an ancient and fatherly show. There were also divers young and comely men attired in white, as were the others: the rest were soldiers.

These canoes were furnished with war-like munition, every man for the most part having his sword and target, with his dagger, besides other weapons, as lances, calivers, darts, bows and arrows.

They rowed about us, one after another, and passing by, did their homage with great solemnity.

The king was a man of tall stature and seemed to be much delighted with the sound of our music, to whom as also to his nobility, our general gave presents.

At length the king craved leave of our general to depart, promising the next day to come aboard, and in the meantime to send us such victuals, as were necessary for our provision: so that the same night we received of them meal, which they call sago, made of the tops of certain trees, tasting in the mouth like sour curds, but melteth like sugar, whereof they make certain cakes, which may be kept the space of ten years, and yet then good to be eaten. We had of them store of rice, hens, unperfect and liquid sugar, sugar canes, with store of cloves.

The king having promised to come aboard, brake his promise, but sent his brother to make his excuse, and to entreat our general to come on shore, offering himself pawn aboard for his safe return. Whereunto the general consented not, upon mislike conceived of the breach of his promise. But to satisfy him, our general sent certain of his gentlemen to the court.

The king at last came in guarded with 12 lances covered over with a rich canopy, with embossed gold. Our men rising to meet him, he graciously did welcome, and entertain them. He was attired after the manner of the country, but more sumptuously than the rest. From his waist down to the ground, was all cloth of gold, and the same very rich: his legs were bare, but on his feet were a pair of shoes made of Cordovan skin. In the attire of his head were finely wreathed hooped rings of gold, and about his neck he had a chain of perfect gold, the links whereof were great, and one fold double. On his fingers he had six very fair jewels, and sitting in his chair of estate, at his right hand stood a page with a fan in his hand, breathing and gathering the air to the king. The fan was in length two foot, and in breadth one foot, set with 8 sapphires, richly embroidered, and knit to a staff 3 foot in length, by which the page did hold, and move it.

This island is the chiefest of all the islands of Molucca. The king with his people are Moors in religion, observing certain new moons, with fasting:[73] during which fasts, they neither eat nor drink in the day, but in the night.

Our general considering the great distance, and how far he was yet off from his country, thought it not best here to linger the time any longer, but weighing his anchors, set out, and sailed to a certain little island to the southwards of Celebes, where we graved our ship, and continued there in that and other businesses 26 days. This island is thoroughly grown with wood of a large and high growth, very straight and without boughs, save only in the head or top, whose leaves are not much differing from our broom in England. Amongst these trees night by night, through the whole land, did show themselves an infinite swarm of fiery worms flying in the air, whose bodies being no bigger than our common English flies, make such a show and light, as if every twig or tree had been a burning candle.

When we had ended our business here, we weighed, and set sail: but having at that time a bad wind, with much difficulty we recovered to the northward of the island of Celebes, where by reason of contrary winds, we were enforced to alter to the southward again, finding that course also to be very hard and dangerous for us, by reason of infinite shoals which lie off, and among the islands. Upon the 9 of January in the year 1579 we ran suddenly upon a rock, where we stuck fast from 8 of the clock at night till 4 of the clock in the afternoon the next day, being indeed out of all hope to escape the danger: but our general showed himself courageous, and of a good confidence in the mercy and protection of God: and we did our best endeavour to save ourselves, which it pleased God so to bless, that in the end we cleared ourselves most happily of the danger.

We lighted our ship upon the rocks of 3 ton of cloves, 8 pieces of ordnance, and certain meal and beans: and then the wind (as it were in a moment by the special grace of God) changing from the starboard to the larboard of the ship, we hoisted our sails, and the happy gale drove our ship off the rock into the sea again, to the no little comfort of all our hearts, for which we gave God such praise and thanks, as so great a benefit required.

The 8 of February following, we fell with the fruitful island of Barateve.[74] The people of this island are comely in body and stature, and of a civil behaviour, just in dealing, and courteous to strangers. The men go naked, saving their heads and privities, every man having something or other hanging at their ears. The women are covered from the middle down to the foot, wearing a great number of bracelets upon their arms, being made some of bone, some of horn, and some of brass, the lightest whereof by our estimation weighed two ounces apiece.

With this people linen cloth is good merchandise, whereof they make rolls for their heads, and girdles to wear about them.

Their island is both rich and beautiful: rich in gold, silver, copper, and sulphur, wherein they seem skilful and expert.

Their fruits be diverse and plentiful, as nutmegs, ginger, long pepper, lemons, cucumbers, cocos, sago, with divers other sorts: since the time that we first set out of our own country of England, we happened on no place, wherein we found more comforts and better means of refreshing.

We set our course for Java, where arriving, we found great courtesy, and honourable entertainment. This island is governed by 5 kings, whom they call Rajah.

Of these five we had four a shipboard at once, and two or three often. They are wonderfully delighted in coloured clothes, as red and green: their upper parts of their bodies are naked save their heads, whereupon they wear a Turkish roll, as do the Moluccans: from the middle downward they wear a pintado of silk, trailing upon the ground, in colour as best they like.

They have an house in every village for their common assembly: every day they meet twice, men, women, and children, bringing with them such victuals as they think good, some fruits, some rice boiled, some hens roasted, some sago, having a table made 3 foot from the ground, whereon they set their meat, that every person sitting at the table may eat, one rejoicing in the company of another.

They boil their rice in an earthen pot, made in form of a sugar loaf, being full of holes, as our pots which we water our gardens withal, and it is open at the great end, wherein they put their rice dry, without any moisture. In the meantime they have ready another great earthen pot, set fast in a furnace, boiling full of water, whereinto they put their pot with rice, by such measure, that they swelling become soft at the first, and by their swelling stopping the holes of the pot, admit no more water to enter, but the more they are boiled, the harder and more firm substance they become, so that in the end they are a firm and good bread, of the which with oil, butter, sugar, and other spices, they make diverse sorts of meats very pleasant of taste, and nourishing to nature.

The French pox is here very common to all, and they help themselves, sitting naked from ten to two in the sun, whereby the venomous humour is drawn out. Not long before our departure, they told us, that not far off there were such great ships as ours, wishing us to beware: upon this our captain would stay no longer.

From Java we sailed for the Cape of Good Hope, which was the first land we fell withal: neither did we touch with it, or any other land, until we came to Sierra Leone, upon the coast of Guinea: we ran hard aboard the Cape, finding the report of the Portuguese to be most false, who affirm, that it is the most dangerous cape of the world, never without intolerable storms and present dangers to travellers.

This cape is a most stately thing, and the fairest cape we saw in the whole circumference of the earth, and we passed by it the 18 of June.

From thence we continued our course to Sierra Leone, on the coast of Guinea, where we arrived the 22 of July, and found necessary provisions, great store of elephants, oysters upon trees of one kind, spawning and increasing infinitely.

We arrived in England the third of November 1580 being the third year of our departure.

XXXV

The second voyage of master Martin Frobisher, made to the west and northwest regions, in the year 1577, with a description of the country, and people.

On Whit Sunday, the six and twentieth of May 1577, Captain Frobisher departed from Blackwall, with one of the Queen's ships, called the *Aide*, of nine score tons: and two other little barks likewise, one called the *Gabriel*, and the other, the *Michael*, accompanied with seven score gentlemen, soldiers, and sailors, well furnished with victuals, and other provisions necessary for one half year, on this his second voyage, for the further discovering of the passage to Cathay, supposed to be on the north and northwest part of America: where through our merchants may have course and recourse with their merchandise, from these our northernmost parts of Europe, to those Oriental coasts of Asia, in much shorter time, and with greater benefit than any others, to their no little commodity and profit.

With a merry wind the 7th of June we arrived at the islands called Orcades, or vulgarly Orkney, where we made provision of fresh water; in the doing whereof our General licensed the gentlemen and soldiers for their recreation, to go on shore. At our landing, the people fled from their poor cottages, with shrieks and alarms, to warn their neighbours of enemies, but by gentle persuasions we reclaimed them to their houses. It seemeth they are often frighted with pirates. Their houses are very simply builded with pebble stone, without any chimneys, the fire being made in the midst thereof. The good man, wife, children, and other of their family eat and sleep on the one side of the house, and the cattle on the other, very beastly and rudely, in respect of civility. They are destitute of wood, their fire is turfs, and cowshards. They have corn, bigg, and oats, with which they pay their king's rent, to their maintenance of his house. They take great quantity of fish, which they dry in the wind and sun. They dress their meat very filthily and eat it without salt. Their church and religion is reformed according to the Scots.

We departed hence the 8th of June and followed our course until the 4th of July: all which time we had no night, but that easily, and without any impediment we had when we were so disposed, the fruition of our books, and other pleasures to pass away time. This benefit endureth in those parts not 6 weeks, while the sun is near the Tropic of Cancer.

We met great islands of ice, of half a mile, some more, some less in compass, showing above the sea, 30 or 40 fathoms, and as we supposed fast on the ground, where with our lead we could scarce sound the bottom for depth.

Here, in place of odoriferous and fragrant smells of sweet gums, and pleasant notes of musical birds, which other countries in more temperate zones do yield, we tasted the most boisterous Boreal blasts mixed with snow and hail, in the months of June and July, nothing inferior to our untemperate winter: a sudden alteration.

All along this coast ice lieth, as a continual bulwark, and so defendeth the country, that those that would land there, incur great danger. Our general 3 days together attempted with the ship-boat to have gone on shore, which for that without great danger he could not accomplish, he deferred it. All along the coast lie very high mountains covered with snow. Four days coasting along this land, we found no sign of habitation. Little birds came flying into our ships, which causeth us to suppose, that the country is both more tolerable and also habitable within, than the outward shore maketh show.

On July 16th, we came with the making of land. Between two islands there is a large entrance or strait, called Frobisher's Strait, after the name of our general, the first finder thereof.

Whilst he was searching the country near the shore, some of

the people of the country showed themselves leaping and dancing, with strange shrieks and cries, which gave no little admiration to our men. Our general desirous to allure them unto him by fair means, caused knives, and other things to be proffered unto them, which they would not take at our hands: but being laid on the ground, and the party going away, they came and took up, leaving some thing of theirs to countervail the same. At length two of them leaving their weapons, came down to our general and master, who did the like to them, commanding the company to stay, and went unto them: who after certain dumb signs, and mute congratulations, began to lay hands upon them, but they deliverly escaped, and ran to their bows and arrows, and came fiercely upon them (not respecting the rest of our company which were ready for their defence), but with their arrows hurt divers of them: we took the one, and the other escaped.

The day following, being the 19th of July, our captain returned to the ship, with report of supposed riches, which showed itself in the bowels of those barren mountains, wherewith we were all satisfied.

Within four days after we had been at the entrance of the straits, the northwest and west winds dispersed the ice into the sea, and made us a large entrance. We entered them, and our general and master with great diligence, sought out and found out a fair harbour for the ship and barks to ride in, and brought the ship, barks, and all their company to safe anchor, except one man, which died by God's visitation.

After the ship rode at anchor, our general, with such company as could well be spared from the ships, in marching order entered the land, having special care that we should all with one voice thank God for our safe arrival: secondly beseech Him, that it would please His divine majesty, long to continue our Queen, for whom he, and all the rest of our company in this order took possession of the country: and thirdly, that by our Christian study and endeavour, those barbarous people trained up in paganism, and infidelity, might be reduced to the knowledge of true religion, and to the hope of salvation in Christ our Redeemer.

We marched through the country, with ensign displayed, so

far as was thought needful, and now and then heaped up stones on high mountains, and other places in token of possession, as likewise to signify unto such as hereafter may chance to arrive there, that possession is taken in the behalf of some other prince, by those that first found out the country.

The stones of this supposed continent with America be altogether sparkled, and glister in the sun like gold: so likewise doth the sand in the bright water, yet they verify the old proverb: all is not gold that glistens.

On this west shore we found a dead fish floating, which had in his nose a horn straight and torqued, of length two yards lacking two inches, being broken in the top, where we might perceive it hollow, into which some of our sailors putting spiders they presently died. I saw not the trial thereof. By the virtue thereof we supposed it to be the sea unicorn.[75]

After our general had found out such store of supposed gold ore as he thought himself satisfied withal, he returned to the *Michael*. Coasting along the west shore not far from whence the ship rode, they perceived a fair harbour, and willing to sound the same, at the entrance thereof, they espied two tents of seal skins, unto which the captain, and other company resorted. At the sight of our men the people fled into the mountains: nevertheless they went to their tents, where leaving certain trifles of ours, as glasses, bells, knives, and such like things they departed, not taking any thing of theirs except one dog. They did in like manner leave behind them a letter, pen, ink, and paper, whereby our men whom the captain lost the year before, and in that people's custody, might (if any of them were alive) be advertised of our presence.

At our coming back again to the place where their tents were before, they had removed their tents further into the said bay or sound, where they might if they were driven from the land, flee with their boats into the sea. We parting ourselves into two companies, and compassing a mountain came suddenly upon them by land, who espying us, without any tarrying fled to their boats, leaving the most part of their oars behind them for haste, and rowed down the bay, where our two pinnaces met them and drove them to shore: but if they had had all their oars, so swift are they in rowing, it had been lost time to have chased them.

When they were landed they fiercely assaulted our men with their bows and arrows, who wounded three of them with our arrows: and perceiving themselves thus hurt, they desperately leapt off the rocks into the sea, and drowned themselves. Two women not being so apt to escape as the men were, the one for her age, and the other being encumbered with a young child, we took. The old wretch, whom divers of our sailors supposed to be either a devil, or a witch, had her buskins plucked off, to see if she were cloven footed, and for her ugly hue and deformity we let her go: the young woman and the child we brought away. We named the place where they were slain, Bloody Point.

Having this knowledge of their fierceness and cruelty, we disposed ourselves, contrary to our inclination, something to be cruel, returned to their tents and made a spoil of the same: where we found an old shirt, a doublet, a girdle, and also shoes of our men, whom we lost the year before.

Their riches are not gold, silver or precious drapery, but their said tents and boats, made of the skins of red deer and seal skins: also dogs like unto wolves.

They are men of a large corporature, and good proportion: their colour is not much unlike the sunburnt country man who laboureth daily in the sun for his living.

They wear their hair something long, and cut before either with stone or knife, very disorderly. Their women wear their hair long, and knit up with two loops, showing forth on either side of their faces, and the rest folded upon a knot. Also some of the women raze their faces proportionally, as chin, cheeks, and forehead, and the wrists of their hands, whereupon they lay a colour which continueth dark azurine.

They eat their meat all raw, both flesh, fish, and fowl, or something parboiled with blood and a little water which they drink. For lack of water they will eat ice that is hard frozen, as pleasantly as we will do sugar candy, or other sugar.

If they for necessity's sake stand in need of the premises, such grass the country yielded they pluck up and eat, not daintily, or saladwise to allure their stomachs to appetite: but for necessity's sake without either salt, oils or washing, like brute beasts devouring the same. They neither use table, stool, nor table cloth for comeliness: but when they are imbrued with blood knuckle deep, and their knives in like sort, they use their tongues as apt instrument to lick them clean: in doing whereof they are assured to lose none of their victuals.

They frank or keep certain dogs not much unlike wolves, which they yoke together, as we do oxen and horses, to a sled or trail: and so carry their necessaries over the ice and snow from place to place: as the captive, whom we have, made perfect signs. And when those dogs are not apt for the same use: or when with hunger they are constrained for lack of other victuals, they eat them.

They apparel themselves in the skins of such beasts as they kill, sewed together with the sinews of them.

Upon their legs they wear hose of leather, with the fur side inward two or three pair at once, and especially the women. In those hose they put their knives, needles, and other things needful to bear about.

They dress their skins very soft and supple with the hair on. In cold weather or winter they wear the fur side inward: and in summer outward. Other apparel they have none.

Those beasts, fishes, and fowls, which they kill, are their meat, drink, apparel, houses, bedding, hose, shoes, thread, and sails for their boats, with many other necessaries whereof they stand in need, and almost all their riches.

Their houses are tents made of seal skins, pitched up with 4 fir quarters foursquare meeting at the top, and the skins sewed together with sinews, and laid thereupon: they are so pitched up, that the entrance into them is always south or against the sun.

Their darts are made of two sorts: the one with many forks of bones in the fore end and likewise in the midst: their proportions are not much unlike our toasting irons but longer: these they cast out of an instrument of wood, very readily. The other sort is greater than the first aforesaid, with a long bone made sharp on both sides not much unlike a rapier, which I take to be their most hurtful weapon.

They have two sorts of boats made of leather, set out on the inner side with quarters of wood, artificially tied together with thongs of the same: the other boat is but for one man to sit and row in with one oar.

What knowledge they have of God, or what idol they adore, we have no perfect intelligence. I think them rather anthropophagi, or devourers of man's flesh than otherwise: for that there is no flesh or fish which they find dead (smell it never so filthily) but they will eat it, as they find it without any other dressing.

The countries on both sides of the straits lie very high with rough stony mountains, and great quantity of snow thereon. There is very little plain ground and no grass, except a little which is much like unto moss that groweth on soft ground, such as we get turfs in. There is no wood at all. To be brief there is nothing fit or profitable for the use of man, which that country with root yieldeth or bringeth forth: howbeit there is great quantity of deer, whose skins are like unto asses, their heads or horns do far exceed, as well in length as also in breadth, any in these our parts or countries: their feet likewise are as great as our oxen's, which we measured to be seven or eight inches in breadth. There are also hares, wolves, fishing bears, and sea fowl of sundry sorts.

The 24th of August, after we had satisfied our minds with freight sufficient for our vessels, though not our covetous desires with such knowledge of the country, people, and other commodities as are before rehearsed, we departed therehence. The 17th of September we fell with the Land's End of England and so sailed to Milford Haven.

In this voyage we lost two men, one in the way by God's visitation, and the other homeward cast overboard with a surge of the sea.

Our general named sundry islands, mountains, capes and harbours after the names of divers noblemen and other gentlemen his friends, as well on the one shore as also on the other.

XXXVI

From a letter written to Mr Richard Staper by John Withal [from] Santos in Brazil, the 26 of June 1578.

Worshipful sir, and well beloved friend Mr Staper, I am about three days ago consorted with an Italian gentleman to marry with his daughter within these four days. He hath but only this child which is his daughter, which he hath thought better bestowed on me than on any Portuguese in all the country, and doth give with her in marriage to me part of an ingenio which he hath, that doth make every year a thousand roves of sugar. This my marriage will be worth to me two thousand ducats, little more or less. Also my father in law doth intend to put into my hands the whole ingenio[76] with sixty or seventy slaves, and thereof to make me factor for us both. I give my living Lord thanks for placing me in such honour and plentifulness of all things.

Also certain days past I talked with the provedor and the captain, and they have certified me, that they have discovered certain mines of silver and gold, and look every day for masters to come to open the said mines: which when they be opened will enrich this country very much.

I have talked with the captain and provedor, and my father in law, who rule all this country, for to have a ship with goods to come from London hither, which have promised me to give me licence, saying that now I am free denizen of this country. To cause a ship to come hither with such commodities as would serve this country, would come to great gains, God sending in safety the profit and gains. In such wares and commodities as you may ship hither from London is for every one commodity delivered here three for one, and then after the proceed may be employed in white sugar at four hundred reis the rove.

This voyage is as good as any Peru voyage. If you and Master Osborne will deal here, I will deal with you before any other, because of our old friendly friendship in time past. If you have any stomach thereto, in the name of God do you spy out a fine bark of seventy or eighty tons, and send her hither with a Portuguese pilot to this port of São Vicente in Brazil, bordering upon the borders of Peru.

Also I herewith write unto you in what form and manner you shall furnish this voyage both in commodities and otherwise.

First you must lade in the said ship certain Hampshire and Devonshire kerseys: for the which you must let her depart from London in October, and to touch in the Canaries, and there to make sale of the said kerseys, and with the proceed thereof to load fifteen tons of wines that be perfect and good, and six dozen of Cordovan skins of these colours to wit, orange, tawny, yellow, red, and very fine black. Three hogsheads of sweet oil for this voyage are very necessary, or a hundred and fifty jars of oil. Also in London you may lade in the said ship these parcels of commodities or wares as followeth:

Four pieces of hollands of middle sort.
One piece of fine holland.
Four dozen of scissors of all sorts.
Twenty dozen of great knives which be made in fardels, of a low price.
Four hundred ells of Manchester-cottons, most black, green, some yellow.
Eight or ten dozen of hats, the one half trimmed with taffeta, the other plain, with the bands of cypress.
Six dozen of coarse shirts.
Six dozen of locks for doors and chests.
Two dozen glasses of divers sorts.
Three dozen of frieze gowns.
Four hundred pound of tin of the use of Portugal, most small dishes and trenchers.
Twenty pounds of spices, cloves, cinnamon, pepper, and saffron.
Two quintals of white soap.
Three pound of thread, white, black, and blue.
Six yards of black velvet.
Two barrels of nails for ships and barks.

Four yards of taffeta red, black, and blue, with some green.
Four tons of iron.

Have you no doubt, but by the help of God I shall put all things in good order according to your contentment and profit.

Thus I commit you with all yours to the Holy Ghost for ever.

By your assured friend
JOHN WITHAL.

XXXVII

The third voyage of Captain Frobisher, pretended for the discovery of Cathay, 1578.

The general being returned from the second voyage, immediately after his arrival in England, repaired with all haste to the Court being then at Windsor, to advertise Her Majesty of his good success in this last voyage, and of the plenty of gold ore, with other matters of importance which he had in these septentrional parts discovered. He was heartily welcomed of many noble men, but especially for his great adventure, commended of Her Majesty, at whose hands he received most gracious countenance. Her Highness also commended the rest of the gentlemen in this service, for their great forwardness in this so dangerous an attempt: finding that the matter of the gold ore had appearance and made show of great riches and profit, and the hope of the passage to Cathay, by this last voyage greatly increased, Her Majesty appointed special commissioners to look thoroughly into the cause. The commissioners after sufficient trial and proof made of the ore, and having understood the possibility and likelihood of the passage, advertised Her Highness, that the cause was of importance. Whereupon preparation was made of ships and all other things necessary, because it was assuredly made account of, that the commodity of mines, there already discovered, would at least countervail in all respects the adventurers' charge: it was thought needful, both for the better guard of those parts already found, and for further discovery of the inland and secrets of those countries, and also for further search of the passage to Cathay that certain numbers of chosen soldiers be assigned to inhabit there. Whereupon there was a strong fort of timber, artificially framed, and cunningly devised in ships to be carried thither, whereby those men that were appointed to winter and stay there the whole year, might as well be defended from the danger of the snow and cold air, as also fortified from the force or offence of those country people. To this great adventure many forward young gentlemen of our country willingly have offered themselves. The whole number of men which had offered were one hundred persons, whereof 40 should be mariners for the use of ships, 30 miners for gathering the gold ore together for the next year, and 30 soldiers for the better guard of the rest, within which last number are included the gentlemen, goldfiners, bakers, carpenters, and all necessary persons. Being therefore thus furnished with all necessaries, there were ready to depart upon the said voyage 15 sail of good ships, whereof the whole number was to return again with their loading of gold ore in the end of the summer, except those 3 ships, which should be left for the use of those captains which should inhabit there the whole year. The general with all the captains came to the Court, then lying at Greenwich, to take their leave of Her Majesty, at whose hands they all received great encouragement, and gracious countenance. Her Highness besides other good gifts, and greater promises, bestowed on the general a fair chain of gold, and the rest of the captains kissed her hand, took their leave, and departed every man towards their charge.

We departed from Harwich the one and thirtieth of May. And sailing along the south part of England westward, we at length came by the coast of Ireland at Cape Clear the sixth of June, and gave chase there to a small bark which was supposed to be a pirate, or rover on the seas, but it fell out indeed that they were poor men of Bristol, who had met with such company of Frenchmen as had spoiled and slain many of them, and left the rest so sore wounded that they were like to perish in the sea. Our general who well understood the office of a soldier and an Englishman, and knew well what the necessity of the sea meaneth, relieved them with salves to heal their hurts, and with meat and drink to comfort their pining hearts.

We sailed about fourteen days without sight of any land, or any other living thing, except certain fowls, as guillemots, noddies, gulls &c.

The twentieth of June, at two of the clock in the morning, the general descried land, and found it to be West Friesland.[77] Here the general, and other gentlemen went ashore, being the first known Christians that we have true notice of, that ever set foot upon that ground: and therefore the general took possession thereof to the use of our sovereign lady the Queen's Majesty, and discovered here a goodly harbour for the ships, where were also certain little boats of that country.

The savage and simple people so soon as they perceived our men coming towards them fled fearfully away. They left in their tents all their furniture where amongst other things were found a box of small nails, whereby it appeareth that they have trade with some civil people.

Our men brought away with them only two of their dogs, leaving in recompense bells, looking glasses, and divers of our country toys behind them. This country, no doubt, promiseth good hope of great commodity and riches, if it may be well discovered.

Having a fair and large wind we departed from thence towards Frobisher's Straits, the three and twentieth of June. Then we bore southerly towards the sea, because to the northwards of this coast we met with much driving ice.

On Monday the last of June, the *Salamander* happened to strike a great whale with her full stem, with such a blow that the ship stood still, and stirred neither forward nor backward. The whale thereat made a great and ugly noise, and cast up his body and tail, and so went under water, and within two days after, there was found a great whale dead swimming above water.

We were forced many times to stem and strike great rocks of ice, and so as it were to make through mighty mountains. By which means some of the fleet, where they found the ice to open, entered in, and passed so far within the danger thereof, that it was the greatest wonder of the world that they ever escaped safe. We missed two of the fleet, that is, the *Judith*, and the *Michael*, whom both we supposed had been utterly lost, having not heard any tidings of them in more than 20 days before.

One of our fleet named the *Bark Dennis*, being of an hundred ton burthen, seeking way in amongst the ice, received such a blow with a rock of ice that she sunk down therewith in the sight of the whole fleet. Howbeit having signified her danger by shooting off a piece of great ordnance, new succour of other ships came so readily unto them, that the men were all saved with boats.

Within this ship that was drowned there was parcel of our house which was to be erected for them that should stay all the winter.

There arose a sudden and terrible tempest at the southeast, which blowing from the main sea, directly upon the place of the straits, brought together all the ice a sea board of us upon our backs, and thereby debarred us of turning back to recover sea room again: being thus compassed with danger on every side, some of the ships, where they could find a place more clear of ice, and get a little berth of sea room, did take in their sails, and there lay adrift. Other some fastened and moored anchor upon a great island of ice, and rode under the lee thereof, supposing to be better guarded thereby from the outrageous winds, and the danger of the lesser fleeting ice. And again some were so fast shut up, and compassed in amongst an infinite number of great countries and islands of ice, that they were fain to submit themselves and their ships to the mercy of the unmerciful ice, and strengthened the sides of their ships with junks of cables, beds, masts, planks and such like, which being hanged overboard on the sides of their ships, might the better defend them from the outrageous sway and strokes of the said ice.

It pleased God to send them next day a more favourable wind at the west northwest, which did not only disperse and drive forth the ice before them, but also gave them liberty of more scope and sea room, where (to their great comfort) they enjoyed again the fellowship one of another. Some in mending the sides of their ships, some in setting up their top masts, and mending their sails and tackling. And now the whole fleet plied off to seaward, resolving there to abide until the sun might consume, or the force of wind disperse these ice from the place of their passage: and being a good berth off shore, they took in their sails, and lay adrift.

The tenth of July, the weather still continuing thick and dark, some of the ships in the fog lost sight of the admiral and the rest of the fleet, and wandering to and fro, with doubtful opinion whether it were best to seek back again to seaward through great store of ice, or to follow on a doubtful course in a sea, bay, or straits they knew not, or along a coast, whereof by reason of the dark mists they could not discern the dangers.

The vice-admiral Captain York having lost sight of the fleet, turned back to sea again, having two other ships in company with him.

Also the captain of the *Anne Francis* held it for best to turn it out to sea again, until they might have clear weather to take the sun's altitude, and with incredible pain and peril got out of the doubtful place, into the open sea again, being so narrowly distressed by the way, by means of continual fog and ice, that they were many times ready to leap upon an island of ice to avoid the present danger, and so hoping to prolong life awhile meant rather to die a pining death.

Some hoped to save themselves on chests, and some determined to tie the hatches of the ships together, and to bind themselves with their furniture fast thereunto, and so to be towed with the ship boat ashore, which otherwise could not receive half of the company, by which means if happily they had arrived, they should either have perished for lack of food to eat, or else should themselves have been eaten of those ravenous, bloody, and men-eating people.

The rest of the fleet following the course of the general which led them the way, passed up above sixty leagues within the said doubtful and supposed straits, having always a fair continent upon their starboard side, and a continuance still of an open sea before them.

The general albeit with the first perchance he found out the error, and that this was not the old straits, yet he persuaded the fleet always that they were in their right course, and known straits. Howbeit I suppose he rather dissembled his opinion therein than otherwise, to induce the fleet to follow him, to see a further proof of that place. And as some of the company reported, he hath since confessed that if it had not been for the charge and care he had of the fleet he both would and could have gone through to the South Sea, and dissolved the large doubt of the passage which we seek to find to the rich country of Cathay.

Long time now the *Anne Francis* had lain beating off and on all alone before the Queen's Foreland, not being able to recover their port for ice, albeit many times they dangerously attempted it, for yet the ice choked up the passage, and would not suffer them to enter. And having never seen any of the fleet since twenty days past, when by reason of the thick mists they were severed in the mistaken straits, they did now this present 23rd of July overthwart a place in the straits called Hatton's Headland, where they met with seven ships of the fleet again.

At their meeting they hailed the admiral after the manner of the sea, and with great joy welcomed one another with a thundering volley of shot.

Upon the seven and twentieth of July, the ship of Bridgewater got out of the ice and met with the fleet. They reported of their marvellous accidents and dangers, declaring their ship to be so leaky that they must of necessity seek harbour, having their stem so beaten within their huddings, that they had much ado to keep themselves above water. They had (as they say) five hundred strokes at the pump in less than half a watch, being scarce two hours; their men being so overwearied therewith, and with the former dangers that they desired help of men from the other ships. Moreover they declared that there was nothing but ice and danger where they had been, and that the straits within were frozen up, and that it was the most impossible thing of the world, to pass up unto the Countess of Warwick's Sound, which was the place of our port.

Some began privily to murmur against the general for this wilful manner of proceeding. Some desired to discover some harbour thereabouts to refresh themselves and reform their broken vessels for a while, until the north and northwest winds might disperse the ice, and make the place more free to pass. Other some forgetting themselves, spoke more undutifully in this behalf, saying: that they had as lief be hanged when they came home, as without hope of safety to seek to pass, and so to perish among the ice.

The general chiefly respecting the accomplishment of the cause he had undertaken and calling to his remembrance the short time he had in hand to provide so great number of ships their loading, determined with this resolution to pass and recover his port, or else there to bury himself with his attempt.

He sought to go in with his pinnaces amongst the islands there, as though he meant to search for harbour, where indeed he meant nothing less, but rather sought if any ore might be found in that place.

In the meantime whilst the fleet lay thus doubtful without any certain resolution what to do, being hard aboard the lee shore, there arose a sudden and terrible tempest at the south southeast, whereby the ice began marvellously to gather about us.

In this storm being the six and twentieth of July, there fell so much snow, with such bitter cold air, that we could not scarce see one another for the same, nor open our eyes to handle our ropes and sails, the snow being above half a foot deep upon the hatches of our ship, which did so wet through our poor mariners' clothes, that he that had five or six shifts of apparel had scarce one dry thread to his back, which kind of wet and coldness, together with the over labouring of the poor men amidst the ice, bred no small sickness amongst the fleet, which somewhat discouraged some of the poor men, who had not experience of the like before, every man persuading himself that the winter there must needs be extreme, where they found so unseasonable a summer.

The general, notwithstanding the great storm, where he saw the ice never so little open, he got in at one gap and out at another, and so himself valiantly led the way through to induce the fleet to follow after, and with incredible pain and peril at length got through the ice, and upon one and thirtieth of July recovered his long wished port after many attempts and sundry times being put back, and came to anchor.

At their arrival here they perceived two ships at anchor within the harbour, whereat they began much to marvel and greatly to rejoice, the *Michael*, and the small bark called the *Gabriel*, who so long time were missing, and never heard of.

Master Wolfall a learned man, appointed by Her Majesty's Council to be their minister and preacher made unto them a godly sermon, exhorting them especially to be thankful to God for their strange and miraculous deliverance. This Master Wolfall being well settled at home in his own country, with a good and large living, having a good honest woman to wife and very towardly children, refused not to take in hand this painful voyage, for the only care he had to save souls, and to reform those infidels if it were possible to Christianity.

The general immediately at his first landing called the chief captains of his council together.

The muster of the men being taken, and the victuals with all other things viewed and considered, every man was set to his charge, as his place and office required. The miners were appointed where to work, and the mariners discharged their ships.

In the meantime the captains sought out new mines, the gold-finers made trial of the ore.

The ninth of August the general began to consider the erecting up of the fort for them that were to inhabit there the whole year. There was arrived only the east side, and the south side of the house, and yet not that perfect and entire: for many pieces thereof were used for fenders in many ships, and so broken in pieces whilst they were distressed in the ice. Also after due examination had, and true account taken, there was found want of drink and fuel to serve one hundred men, which was the number appointed first to inhabit there, because their greatest store was in the ships which were not yet arrived. The Captain Fenton seeing the scarcity of the necessary things aforesaid, was contented, and offered himself to inhabit there with sixty men. Whereupon they caused the carpenters and masons to come before them, and demanded in what time they would take upon them to erect a less house for sixty men. They required eight or nine weeks, if there were timber sufficient, whereas now they had but six and twenty days in all to remain in that country. Wherefore it was fully agreed upon, and resolved by the general and his council, that no habitation should be there this year.

The thirtieth of August the masons finished a house which Captain Fenton caused to be made of lime and stone upon the Countess of Warwick's island, to the end we might prove against the next year, whether the snow could overwhelm it, the frost break it up, or the people dismember the same. And the better to allure those brutish and uncivil people to courtesy against other times of our coming, we left therein divers of our country toys, as bells, and knives, wherein they specially delight, one for the necessary use, and the other for the great pleasure thereof. Also pictures of men and women in lead, men on horseback, looking glasses, whistles, and pipes. Also in the house was made an oven, and bread left baked therein for them to see and taste.

We buried the timber of our pretended fort. Also here we sowed pease, corn, and other grain, to prove the fruitfulness of the soil against the next year.

The fleet now being in some good readiness for their lading, the general calling together the gentlemen and captains to consult, told them that he was very desirous that some further discovery should be attempted, and that he would not only by God's help bring home his ships laden with ore, but also meant to bring some certificate of a further discovery of the country. The captains were contented and willing, as the general should appoint and command, to take any enterprise in hand. Which after long debating was found a thing very impossible. First the dark foggy mists, the continual falling snow and stormy weather which they commonly were vexed with, and now daily ever more and more increased, have no small argument of the winter's drawing near. Again, drink was scant throughout all the fleet by means of the great leakage. Yet notwithstanding these reasons alleged the general himself went in a pinnace and discovered further northward in the straits, and found that by Bear's Sound and Hall's Island, the land was not firm, as it was first supposed, but all broken islands in manner of an archipelago, and so with other secret intelligence to himself, he returned to the fleet.

The last day of August the whole fleet departed from the Countess's Sound, excepting the *Judith*, and the *Anne Francis*, who stayed for the taking in of fresh water, and came the next day.

Thanks be to God, all the fleet arrived safely in England about the first of October, some in one place and some in another.

There died in the whole fleet in all this voyage not above forty persons, which number is not great, considering.

XXXVIII

Certain directions given by Mr Richard Hakluyt of the Middle Temple, to Mr Morgan Hubblethorne, Dyer, sent into Persia, 1579.

For that England hath the best wool and cloth of the world, and for that the cloths of the realm have no good vent, if good dyeing be not added: therefore it is much to be wished, that the dyeing of foreign countries were seen, for thereof will follow honour to the realm, and great and ample vent of our cloths: and of the vent of cloths, will follow the setting of our poor on work, in all degrees of labour in clothing and dyeing: for which cause most principally you are sent over at the charge of the city: and therefore for the satisfying the lords, and of the expectation of the merchants and of your company, it behoves you to have care to return home with more knowledge than you carried out.

The price of a cloth, for a fifth, sixth and seventh part riseth by the colour and dyeing: and therefore to devise to dye as good colours with the one half of the present price were to the great commodity of the realm, by saving of great treasure in time to come. And therefore you must have great care to have knowledge of the materials of all the countries that you shall pass through, that may be used in dyeing, be they herbs, weeds, barks, gums, earths, or what else soever.

In Persia you shall find carpets the best of the world, and excellently coloured: you must use means to learn all the order of the dyeing which are so dyed as neither rain, wine, nor yet vinegar can stain.

For that in Persia they have great colouring of silks, it behoves you to learn that also, for that cloth dyeing and silk dyeing have a certain affinity, and your merchants mind to bring much raw silk into the realm, and therefore it is more requisite you learn the same.

In Persia there are that stain linen cloth: it is not amiss you learn it if you can: it hath been an old trade in England, whereof some excellent cloths yet remain: but the art is now lost, and not to be found in the realm.

They have a cunning in Persia to make in buskins of Spanish leather flowers of many kinds, in most lively colours, and these the courtiers do wear there: to learn which art were no harm.

You shall find anil[78] there, if you can procure the herb that it is made of, either by seed or by plant, to carry into England, you may do well to endeavour to enrich your country with the same: but withal learn you the making of the anil, and if you can get the herb, you may send the same dry into England, for possibly it groweth here already.

Return home with you all the materials and substances that they dye withal in Russia, and also in Persia, that your company may see all.

In some little pot in your lodging, I wish you to make daily trials in your art, as you shall from time to time learn aught among them.

Set down in writing whatsoever you shall learn from day to day, lest you should forget, or lest God should call you to his mercy: and by each return I wish you to send in writing whatsoever you have learned, or at the least keep the same safe in your coffer, that come death or life your country may enjoy the thing that you go for, and not lose the charge, and travel bestowed in this case.

Learn you there to fix and make sure the colour to be given by logwood:[79] so shall we not need to buy wood so dear, to the enriching of our enemies.

If before you return you could procure a singular good workman in the art of Turkish carpet making, you should bring the art into this realm, and also thereby increase work to your company.

XXXIX

A letter of Gerardus Mercator, written to Mr Richard Hakluyt of Oxford, touching the intended discovery of the Northeast Passage, An. 1580.

Sir, I wish Arthur Pet had been informed before his departure of some special points. The voyage to Cathay by the east is doubtless very easy and short, and I have oftentimes marvelled, that being so happily begun, it hath been left off, and the course changed into the west, after that more than half of your voyage was discovered. For beyond Novaya Zemlya there followeth presently a great bay, which on the left side is enclosed with the mighty promontory Tabin.[80] Into the midst hereof fall great rivers, which passing through the whole country of Serica, and being as I think navigable with great vessels into the heart of the continent, may be an easy means whereby to traffic for all manner of merchandise, and transport them out of Cathay, and other kingdoms thereabouts into England. But considering with myself that that navigation was not intermitted, but upon great occasion, I thought that the Emperor of Russia and Muscovy had hindered the proceeding thereof. If so be that with his grace and favour a further navigation may be made, I would counsel them certainly to search this bay and rivers aforesaid, and in them to pick and choose out some convenient port and harbour for the English merchants, from whence afterward with more opportunity and less peril all the coast of Cathay may be discovered. That the pole of the loadstone[81] is not far beyond I have learned by the certain observations of the loadstone: about which pole I think there are very many rocks, and very hard and dangerous sailing: and yet a more hard and difficult passage I think it to be this way which is now attempted by the west, for it is nearer to the pole of the loadstone, to the which I think it not safe to approach. And because the loadstone hath another pole than that of the world, the nearer you come unto it, the more the needle of the compass doth vary from the north, sometimes to the west, and sometimes to the east, according as a man is to the eastward or to the westward of that meridian.

This is a strange alteration and very apt to deceive the sailor, unless he know the unconstancy and variation of the compass, and take the elevation of the pole sometimes with his instruments. If Master Arthur be not well provided in this behalf, or of such dexterity, that perceiving the error he be not able to correct the same, I fear lest in wandering up and down he lose his time, and be overtaken with the ice in the midst of the enterprise. For that gulf, as they say, is frozen every year very hard. Which if it be so, the best counsel I could give for their best safety, were to seek some harbour in that bay, and those rivers whereof I have spoken, and by some ambassador to make friendship and acquaintance with the Great Khan,[82] in the name of the Queen's Majesty. I think from the mouths of the mighty rivers to Cambalu[83] the chiefest seat of the Khan there are not past 300 German miles.

I would gladly know how high the sea doth flow commonly in the port of Moscovy where your men do harbour. And also whether the sea in this strait do flow always one way to the east or to the west, or whether it do ebb and flow according to the manner of the tides in the middle of the Channel, that is to say, whether it flows there six hours into the west, and as many back again to the east, for thereupon depend other speculations of importance. I would wish Mr Frobisher to observe the same westwards. Concerning Canada, and New France which are in my maps, they were taken out of a certain sea card drawn by a certain priest out of the description of a Frenchman, a pilot very skilful in those parts: for the trending of the coast, and the elevation of the pole, I doubt not but they are very near the truth: for the chart had, beside a scale of degrees of latitude passing through the midst of it, another particularly annexed to the coast of New France, wherewith the error of the latitudes committed by reason of the variation of the compass might be corrected.

If there be anything else that you would require of me, I will most willingly communicate it with you, craving this likewise of your courtesy, that whatsoever observations of both these voyages shall come to your hands, you would impart them to me. Whatsoever I gather of them, I will faithfully signify unto you by letters, if happily they may yield any help or light unto this most excellent enterprise of navigation, and most profitable to our Christian commonwealth. Fare you well most learned friend. At Duisburg in Cleveland,[84] 28 of July, the year, 1580.

Yours wholly to my power to be commanded,

GERARDUS MERCATOR.

XL

Notes given 1580: to Mr Arthur Pet, and to Mr Charles Jackman, sent by the merchants of the Moscovy Company for the discovery of the Northeast Strait.

Whereas the Portuguese have in their course to their Indies in the southeast, certain ports and fortifications to thrust into by the way, so you are to see what islands, and what ports you had need to have in your course to the northeast. For which cause I wish you to note all the islands, and to set them down in plat, to two ends: that we may devise to take the benefit by them, and also foresee how by them the savages or civil princes may in any sort annoy us in our purposed trade that way.

And for that the people be no Christians, it were good that the mass of our commodities were always in our own disposition, and not at the will of others. Therefore it were good that we did seek out some small island in the Scythian sea, where we might plant, fortify, and staple safely, from whence (as time should serve) we might feed those heathen nations with our commodities.

And if no such islands may be found in the Scythian sea toward the firm of Asia, then you are to search out the ports that be about Novaya Zemlya, to the end you may winter there the first year, if we may in short time come into Cambalu, and unload and set sail again for return, you may on your way come as far in return as a port about Novaya Zemlya: that the summer following, you may the sooner be in England for the more speedy vent of your East commodities, if you cannot go forward and back in one self same summer.

If you find the soil planted with people, it is like that in time an ample vent of our warm woollen cloths may be found. And if there be no people at all there to be found, then you shall specially note what plenty of whales, and of other fish is to be found there.

And if the air may be found upon that tract temperate, and the the soil yielding wood, water, land and grass, and the seas fish, then we may plant on that main the offals of our people, as the Portuguese do in Brazil, and so they may yield commodity to England by harbouring and victualling us.

And it may be, that the inland there may yield masts, pitch, tar, hemp, and all things for the Navy.

And if there be a strait in the passage into the Scythian seas, the same is specially and with great regard to be noted, for what prince soever shall possess the same, as the King of Denmark doth possess the strait of Denmark, he only shall have the trade out of these regions into the northeast parts of the world for himself, and for his private profit, or for his subjects only, or to enjoy wonderful benefit of the toll of the same. If any such strait be found, the havens near, the length of the straits, and all other such circumstances are to be set down and all the mariners in the voyage are to be sworn to keep close all such things.

You must have great care to preserve your people, since your number is so small, and not to venture any one man in any wise.

Bring home with you (if you may) from Cambalu or other civil place, one young man, although you leave one for him.

Also the fruits of the countries if they will not of themselves dure, dry them and so preserve them.

Also the seeds of all strange herbs and flowers, for such seeds of fruit and herbs coming from another part of the world, and so far off, will delight the fancy of many for the strangeness, and for that the same may grow, and continue the delight long time.

If you arrive at Cambalu, bring thence the map of that country. Bring thence some old printed book, to see whether they have had print there before it was devised in Europe as some write. If you arrive in Cambalu, take special view of their navy, and to note the force, greatness, manner of building of them, the sails, the tackles, the anchors, with ordnance, armour, and munition.

Also to note the force of the walls and bulwarks of their cities, their ordnance, and whether they have any calivers, and powder and shot.

And so throughout to note the force of the country both by sea and by land.

Take a special note of their apparel and furniture, and of the substance that the same is made of, of which a merchant may make a guess as well of their commodity, as also of their wants.

To note their shops and warehouses, and with what commodities they abound, the price also.

Things to be carried with you: kerseys of all orient colours, frizadoes, motleys, bristow friezes, Spanish blankets, bays of all colours, felts of divers colours. Taffeta hats. Deep caps for mariners, whereof if ample vent may be found, it would turn to an infinite commodity of the common poor people by knitting.

Quilted caps of Levant taffeta of divers colours, for the night.

Knit stocks of silk of orient colours.

Garters, girdles of buff and all other leather, with gilt and ungilt buckles.

Gloves of all sorts knit, and of leather.

Gloves perfumed.

Shoes of Spanish leather.

Shoes of other leather.

Velvet shoes and pantoufles.

Purses knit, and of leather.

Nightcaps knit.

A garnish of pewter for a show of a vent of that English commodity, bottles, flagons, spoons etc. of that metal.

Glasses of English making.

Looking glasses for women, great and fair.

Spectacles of the common sort.

Hour glasses.

Combs of horn.

Linen of divers sorts.

Handkerchiefs with silk of several colours wrought.

Glazen eyes to ride with against dust.

Knives in sheaths both single and double, of good edge.

Needles great and small of every kind.

Buttons.

All the several silver coins of our English monies, to be carried with you to be showed to the governors at Cambalu, which is a thing that shall in silence speak to wise men more than you imagine.

Locks and keys, hinges, bolts, hasps, etc. great and small of excellent workmanship, whereof if vent may be, we shall set our subjects in work, which you must have in great regard. For in finding ample vent of any thing that is to be wrought in this realm, is more worth to our people besides the gain of the merchant, than Christchurch, Bridewell, the Savoy, and all the hospitals of England.

Take with you the map of England set out in fair colours, one of the biggest sort I mean, to make show of your country from whence you come. And also the large map of London to make show of your city. And let the river be drawn full of ships of all sorts, to make the more show of your great trade and traffic in trade of merchandise.

Rolls of parchment, for that we may vent much without hurt to the realm, and it lieth in small room.

Carry glue, for that we have plenty.

Red ochre for painters. We have great mines of it, and have no vent.

Try what vent you may have of saffron, because this realm yields the best of the world, and the tillage may set the poor greatly in work to their relief.

Black coneys' skins. Try the vent at Cambalu, for we abound with the commodity.

Before you offer your commodities to sale, endeavour to learn what commodities the country there hath. For if you bring thither velvet, taffeta, spice, or any such commodity that you yourself desire to lade yourself home with, you must not sell yours dear, lest hereafter you purchase theirs not so cheap as you would.

Antimony. See whether they have any ample use there for it, for that we may load whole navies of it and have no use for it unless it be for some small portion in founding of bells.

A painted bellows. For that perhaps they have not the use of them.

A pot of cast iron. It is a natural commodity of this realm.

Note especially what excellent dyeing they use in these regions: see their dye houses and the materials and simples that they use about the same, and bring musters and shows of the colours and of the materials.

Take with you for your own use. All manner of engines to take fish and fowl.

Take with you those things that be in perfection of goodness. For false and sophisticated commodities shall draw you and all your commodities into contempt and ill opinion.

XLI

Reports of the voyage into the parts of Persia and Media, for the company of English merchants for the discovery of new trades, in the years 1579 1580 and 1581 gathered out of sundry letters written by Christopher Burrough.

The ships for the voyage to St Nicholas in Russia, in which the factors and merchandise for the Persian voyage were transported, departed from Gravesend the 19 of June 1579, arrived at St Nicholas in Russia the 22 of July, where the factors and merchants landed, and the merchandise were discharged and laden into barks of the country, to be carried from thence up by river unto Vologda. By continual sailing, rowing, setting with poles or drawing of men, they came to the city of Vologda the 19 of August, where they landed their goods and stayed at that place till the 30. Having provided at Vologda, telegas, or wagons, whereupon they laded their goods, they departed thence towards Yaroslavl, and came to the east side of the river Volga over against Yaroslavl, with 25 telegas laden with goods the seventh of September at five of the clock afternoon.

Then the three stroogs or barks provided to transport the goods to Astrakhan (where they should meet the ship that should carry the same into Persia) came over unto the same side of the River Volga, and there took in the said goods. They arrived at Nijni Novgorod the 17 day [of September] at three of the clock afternoon, where they shewed the Emperor's letters to pass free without paying any custom and then departing, arrived at Kazan (or near the same town) on the 22 of September at five of the clock afternoon, departing thence came to Uvek[85] the fifth of October about five of the clock in the morning. This place is accounted half the way between Kazan and Astrakhan: and here there groweth great store of liquorice: the soil is very fruitful: they found there apple trees, and cherry trees. The latitude of Uvek is 51 degrees 30 minutes. At this place had been a very fair stone castle, and adjoining the same was a town called by the Russians, Sodom: this town and part of the castle (by report of the Russians) was swallowed into the earth by the justice of God, for the wickedness of the people that inhabited the same. There remaineth at this day to be seen a part of the ruins of the castle, and certain tombs, wherein as it seemeth have been laid noble personages: for upon a tomb stone might be perceived the form of a horse and a man sitting on it with a bow in his hand, and arrows girt to his side: there was a piece of a scutcheon also upon one of the stones, which had characters graven on it, whereof some part had been consumed with the weather, and the rest left unperfect but by the form of them that remained, we judged them to be characters of Armenia.

They departed from Uvek the said fifth of October at five of the clock after noon, and came to Peravolok the 10 day, making no abode at that place but passed along by it. Seven versts beneath, upon an island called Tsaritsna the Emperor of Russia hath fifty gunners all the summer time to keep watch, called by the Tartar name *carawool* Between this place and Astrakhan are five other *carawools* or watches.

The 16 of October they arrived at Astrakhan in safety, where they found the ship provided for the Persia voyage in good order and readiness. The 17 day the four principal factors of the company, Arthur Edwards, William Turnbull, Matthew Talbois, and Peter Garrard, were invited to dine with the chief secretary of Astrakhan (Vasili Pheodorovich Shelepin) who declared then unto them the troubles that were in Media and Persia: and how the Turk with the help of the Crims had conquered, and did possess the greatest part of Media: winter was at hand, and if they should put out with their ship to the sea, they should be constrained to take what hazards might happen them: whereupon the factors determined to stay there all winter to learn further of the state of those countries.

The 19 of November the wind being northerly, there was a great frost, and much ice in the river: the next day being the 20 of

November the ice stood in the river, and so continued until Easter Day.

The 22 of December departed this life John Moore the gunner of the ship.

The 6 of January being Twelfth Day the Russians of Astrakhan broke a hole in the ice upon the river Volga, and hallowed the water with great solemnity according to the manner of their country, at which time all the soldiers of the town shot off their small pieces upon the ice, to gratify the captain of the castle, who stood hard by the ship, and afterwards the great ordnance of the castle was shot off.

On the 31 of January there happened a great eclipse of the moon, which continued before she was clear an hour and a half by estimation: she was wholly darkened the space of half an hour.

The 7 of March 1580 the Nogaians and Crims[86] came before Astrakhan to the number of one thousand four hundred horsemen, which encamped round about: some of them lay on the Crims' side of Volga, and some on the Nogaian side, but none of them came upon the island that Astrakhan standeth on. It was said that two of the prince of the Crims his sons were amongst them. They sent a messenger on the eighth day to the captain of Astrakhan, to signify that they would come and visit him: who answered, he was ready to receive them: and taking a great shot or bullet in his hand, willed the messenger to tell them that they should not want of that gear, so long as it would last. The ninth day news was brought that the Crims determined to assault the town or castle, and were making of faggots of reed, to bring with them for that purpose. The 13 day they broke up their camps, and marched to the northwards into the country of Nogaia.

The 17 of April the variation of the compass observed in Astrakhan was 13 deg. 40 min. from north to west. This spring there came news to Astrakhan that the Queen of Persia (the King being blind) had been with a great army against the Turks that were left to possess Media, and had given them a great overthrow: yet nothwithstanding Derbent,[87] and the greatest part of Media were still possessed and kept by the Turks. The factors of the company consulting upon their affairs, determined to leave at

Astrakhan the one half of their goods with Arthur Edwards, and with the other half the other three factors would proceed in the ship on their purposed voyage to the coast of Media, to see what might be done there.

The 29 of April Amos Riall, and Anthony Marsh, the company's servants were sent from Astrakhan up the river to Yaroslavl, with letters of advice to be sent for England, and had order for staying the goods in Russia that should come that year out of England for maintaining the trade purposed for Persia, until further trial were made of what might be done in those parts.

The first day of May in the morning, having the ship in readiness to depart, the factors invited the Duke and the principal secretary to a banquet aboard the ship, where they were entertained to their good liking, and at their departure was shot off all the ordnance of the ship, and about nine of the clock at night they weighed anchor, and departed with their ship from Astrakhan. The seventh of May in the morning they passed by a tree that standeth on the left hand of the river as they went down, which is called Mahomet's Tree, and about three versts to the southward of the said tree is a place called Uchoog, where are certain cottages and the Emperor hath lying at that place certain gunners to guard his fishermen that keep the weir. The ninth and tenth days they met with shoaled water, and were forced to lighten their ship. The 15 day by great industry and travail they got their ship clear off the shoals and flats. The 21 having the wind at northwest they set sail, and steered thence south by west, and south until eleven of the clock, and had then nine foot water: and at noon observed the latitude, and found it to be 44 degrees 47 minutes: then they had three fathoms and a half water, being clear of the flats.

From thence they sailed until the 27 day two of the clock in the morning westsouthwest eight leagues, the wind blowing at north very much. They came to Bildih[88] in the country of Media, against which place they anchored in 9 foot water. Presently there came aboard of them a boat, wherein were seven or eight persons, two Turks, the rest Persians, the Turks' vassals, which bade them welcome, and seemed to be glad of their arrival, who told the factors that the Turks had conquered all Media, and how that the Turks' Pasha[89] remained in Derbent with a garrison of Turks, and that Shemakha was wholly spoiled, and had few or no inhabitants left in it. The factors then being desirous to come to the speech of the Pasha, sent one of the company's servants Robert Golding, with those soldiers, to the captain of Baku,[90] which place standeth hard by the sea, to certify him of their arrival, and what commodities they had brought, and to desire friendship to have quiet and safe traffic for the same.

When the messenger came to the captain of Baku, the said captain gave him very friendly entertainment, and after he understood what they were that were come in the ship, and what they had brought, he seemed to rejoice much thereat: the said Golding returned and came to the ship. The factors caused a tent to be set up at shore near the ship, against the coming of the captain: who came thither about three of the clock afternoon, and brought about thirty soldiers, in shirts of mail, and some of them had gauntlets of silver, others of steel, and very fair. The factors met him at their tent, and after very friendly salutations passed between them, they gave him for a present a garment of cloth of velvet, and another of scarlet. The factors made request that he would help them to the speech of the Pasha, who answered that their demand was reasonable, and said, because the way to Derbent, where the Pasha remained, was dangerous, he would send thither, and certify him of their arrival, and such commodities as they would desire to exchange or barter he would procure the said Pasha to provide for them: and therefore willed the factors to consult together, and certify him what they most desired, and what quantity they would have provided.

That night they rode from the seaside, to a village about ten miles off, where at suppertime the captain had much talk with Mr Garrard of our country, demanding where about it did lie, what countries were near unto it, and with whom we had traffic: but when by the situation he perceived we were Englishmen, he demanded if our prince were a maiden Queen: which when he was certified of, then (quoth he) your land is called *Ingiltere*, is it not?

Supper being brought in, he requested them to eat. After their pottage (which was made of rice) and likewise their boiled meat, there came in platters of rice sodden thick, and honey mingled withal: after all which, came a sheep roasted whole. Divers questions he had with Mr Garrard and Christopher Burrough at suppertime, about their diet, inquiring whether they eat fish or flesh voluntarily, or by order. Their drink in those parts is nothing but water. After supper (walking in the garden) the captain demanded of Mr Garrard, whether the use was in England to lie in the house or in the garden, and which he had best liking of: he answered, where it pleased him, but their use was to lie in houses: whereupon the captain caused beds to be sent into the house for them, and caused his kinsman to attend on them in the night.

In the morning very early he sent horse for the rest of the company which should go to Derbent, sending by them that went ten sheep for the ship. In that village there was a Stove,[91] into which the captain went in the morning, requesting Mr. Garrard to go also to the same to wash himself, which he did. Shortly after their coming out of the Stove, whilst they were at breakfast, Mr Turnbull, Mr Talbois, and Thomas Hudson the master of the ship, came thither, and when they had all broken their fasts, they went to Baku: but Christopher Burrough returned to the ship, for that he had hurt his leg, and could not well endure that travel. In their journey to Derbent they forsook the ordinary ways, being very dangerous, and travelled through woods till they came almost to the town of Derbent: the gentleman rode before with the captain's letters to the Pasha, to certify him of the English merchants' coming, who was very glad of the news, and sent forth to receive them certain soldiers, who met them about two miles out of the town, saluting them with great reverence, and then came forth noble men, captains, and gentlemen, to receive them into the castle and town. As they entered the castle, there was a shot of twenty pieces of great ordnance, and the Pasha sent Mr Turnbull a very fair horse with furniture to mount on, esteemed to be worth an hundred marks,[92] and so they were conveyed to his presence: who after he had talked with them, sent for a coat of cloth of gold, and caused it to be put on Mr Turnbull's back, and then willed them all to depart, and take their ease, for that they were weary of their journey, and on the morrow he would talk further with them. The next day they requested that he would grant them his privilege, whereby they might traffic safely in any part and place of his country, offering him, that if it pleased His Majesty to have any of the commodities that they had brought, and to write his mind thereof to the captain of Baku, it should be delivered him accordingly. The Pasha's answer was, that he would willingly give them his privilege: yet for that he regarded their safety, having come so far, and knowing the state of his country to be troublesome, he would have them to bring their commodity thither, and there to make sale of it, promising he would provide such commodities as they needed, and that he would be a defence unto them: whereupon the factors sent Thomas Hudson back for the ship to bring her to Derbent, and the Pasha sent a gentleman with him to the captain of Baku, to certify him what was determined.

The 29 day their goods were unladen and carried to the Pasha's garden, where he made choice of such things as he liked, taking for custom of every five and twenty kerseys, or whatsoever, one, or after the rate of four for the hundred. The factors determined to send a part of the rest of the goods to Baku, for the speedier making sale thereof, for which cause they obtained the Pasha's letter to the captain of Baku, written very favourably in their behoof: and thereupon was laden and sent in a small boat of that country in merchandise, to the value (very near) of one thousand pound sterling: videlicet, one hundred pieces of kerseys, seven broadcloths, two barrels of cochineal,[93] two barrels of tin, four barrels of shaffe.[94] There went with the same of the company's servants William Winckle, Robert Golding, and Richard Relfe, with two Russians, whereof one was an interpreter, besides four barkmen. Their passage and carriage of their goods to Baku was chargeable, although their sales when they came thither were small: they had great friendship shewed them of the captain of Baku, as well for the Pasha's letter as also for the factors' sakes, who had dealt friendly with him, as before is declared.

Robert Golding desirous to understand what might be done at Shemakha, which is a day's journey from Baku, went thither, from whence returning, he was set on by thieves, and was shot into the knee with an arrow, who had very hardly escaped with his life and goods, but that by good hap he killed one of the thieves' horses with his caliver, and shot a Turk through both cheeks with a dag. On the sixth day of August the factors being advertised at Derbent that their ship was so rotten and weak, that it was doubtful she would not carry them back to Astrakhan, did thereupon agree and bargain at that place with an Armenian, whose name was Jacob, for a bark called a buss, being of burden about 35 tons, which came that year from Astrakhan. When all their goods were laden aboard the said buss, being ready to have departed for Derbent, there arose a great storm with the wind out of the sea, by force whereof the cables and hawsers were broken, and their vessel put ashore, and broken to pieces against the rocks: every of them that were in her saved their lives, and part of the goods. But there was a chest wherein were dollars, and gold, which they had received for the commodities of the company, which they sold at Baku, which at the taking out of the buss, fell by the bark's side into the water amongst the rocks, and so was lost. The packs of cloth which they could not well take out of the buss were also lost, other things that were more profitable they saved.

The 18 of August, the factors received from the Pasha 500 batmans[95] of raw silk, parcel of the bargain made with him, who bade them come the next day for the rest of the bargain.

The 19 day the factors went to the Pasha according to his appointment, but that day they could not speak with him, but it was delivered them as from him, that they should look and consider whether anything were due unto him or not, which grieved the factors: and thereupon Mr Turnbull answered, that their heads and all that they had were at the Pasha's pleasure: but then it was answered that there was no such matter in it: but that they should cast up their reckonings, to see how it stood between them. The 20 day they cast up their reckonings. The 21 they went to have spoken with the Pasha, but were denied audience.

The 22 day they heard news by a buss that came from Astrakhan, that Arthur Edwards (whom the factors left at Astrakhan with the moiety of the goods) was dead.

The 23 day the factors received more from the Pasha 500 batmans of silk.

The 5 Tobias Atkins the gunner's boy died of the flux, who was buried the 6 day 2 miles to the southward of the castle of Derbent, where the Armenian Christians do usually bury their dead.

The 26 of September was laden aboard the ship 40 bales of silk. From the 26 till the 2 of October, they took into the ship, bread, water, and other necessary provision for their sea store: the said 2 day of October, the factors were commanded upon the sudden to avoid their house, and get them with their provision out of the town: whereupon they were constrained to remove and carry their things to the seaside against the ship, and remained there all night. The cause of this sudden avoiding of them out of the town (as afterwards they perceived) was for that the Pasha had received news of a supply with treasure that the Turk had sent, which was then near at hand coming towards him.

The 3 day of October all things were brought from the shore aboard the ship: and the factors went to the Pasha to take their leave of him, unto whom they recommended those the company's servants which they had sent to Baku, who caused their names to be written, and promised they should want nothing, nor be injured of any. After this leave taken the factors departed towards Astrakhan, the wind serving well for that purpose at south southeast: and as they were ready to set sail, there came against the ship a man, who waved: whereupon the boat was sent ashore to him, who was an Armenian sent from William Winckle, with his writing tables, wherein Winckle had written briefly, the mishap of the loss of the buss, and that they were coming from Bildih towards Derbent with a small boat, forced to put ashore in a place by the seaside called the Armenian village: whereupon the factors caused the ship to stay, which stay and loss of those southerly winds, was a cause of great troubles, that they afterwards sustained through ice, &c. entering the Volga as shall be declared.

The 4 day the wind south southeast Christopher Burrough was sent to shore to Derbent to provide some necessaries for the voyage, and being on shore he saw there the coming in of the

Turks' treasure, being accompanied with 200 soldiers, and one hundred pioneers, besides captains and gentlemen: the Pasha with his captains and soldiers very gallantly apparelled and furnished went out from Derbent about three or four miles, to meet the said treasure, and received the same with great joy and triumph. Treasure was the chief thing they needed, for not long before the soldiers were ready to break into the court against the Pasha for their pay. The treasure came in seven wagons, and with it were brought ten pieces of brass.[96]

In Media there was no commodity to be brought of any value, but raw silk, neither was that to be had but at the Pasha's hands: who shortly after their coming thither taxed the country for that commodity. His dealing with our merchants as it was not with equity in all points according to his bargain, so it was not extreme ill. Of the commodities they carried he took the chiefest part, for which he gave but a small price in respect of the value, and because he provided such quantity of commodity for them, which otherwise they could not have had, the country being so troublesome, and travel by land so dangerous, he used them at his pleasure.

The news that was reported unto them at Astrakhan touching the wars between the Turks and Persians differed little from the truth: for the Turks' army with the aid of the Crims invaded and conquered the country of Media in anno. 1577. The Great Turk appointed Osman Pasha (the said Pasha, and now captain of Derbent) governor of the whole country, who settled himself in Shemakha the chief city of Media, and principal place of traffic.

After the Pasha had brought the country into order according to his liking, and placed garrisons where he thought convenient, the army was dissolved and sent back. When the Persians understood that the Turks' army was dissolved and returned, they gathered a power together, and with the Queen of their country as chief, they entered the country of Media, and overran the same with fire and sword, destroying whatsoever they found, as well people, cattle, as whatsoever else, that might be commodious to the Turks. And after they had so overrun the country, they came to Shemakha, where the said Pasha lieutenant-general of the Great Turk was settled, and besieged it: whereupon the Pasha seeing he could not long endure to withstand them, fled thence to Derbent where he now remaineth.

Derbent is a strong castle which was built by Alexander the Great, the situation whereof is such, that the Persians being without ordnance, are not able to win it but by famine. When the Turks were fled from Shemakha, the Persians entered and spoiled it, leaving therein neither living creature nor any commodity, and so returned back into Persia.

The latitude of Derbent (by divers observations exactly there made) is 41 deg 52 min. The variation of the compass at that place about 11 degrees from north to west. From Derbent to Bildih by land 46 leagues. From Derbent to Shemakha by land 45 leagues. From Shemakha to Baku about 10 leagues, which may be 30 miles. From the Castle Derbent eastwards, there reach two stone walls to the border of the Caspian Sea, which is distant one English mile. Those walls are 9 foot thick, and 28 or 30 foot high, and the space between them is 160 geometrical paces, that is 800 foot.

The 5 of October about noon the wind north northeast they weighed anchor, and set sail from Derbent, being alongst the coast to the southwards to seek their men: but as they had sailed about four leagues the wind scanted easterly, so that they were forced to anchor in three fathom water.

The 6 day they weighed anchor, and bare further off into the sea, where they anchored in seven fathom water, the ship being very leak, and so rotten abaft the mainmast, that a man with his nails might scrape through her side.

The 7 day about 7 of the clock in the morning they set sail, the wind southwest. They considered the time of the year was far spent, the ship weak, leak and rotten, and therefore determining not to tarry any longer for Winckle and his fellows, but to leave them behind, bent themselves directly towards Astrakhan: and sailing north northeast until midnight about 16 leagues, the wind then came to the north northwest, and blew much, a very storm, which caused them to take in all their sails, saving the fore course, with which they were forced to steer before the sea. And on the 8 day about two of the clock in the morning their great boat sunk at the ship's stern, which they were forced to cut from the ship to their great grief and discomfort: for in her they hoped to save their lives if the ship should have miscarried. About 10 of the clock before noon they had sight of the land about 5 leagues to the south of Derbent, where they came at an anchor in three fathoms, and black ooze, good anchor hold, whereof they were glad. Winckle and the rest of his fellows being in the Armenian village, saw the ship as she passed by that place, and sent a man in the night following along the coast after her, who came against the ship where she rode, and with a firebrand in the top of a tree made signs, which was perceived by them in the ship, whereupon they hoisted out their skiff, and sent her ashore, which returned a letter from Winckle, wherein he wrote that they were with such goods as they had at the Armenian village, and prayed that there they might with the same goods be taken into the ships. The 9 day it was little wind, they weighed and bare a little further off into the sea towards the village, and anchored.

The 12 day the wind southeast they weighed anchor, and bare against and near to the Armenian village where they anchored, and then the skiff came aboard and told them that our people at shore were likely to be spoiled of the Tartars, were it not that the gunners defended them: then was the skiff sent back again to charge them at any hand they should hasten aboard the ship whatsoever it cost them. Whereupon, all the company came aboard the same day saving Richard Relfe and two Russians, but as soon as the skiff was returned aboard the ship, the wind blew at southeast, and the sea was growing, so as they were forced to take in their skiff into the ship, and rode still till the 13 day, and then being fair weather, early in the morning the skiff was hoisted out of the ship, and sent to shore to fetch the said Relfe and the two Russians, which were ready at the shore side, and with them two Spaniards that were taken captives at the Goletta in Barbary,[97] which served the Turk as soldiers. Those Spaniards (of Christian charity) they brought also aboard the ship to redeem them from their captivity, which were brought over into England, and set free and at liberty here in London, in September 1581. The 16 day the wind east southeast, they weighed anchor and set sail, bearing northwards towards Astrakhan.

The 30 day the wind southeast, they set sail to the northeastwards: but the ship fell so on the side to the shorewards, that they were forced eftsoons to take in their sail, and anchor again, from whence they never removed her. That day they shared their bread: but in their want God sent them two coveys of partridges, that came from the shore, and lighted in and about their ships, whereby they were comforted, and one that lay sick, of whose life was small hope, recovered his health.

The 13 day [of November] in the morning Amos Riall was sent away in a small boat towards Astrakhan, to provide victuals and carriages to relieve and help them, who could pass no further than the Four Islands, but was there overtaken with ice, and was forced to leave his boat, and from thence passed post to Astrakhan. The same day they departed in lighters with the goods, leaving the ship at anchor, and in her two Russians, which with three more had offered to undertake for twenty roubles in money to carry the ship into some harbour, where she might safely winter, or else to keep her where she rode all winter, which was promised to be given them if they did it: the wind then at northeast, did freeze the sea so as they could not row, guide, stir or remove the lighters, but as the wind and ice did force them. And so they continued driving with the ice, southeast into the sea by the space of forty hours, and then being the sixteenth day the ice stood. Whiles they drove with the ice, the dangers which they incurred were great: for oftentimes when the ice with force of wind and sea did break, pieces of it were tossed and driven one upon another with great force, terrible to behold, and the same happened at some times so near unto the lighters, that they expected it would have overwhelmed them to their utter destruction: but God who had preserved them from many perils before, did also save and deliver them then.

Within three or four days after the first standing of the ice, when it was firm and strong, they took out all their goods, being forty and eight bales or packs of raw silk, &c., laid it on the ice, and covered the same with such provisions as they had. Then for want of victuals they agreed to leave all the goods there upon the ice, and to go to the shore: and thereupon brake up their chests, and with such other things as they could get, they made sleds for every one of them to draw upon the ice, whereon they laid their clothes to keep them warm, and such victuals as they had, and such other things as they might conveniently carry, and so they departed very early about one of the clock in the morning, and travelling on the ice, directed their way north, as near as they could judge.

The next morning very early they lost their way through the persuasion of the Russians which were with them, taking too much towards the left hand (contrary to the opinion of Mr Hudson) whereby wandering upon the ice four or five days, not knowing whether they entered into the Crim Tartars' land or not, at length it fortuned they met with a way that had been travelled, which crossed backwards towards the sea: that way they took, and following the same, within two days' travel it brought them to a place called (in the English tongue) Red cliff, which divers of the company knew.

There they remained that night, having nothing to eat but one loaf of bread, which they happened to find with the two Russians that were left in the ship to keep her all the winter (as is aforesaid) who certified them that the ship was cut in pieces with the ice, and that they had hard scaping with their lives.

In the morning they departed early, and about 9 of the clock they met Amos Riall, with the carpenter, and a gunner newly come out from England, and also 65 horses with so many cossacks to guide them, and 50 gunners for guard, which brought provision of victuals, and were sent by the Duke to fetch the goods to Astrakhan. The meeting of that company was much joy unto them.

The factors sent back with Amos Riall and the company to fetch the goods, Thomas Hudson the master, Tobias Paris his mate, and so they the factors and their company proceeded on towards Astrakhan, where they arrived the last day of November.

The 3 day [of December] early in the morning they departed all towards the said goods, and the same day did lade all the goods they could find upon the sleds, and with all convenient speed returned back towards Astrakhan. And when they rested the night, in the morning very early before the break of day, they were assaulted by a great company of the Nogaian Tartars' horsemen, which came shouting and hallooing with a great noise, but our people were so environed with the sleds, that they durst not enter upon them, but ran by, and shot their arrows amongst them, and hurt but one man in the head, who was a Russian. Yet when it was day, they showed themselves a good distance off from our men, being a very great troop of them, but did not assault them any more. The same day our men departed from thence towards Astrakhan, where they arrived in safety the 4 of December, about 3 of the clock in the afternoon, where our people greatly rejoiced of their great good hap to have escaped so many hard events, troubles and miseries, as they did in that voyage, and had great cause therefore to praise the Almighty, who had so mercifully preserved and delivered them. They remained the winter at Astrakhan, where they found great favour and friendship of the Duke, captain, and other chief officers of the place: but that winter there happened no great matter worth the noting.

In the spring of the year 1581, about the midst of March, the ice was broken up, and clear gone before Astrakhan, and the ninth of April, having all the goods that were returned from the parts of Media, laden into a stroog, the factors, William Turnbull, Matthew Talbois, Giles Crow, Christopher Burrough, Michael Lane, Laurence Prouse gunner, Randolfe Foxe, Tho. Hudson, Tobias Paris, Morgan Hubblethorne the dyer, Rich the surgeon, Rob. Golding, John. Smith, Edw. Reding carpenter, and William Perrin gunner, having also 40 Russians, whereof 36 were cossacks to row, the rest merchants passengers, departed from Astrakhan up the Volga towards Yaroslavl. They left behind them at Astrakhan, with the English goods and merchandise there remaining, Amos Riall, W. Winckle and Richard Relfe, and appointed them to sell and barter the same, or so much thereof as they could, and with such as they should take in exchange to return up to Yaroslavl that summer when the Emperor's carriage should pass up the Volga.

The 26 day they arrived with their stroog at Kazan, where they remained till the fourth of June: the factors sent Giles Crow from Kazan to Moscow, with their letters the 30 of May. The 29 day they came to Vologda, with all their goods in safety, and good order. The same 29, William Turnbull and Peter Garrard departed from Vologda post by water towards Kholmogory. They arrived in safety the 16 of July, and found there the Agents of Russia, and in the road the ships sent out of England, almost laden ready to depart.

The 25 day departed for England (out of the road of St Nicholas) the ship *Elizabeth.*

The 26 day departed thence the *Thomas Allen* and *Mary Susan,* and in the *Thomas Allen* went William Turnbull, Matthew Talbois, Thomas Hudson and others. The goods returned of the Persia voyage were laden into the ship, *William and John,* whereof was master, William Bigat, and in her with the same goods came Peter Garrard and Tobias Paris.

In their return homeward they had some foul weather, and were separated at sea, the *William and John* put into Newcastle the 24 of September: from whence the said Peter Garrard and Tobias Paris came to London by land, and brought news of the arrival of the ship.

The 25 of September both the said ships arrived at the port of London in safety, and anchored before Limehouse and Wapping, where they were discharged, 1581.

XLII

Of the tree which beareth bombasine cotton, or gossampine.

In Persia is great abundance of bombasine cotton, and very fine: this groweth on a certain little tree or briar, not past the height of a man's waist, or little more: the tree hath a slender stalk like unto a briar, or to a carnation gillyflower, with very many branches, bearing on every branch a fruit or rather a cod, growing in round form, containing in it the cotton: and when this bud or cod cometh to the bigness of a walnut, it openeth and showeth forth the cotton, which groweth still in bigness until it be like a fleece of wool as big as a man's fist, and beginneth to be loose, and then they gather it as it were the ripe fruit. The seeds of these trees are as big as peas, and are black, and somewhat flat, and not round; they sow them in plowed ground, where they grow in the fields in great abundance in many countries in Persia, and divers other regions.

XLIII

A report of the voyage attempted in the year of our Lord 1583 by Sir Humphrey Gilbert, intended to discover and to plant Christian inhabitants in place convenient, upon those large and ample countries extended northward from the cape of Florida, lying under very temperate climes, esteemed fertile and rich in minerals, yet not in the actual possession of any Christian prince, written by Mr Edward Haye who alone continued unto the end, and by God's special assistance returned home.

The first discovery of these coasts (never heard of before) was well begun by John Cabot the father, and Sebastian his son, who were the first finders out of all that great tract of land stretching from the cape of Florida unto those islands which we now call the Newfoundland: all which they brought and annexed unto the crown of England.

Not long after that Christopher Columbus had discovered the islands and continent of the West Indies for Spain, John and Sebastian Cabot made discovery also of the rest from Florida northwards to the behoof of England.

Whensoever afterwards the Spaniards (very prosperous in all

their southern discoveries) did attempt any thing into Florida and those regions inclining towards the north, they proved most unhappy, and were at length discouraged utterly as if God had prescribed limits unto their Spanish nation which they might not exceed.

The French, as they can pretend less title unto these northern parts than the Spaniard, did but review that discovered by the English nation, usurping upon our right, and imposing names upon countries, rivers, bays, capes, or headlands, as if they had been the first finders of those coasts; even so God hath not hitherto permitted them to establish a possession permanent, notwithstanding their manifold attempts.

When first Sir Humphrey Gilbert undertook the western discovery of America, and had procured from Her Majesty a very large commission to inhabit and possess at his choice all remote and heathen lands not in the actual possession of any Christian prince, very many gentlemen of good estimation drew unto him, so that the preparation was expected to grow into a puissant fleet, able to encounter a king's power by sea: nevertheless, amongst a multitude of voluntary men, their dispositions were divers. And when the shipping was in a manner prepared, and men ready upon the coast to go aboard: at that time some failed their promises contracted, leaving the general with few of his assured friends, with whom he adventured to sea.

Having buried only in a preparation a great mass of substance whereby his estate was impaired, his mind yet not dismayed, he continued his former designment.

In furtherance of his determination, amongst others, Sir George Peckham knight showed himself very zealous to the action, greatly aiding him both by his advice and in the charge. Other gentlemen to their ability joined unto him, resolving to adventure their substance and lives in the same cause. Who beginning their preparation from that time, both of shipping, munitions, victual, men, and things requisite, such were the difficulties and cross accidents opposing these proceedings, which took not end in less than two years.

We resolved to begin our course northwards, and to follow

directly as we might, the trade way unto Newfoundland: from whence after our refreshing and reparation of wants, to proceed into the south, not omitting any river or bay which in all that large tract of land appeared to our view to be worthy of search.

We began our voyage upon Tuesday the eleventh day of June, in the year of our Lord 1583, having in our fleet these ships: *The Delight*, 120 tons, the *Bark Raleigh*, 200 tons, set forth by Mr Walter Raleigh, *The Golden Hind*, 40 tons, the *Swallow*, 40 tons, the *Squirrel*, 10 tons.

We were in number in all about 260 men: among whom we had of every faculty good choice, as shipwrights, masons, carpenters, smiths: also mineral men and refiners. Besides, for solace of our people and allurement of the savages, we were provided of music in good variety: not omitting the least toys, as morris dancers, hobby horses, and Maylike conceits to delight the savage people, whom we intended to win by all fair means possible. And to that end we were indifferently furnished of all petty haberdashery wares to barter with those simple people.

Thursday following about midnight the [*Bark Raleigh*] forsook us, notwithstanding we had the wind east, fair and good. But it was after credibly reported, that they were infected with a contagious sickness, and arrived greatly distressed at Plymouth: the reason I could never understand. Sure I am, no cost was spared by their owner Master Raleigh in setting them forth.

By this time we were in 48 degrees of latitude not a little grieved with the loss of the most puissant ship in our fleet: we were encumbered with much fog and mists in manner palpable, in which we could not keep so well together, but were dissevered, losing the company of the *Swallow* and the *Squirrel* upon the 20th day of July.

The 27th of July, we might descry not far from us, as it were mountains of ice driven upon the sea.

Before we came to Newfoundland about 50 leagues on this side, we passed the Bank. The Portuguese and French chiefly, have a notable trade of fishing upon this Bank, where are sometimes a hundred or more sails of ships: who commonly begin the fishing in April, and have ended by July.

During the time of fishing, a man shall know without sounding when he is upon the Bank, by the incredible multitude of sea fowl hovering over the same, to prey upon the offals and garbage of fish thrown out by fishermen.

We had sight of an island named Penguin, of a fowl there breeding in abundance, almost incredible, which cannot fly, their wings not able to carry their body, being very large (not much less than a goose) and exceeding fat: which the Frenchmen use to take without difficulty upon that island, and to barrel them up with salt.

Trending this coast, we came to the island called Bacalaos, being not past two leagues from the main. Here we met with the *Swallow* again, whom we had lost in the fog, and all her men: they spared not to cast up into the air and overboard, their caps and hats in good plenty.

We held on our course southward, until we came against the harbour called St John's: where before the entrance into the harbour, we found also the *Squirrel* lying at anchor. Whom the English merchants (that were and always be admirals by turns interchangeably over the fleets of fishermen within the same harbour) would not permit to enter into the harbour. We made ready our fights, and prepared to enter the harbour, any resistance to the contrary notwithstanding there being within of all nations, to the number of 36 sails. But first the general dispatched a boat to give them knowledge of his coming for no ill intent, having commission from Her Majesty for his voyage he had in hand.

Having taken place convenient in the road, we let fall anchors, the captains and masters repairing aboard our admiral: whither also came immediately the masters and owners of the fishing fleet of Englishmen, to understand the general's intent and cause of our arrival there. They were all satisfied when the general had showed them his commission, and purpose to take possession of those lands to the behalf of the crown of England, requiring but their lawful aid for repairing of his fleet, and supply of some necessaries, so far as conveniently might be afforded him, both out of that and other harbours adjoining.

It was further determined that every ship of our fleet should deliver unto the merchants and masters of that harbour a note of all their wants: which done, the ships as well English as strangers, were taxed at an easy rate to make supply. Whereunto the Portuguese (above other nations) did most willingly and liberally contribute. Insomuch as we were presented (above our allowance) with wines, marmalades, most fine rusk or biscuit, sweet oils and sundry delicacies. Also we wanted not of fresh salmon, trout, lobster and other fresh fish brought daily unto us.

Monday following, the general had his tent set up, who being accompanied with his own followers, summoned the merchants and masters, both English and strangers to be present at his taking possession of those countries. Before whom openly was read and interpreted unto the strangers his commission: by virtue whereof he took possession in the same harbour of St John's and 200 leagues every way, invested the Queen's Majesty with the title and dignity thereof, had delivered unto him (after the custom of England) a rod and a turf of the same soil, entering possession also for him, his heirs and assigns for ever. He proposed and delivered three laws to be in force immediately. That is to say: the first for religion, which in public exercise should be according to the Church of England. The 2nd for maintenance of Her Majesty's right and possession of those territories, against which if any thing were attempted prejudicial, the party or parties offending should be adjudged and executed as in case of high treason, according to the laws of England. The 3rd if any person should utter words sounding to the dishonour of Her Majesty, he should lose his ears, and have his ship and goods confiscate.

These contents published, obedience was promised by general voice and consent of the multitude as well of Englishmen as strangers. Afterwards were erected not far from that place the arms of England engraven in lead, and infixed upon a pillar of wood.

That which we do call the Newfoundland, and the Frenchmen Bacalaos, is an island, or rather (after the opinion of some) sundry islands and broken lands, situate in the north regions of America, upon the gulf of the great river called St Lawrence in Canada. The land with south and north, containing in length between three and 400 miles, accounting from Cape Race, unto the Grand Bay. The island round about hath very many goodly harbours, safe roads for ships.

In the months of June, July, August and September, the heat is somewhat more than in England at those seasons. Those which have arrived there after November and December, have found the snow exceedingly deep. But admitting extraordinary cold in those south parts, above that with us here: it can not be so great as in Sweden, much less in Moscovy or Russia: yet are the same countries very populous, and the rigour of cold is dispersed with by the commodities of stoves, warm clothing, meats and drinks: all which need not be wanting in the Newfoundland, if we had intent there to inhabit.

In the north are savages altogether harmless. Nature hath recompensed that incommodity of some sharp cold, by many benefits, viz. with incredible quantity, and no less variety of kinds of fish in the sea and fresh waters, as trouts, salmons, and other fish to us unknown: also cod, which alone draweth many nations thither, and is become the most famous fishing of the world. Abundance of whales, for which also is a very great trade in the bays of Placentia and the Grand Bay, where is made train oils of the whale: herring the largest that have been heard of, and exceeding the malstrond herring of Norway: but hitherto was never benefit taken of the herring fishing. There are sundry other fish very delicate, namely the bonito, lobsters, turbot, with others infinite not sought after: oysters having pearl but not orient in colour.

Concerning the island commodities, as well to be drawn from this land, as from the exceeding large countries adjoining: there is resin, pitch, tar, soapashes, deal board, masts for ships, hides, furs, flax, hemp, corn, cables, cordage, linen-cloth, metals and many more.

The trees for the most are firtrees, pine and cypress, all yielding gum and turpentine.

The grass and herb doth fat sheep in very short space, proved by English merchants which have carried sheep thither for fresh victual and had them raised exceeding fat in less than three weeks.

Peas which our countrymen have sown in the time of May, have come up fair, and been gathered in the beginning of August.

Upon the land divers sorts of hawks, as falcons, and others by report: partridges most plentiful larger than ours, grey and white of colour, and rough footed like doves, which our men after one flight did kill with cudgels, they were so fat and unable to fly. Birds some like blackbirds, linnets, canary birds, and other very small. Beast of sundry kinds, red deer, buffaloes or a beast, as it seemeth by the track and foot very large in manner of an ox. Bears, ounces or leopards, some greater and some lesser, wolves, foxes, which to the northward a little further are black, whose fur is esteemed in some countries of Europe very rich. Otters, beavers, martens: and in the opinion of most men that saw it, the general had brought unto him a sable alive. We could not observe the hundredth part of creatures in those uninhabited lands: but these mentioned may induce us to glorify the magnificent God, who hath superabundantly replenished the earth with creatures serving for the use of man.

Iron very common, lead, and somewhere copper, I will not aver of richer metals.

The general was most curious in the search of metals, commanding the mineral man and refiner, especially to be diligent. Who after search, brought at first some sort of ore, seeming rather to be iron than other metal. The next time he found ore, which with no small show of contentment he delivered unto the general, using protestation, that if silver were the thing which might satisfy the general and his followers, there it was.

Myself could not follow this confident opinion of our refiner to my own satisfaction: but afterward demanding our general's opinion therein, and to have some part of the ore, he replied: content yourself, I have seen enough. Touching the ore, I have sent it aboard, whereof I would have no speech to be made so long as we remain within harbour: here being both Portuguese, Biscayans and Frenchmen not far off, from whom must be kept any bruit or muttering of such matter. When we are at sea proof shall be made: if it be to our desire, we may return the sooner hither again. Whose answer I judged reasonable.

While the better sort of us were seriously occupied in contriving of matters for the commodity of our voyage: others of another sort and disposition were plotting of mischief. Some casting to steal away our shipping by night: whose conspiracies discovered, they were prevented. Others drew together in company, and carried away out of the harbours adjoining, a ship loaded with fish, setting the poor man on shore. A great many more of our people stole into the woods to hide themselves, attending time and means to return home by such shipping as daily departed from the coast. Some were sick of fluxes, and many dead: and in brief, by one means or other our company was diminished, and many by the general licensed to return home. Insomuch as after we had reviewed our people, resolved to see an end of our voyage, we grew scant of men to furnish all our shipping: it seemed good therefore unto the general to leave the *Swallow* with such provision as might be spared for transporting home the sick people.

The general made choice to go in his frigate the *Squirrel* being most convenient to discover upon the coast, and to search into every harbour or creek, which a great ship could not do. Therefore the frigate was prepared and overcharged with small ordnance, more to give a show, than with judgement to foresee unto the safety of her and the men.

Now having made ready our shipping, the *Delight*, the *Golden Hind*, and the *Squirrel*, and put aboard our provision, which was wines, bread or rusk, fish wet and dry, sweet oils: besides marmalades, figs, lemons barreled, and suchlike: in brief, we were supplied as if we had been in some city populous and plentiful of all things.

We departed from this harbour of St John's upon Tuesday the twentieth of August.

Upon Tuesday the 27 of August, toward the evening, our general caused them in his frigate to sound, who found white sand at 35 fathoms, being then in latitude about 44 degrees.

The evening was fair and pleasant, yet not without token of storm to ensue, and most part of this Wednesday night, like the swan that singeth before her death, they in the *Delight*, continued in sounding of trumpets, with drums, and fifes: also winding the cornets.

Thursday, the 29 of August, the wind arose, and blew vehemently at south and by east, bringing withal rain, and thick mist, so that we could not see a cable length before us. And betimes in the morning we were altogether run in amongst flats and sands, amongst which we found shoal and deep in every three or four ship's length, after we began to sound: Master Cox looking out discerned (in his judgement) white cliffs, crying (land) withall, though we could not afterward descry any land, it being very likely the breaking of the sea white, which seemed to be white cliffs, through the haze and thick weather.

Immediately tokens were given unto the *Delight*, to cast about to seaward, which, being the greater ship, was yet foremost upon the breach, keeping so ill watch, that they knew not the danger, before they felt the same: presently struck a ground, and soon after her stern and hinder parts beaten in pieces: whereupon the rest (that is to say, the frigate in which was the general and the *Golden Hind*) cast about east southeast, bearing to the south, even for our lives into the wind's eye, because that way carried us to the seaward. Making out from this danger, we sounded one while seven fathom, then five fathom, then four fathom and less, again deeper, immediately four fathom, then but three fathom, the sea going mightily and high. At last we recovered (God be thanked) in some despair, to sea room enough.

In this distress, we desired to save the men by every possible means. But all in vain, sith God had determined their ruin: yet all that day, and part of the next, we beat up and down as near unto the wreck as was possible for us, looking out, if by good hap we might espy any of them.

This was a heavy and grievous event, to lose at one blow our chief ship freighted with great provision. But more was the loss of our men, which perished to the number almost of a hundred souls.

Those in the frigate were already pinched with spare allowance, and want of clothes chiefly: whereupon they besought the general to return for England, before they all perished. And to them of the *Golden Hind,* they made signs of their distress, pointing to their mouths, and to their clothes thin and ragged: then immediately they also of the *Golden Hind,* grew to be of the same opinion and desire to return home.

The general calling the captain and master of the *Hind,* yielded them many reasons, enforcing this unexpected return.

Reiterating these words, be content, we have seen enough, and take no care of expense past: I will set you forth royally the next spring, if God send us safe home. Therefore I pray you let us no longer strive here, where we fight against the elements.

So upon Saturday in the afternoon the 31 of August, we changed our course, and returned back for England.

The wind was large for England at our return, but very high, and the sea rough, insomuch as the frigate wherein the general went was almost swallowed up.

The general came aboard the *Hind* to have the surgeon of the *Hind* to dress his foot, which he hurt by treading upon a nail. So agreeing to carry out lights always by night, that we might keep together, he departed into his frigate, being by no means to be entreated to tarry in the *Hind,* which had been more for his security. Immediately after followed a sharp storm, which we overpassed for that time. Praised be God.

The weather fair, the general came aboard the *Hind* again, to make merry together with the captain, master, and company, which was the last meeting, and continued there from morning until night. Lamenting greatly the loss of his great ship, more of the men, but most of all of his books and notes, and what else I know not: yet by circumstances I gathered the same to be the ore brought unto him in Newfoundland. Whatsoever it was, the remembrance touched him so deep, as not able to contain himself, he beat his boy in great rage, even at the same time, so long after the miscarrying of the great ship, because upon a fair day, when we were becalmed upon the coast of Newfoundland, he sent his boy aboard to fetch certain things: amongst which, this being chief, was yet forgotten and left behind. After which time he could never conveniently send again aboard the great ship, much less he doubted her ruin so near at hand.

Herein my opinion was better confirmed diversely, and by sundry conjectures, which maketh me have the greater hope of this rich mine. For now his mind was wholly fixed upon the Newfoundland. And as before he refused not to grant assignments liberally, now he became contrarily affected, refusing to make any so large grants, especially of St John's, which certain English merchants made suit for.

Last, being demanded what means he had at his arrival in England, to compass the charges of so great preparation as he intended to make the next spring: having determined upon two fleets, one for the south, another for the north: leave that to me (he replied) I will ask a penny of no man. I will bring good tidings unto Her Majesty, who will be so gracious, to lend 10,000 pounds, willing us therefore to be of good cheer. And these last words he would often repeat, with demonstration of great fervency of mind, being himself very confident, and settled in belief of inestimable good by this voyage.

The vehement persuasion and entreaty of his friends could nothing avail, to divert him from a wilful resolution of going through in his frigate, which was overcharged upon their decks, with fights, nettings, and small artillery, too cumbersome for so small a boat, that was to pass through the ocean sea at that season of the year.

When he was entreated by the captain of the *Hind*, not to venture in the frigate, this was his answer: I will not foresake my little company going homeward, with whom I have passed so many storms and perils. And in very truth, he was urged to be so over hard, by hard reports given of him, that he was afraid of the sea, albeit this was rather rashness, than advised resolution.

Monday the ninth of September, in the afternoon, the frigate was nearly cast away, oppressed by waves, yet at that time recovered: and giving forth signs of joy, the general sitting abaft with a book in his hand, cried out unto us in the *Hind* (so oft as we did approach within hearing) we are as near to heaven by sea as by land.

The same Monday night, about twelve of the clock, or not long after, the frigate being ahead of us in the *Golden Hind*, suddenly her lights were out, whereof as it were in a moment, we lost the sight, and withal our watch cried, the general was cast away, which was too true. For in that moment, the frigate was devoured and swallowed up of the sea. Yet still we looked out all that night, and ever after, until we arrived upon the coast of England.

In great torment of weather, and peril of drowning, it pleased God to send safe home the *Golden Hind*, which arrived in Falmouth, the 22 day of September, being Sunday.

All the men tired with the tediousness of so unprofitable a voyage to their seeming: much toil and labour, hard diet and continual hazard of life was unrecompensed.

Even so, amongst the very many difficulties, discontentments, mutinies, conspiracies, sicknesses, mortality, spoilings, and wrecks by sea, it pleased God to support this company in reasonable contentment and concord.

Thus have I delivered the contents of the enterprise and last action of Sir Humphrey Gilbert knight: wherein may always appear (though he be extinguished), some sparks of his virtues, he remaining firm and resolute in a purpose by all pretence honest and godly, to discover, possess, and to reduce unto the service of God, and Christian piety, those remote and heathen countries of America, not actually possessed by Christians, and most rightly appertaining unto the crown of England.

Besides that fruit may grow in time of our travelling into those northwest lands, the crosses, turmoils, and afflictions, both in the preparation and execution of this voyage, did correct the intemperate humours, which before we noted to be in this gentleman, and made unsavoury, and less delightful his other manifold virtues.

Then as he was refined, and made nearer drawing unto the image of God: so it pleased the divine will to resume him unto himself.

XLIV

The voyage made to Tripolis in Barbary, in the year 1583 with a ship called the Jesus, *wherein the adventures and distresses of some English men are truly reported, by Thomas Sanders.*

This voyage was set forth by the right worshipful Sir Edward Osborne knight, chief merchant of all the Turkish Company, and one Master Richard Staper, the ship being of the burden of one hundred tons, called the *Jesus*. The purser was one William Thomson, our owner's son: the merchant's factors were Romane Sonnings a Frenchman, and Richard Skegs servant unto the said Master Staper. The owners were bound unto the merchants by charter party thereupon, in one thousand marks, that the said ship by God's permission should go for Tripolis in Barbary, that is to say, first from Portsmouth to Newhaven in Normandy, from thence to San Lucar, otherwise called Saint Lucas in Andalusia, and from thence to Tripolis, and so return to London. The owner did send down one Richard Diamond, and shipped him for master, who did choose for his mate one Andrew Dier, and so the said ship departed on her voyage accordingly: that is to say, about the 16 of October in An. 1583. By force of weather we were driven to Falmouth, where we remained until the first day of January: at which time the wind coming fair, we departed thence, and about the 20 day of the said month we arrived safely at San Lucar. And about the 9 day of March next following, we made sail from thence, and about the 18 day of the same month we came to Tripolis in Barbary, where we were very well entertained by the king of that country, and also of the commons. The commodities of that place are sweet oils: the king there is a merchant, and the factors bought all their oils of the king custom free, and so laded the same aboard.

There was a man in the said town a pledge, whose name was

Patrone Norado, indebted unto a Turk of that town, in the sum of four hundred and fifty crowns, for certain goods sent by him unto Christendom in a ship of his own, and by his own brother, and himself remained in Tripolis as pledge until his said brother's return: and as the report went there, after his brother's arrival into Christendom, he came among lewd company, and lost his brother's said ship and goods at dice, and never returned unto him again.

The said Patrone Norado being void of all hope, and finding now opportunity, consulted with the said Sonnings for to swim a seaboard the islands, and the ship being then out of danger, should take him in and so go to Toulon in the province of Marseilles with this Patrone Norado.

The ship being ready the first day of May, and having her sails all aboard, our factors did take their leave of the king, who very courteously bid them farewell, and when they came aboard, they commanded the master and the company hastily to get out the ship: the master answered that it was impossible, for that the wind was contrary and overblowed. Then went we to warp out the ship, and presently the king sent a boat aboard of us, with three men in her, commanding Sonnings to come ashore: at whose coming, the king demanded of him custom for the oils; Sonnings answered him that his highness had promised to deliver them custom free. But notwithstanding the king weighed not his said promise, and as an infidel that hath not the fear of God before his eyes, nor regard of his word, albeit he was a king, he caused the said Sonnings to pay the custom to the uttermost penny. And afterwards willed him to make haste away, saying that the janissaries would have the oil ashore again.

These janissaries are soldiers there under the Great Turk, and their power is above the king's. And so the said factor departed from the king, and came to the waterside, and called for a boat to come aboard, and he brought with him the foresaid Patrone Norado. The company inquisitive to know what man that was, Sonnings answered, that he was his countryman, a passenger: I pray God said the company, that we come not into trouble by this man. Then said Sonnings angrily, what have you to do with any matters of mine? if any thing chance otherwise than well, I must answer for all.

Now the Turk unto whom this Patrone Norado was indebted, missing him (supposed him to be aboard of our ship) presently went unto the king, and told him that he thought that his pledge Patrone Norado was aboard of the English ship, whereupon the king presently sent a boat aboard of us, with three men in her, commanding Sonnings to come ashore, and not speaking any thing as touching the man, he said that he would come presently in his own boat, but as soon as they were gone he willed us to warp forth the ship, and said that he would see the knaves hanged before he would go ashore. And when the king saw that he came not ashore, but still continued warping away the ship, he straight commanded the gunner of the bulwark next to us, to shoot three shots without ball. Then came we all to Sonnings, and asked of him what the matter was that we were shot at, he said that it was the janissaries who would have the oil ashore again, and willed us to make haste away, and after that he had discharged three shots without ball, he commanded all the gunners in the town to do their endeavour to sink us, but the Turkish gunners could not once strike us, wherefore the king sent presently to the banio: (this banio is the prison where all the captives lay at night) and promised if that there were any that could either sink us, or else cause us to come in again, he should have a hundred crowns, and his liberty. With that came forth a Spaniard called Sebastian, which had been an old servitor in Flanders, and he said, that upon the performance of that promise, he would undertake either to sink us, or to cause us to come in again, and thereto he would gage his life, and at the first shot he split our rudder's head in pieces, and the second shot he struck us under water, and the third shot he shot us through our foremast with a culverin shot, and thus he having rent both our rudder and mast, and shot us under water, we were enforced to go in again.

This Sebastian for all his diligence herein, had neither his liberty, nor an hundred crowns, so promised by the said king, but after his service was done was committed again to prison, whereby may appear the regard that the Turk or infidel hath of his word, although he be able to perform it, yea more though he be a king.

Then our merchants seeing no remedy, they together with five of our company went ashore, and then they ceased shooting: they shot unto us in the whole, nine and thirty shots, without the hurt of any man.

And when our merchant came ashore, the king commanded presently that they with the rest of our company that were with them, should be chained four and four, to a hundredweight of iron, and when we came in with the ship, there came presently above an hundred Turks aboard of us, and they searched us, and stripped our very clothes from our backs, and brake open our chests, and made a spoil of all that we had: and the Christian caitiffs likewise that came aboard of us made spoil of our goods and used us as ill as the Turks did. And our master's mate having a Geneva Bible in his hand, there came the king's chief gunner, and took it out from him, who showed me of it, and I having the language, went presently to the king's treasurer, and told him of it, saying that since it was the will of God that we should fall into their hands, yet that they should grant us to use our consciences to our own discretion, as they suffered the Spaniards and other nations to use theirs, and he granted us: then I told him that the master gunner had taken away a Bible from one of our men: the treasurer went presently and commanded him to deliver up the Bible again, which he did.

Then came the guardian pasha, which is the keeper of the king's captives, to fetch us all ashore, and then I remembering the miserable estate of poor distressed captives, in the time of their bondage to those infidels, went to mine own chest, and took out thereof a jar of oil and filled a basket full of white rusk to carry ashore with me, but before I came to the banio, the Turkish boys had taken away almost all my bread, and the keeper said, deliver me the jar of oil, and when thou comest to the banio thou shalt have it again, but I never had it of him any more.

But when I came to the banio, and saw our merchants and all the rest of our company in chains, and we all ready to receive the same reward, what heart in the world is there so hard, but would have pitied our cause, hearing or seeing the lamentable greeting there was betwixt us: all this happened the first of May, 1584.

And the second day of the same month, the king with all his council sat in judgement upon us. The first that were had forth to be arraigned, were the factors, and the masters, and the king asked them wherefore they came not ashore when he sent for them. And Romane Sonnings answered, if any offence be, the fault is wholly in myself, and in no other. Then forthwith the king gave judgement, that the said Romane Sonnings should be hanged over the northeast bulwark: from whence he conveyed the forenamed Patrone Norado, and then he called for our master Andrew Dier, and used few words to him, and so condemned him to be hanged over the walls of the westernmost bulwark.

Then fell our other factor (named Richard Skeggs) upon his knees before the king, and said, I beseech your highness either to pardon our master, or else suffer me to die for him, for he is ignorant of this cause. And then the people of that country favouring the said Richard Skeggs besought the king to pardon them both. So then the king spake these words: behold, for thy sake, I pardon the master. Then presently the Turks shouted and cried, saying: away with the master from the presence of the king. And then he came into the banio whereas we were, and told us what had happened, and we all rejoiced at the good hap of Master Skeggs, that he was saved, and our master for his sake.

But afterward our joy was turned to double sorrow, for in the meantime the king's mind was altered: for that one of his council advised him, that unless the master died also, by the law they could not confiscate the ship nor goods, neither captive any of the men: whereupon the king sent for our master again, and gave him another judgement after his pardon for one cause, which was that he should be hanged.

When that Romane Sonnings saw no remedy but that he should die, he protested to turn Turk, hoping thereby to save his life. Then said the Turk, if thou wilt turn Turk, speak the words that thereunto belong: and he did so. Then said they unto him, now thou shalt die in the faith of a Turk, and so he did, as the Turks reported that were at his execution. And Patrone Norado, whereas before he had liberty and did nothing, he then was condemned slave perpetual.

The king condemned all us, slaves perpetually unto the Great Turk, and the ship and goods were confiscated to the use of the Great Turk: and then we all fell down upon our knees, giving God thanks for this sorrowful visitation, and giving ourselves wholly to the almighty power of God.

But first to show our miserable bondage and slavery, and unto what small pittance and allowance we were tied, for every five men had allowance but five aspers of bread in a day, which is but twopence English: and our lodging was to lie on the bare boards, with a very simple cape to cover us, we were also forcibly and most violently shaven, head and beard, and within three days after, I and six more of my fellows, together with four score Italians and Spaniards were sent forth in a galliot to take a Greek carmosell, which came into Africa to steal negroes, and went out of Tripolis unto that place, which was two hundred and forty leagues thence, but we were chained three and three to an oar, and we rowed naked above the girdle, and the boatswain of the galley walked abaft the mast, and his mate afore the mast, and each of them a bull's pizzle dried in their hands, and when their devilish choler rose, they would strike the Christians for no cause: and they allowed us but half a pound of bread a man in a day without any other kind of sustenance, water excepted. And when we came to the place whereas we saw the carmosell, we were not suffered to have neither needle, bodkin, knife, or any other weapon about us, nor at any other time in the night, upon pain of one hundred bastinados: we were then also cruelly manacled in such sort, that we could not put our hands the length of one foot asunder the one from the other, and every night they searched our chains three times to see if they were fast riveted: we continued fight with the carmosell three hours, and then we took it, and lost but two of our men in that fight, but there were slain of the Greeks five, and fourteen were cruelly hurt, and they that were sound, were presently made slaves and chained to the oars: and within fifteen days after we returned again into Tripolis, and then we were put to all manner of slavery. I was put to hew stones, and other to carry stones, and some to draw the cart with earth, and some to make mortar, and some to draw stones (for at that time the Turks builded a church:) and thus we were put to all kinds of slavery.

Shortly after our apprehension, I wrote a letter into England unto my father dwelling in Tavistock in Devonshire, signifying unto him the whole estate of our calamities: and I wrote also to Constantinople to the English ambassador, both which letters were faithfully delivered. But when my father received my letter, and understood the truth of our mishap, and the occasion thereof, and what had happened to the offenders, he certified the right honourable the Earl of Bedford thereof, who in short space acquainted Her Highness with the whole cause thereof, and Her Majesty like a most merciful princess took order for our deliverance. Whereupon the right worshipful Sir Edward Osborne knight directed his letters with all speed to the English ambassador in Constantinople, to procure our delivery: and he obtained the Great Turk's commission, and sent it forthwith to Tripolis, by one Master Edward Barton, together with a Justice of the Great Turk's.

The King having intelligence of their coming, sent word to the keeper, that none of the Englishmen (meaning our company) should go to work. Then he sent for Master Barton and the other commissioners, and demanded of the said Master Barton his message: the Justice answered, that the Great Turk his Sovereign had sent them unto him, signifying that he was informed that a certain English ship called the *Jesus*, was by him the said king confiscated about twelve months since, and now my said sovereign hath here sent his especial commission by us unto you, for the deliverance of the said ship and goods, and also the free liberty and deliverance of the Englishmen of the same ship, whom you have taken and kept in captivity. And so did the Justices deliver unto the king the Great Turk's commission: after the perusing of the same, he forthwith commanded all the English captives to be brought before him, and then willed the keeper to strike off all our irons, which done, the king said, you Englishmen, for that you did offend the laws of this place, by the same laws therefore some of your company were condemned to die as you know, and you to be perpetual captives during your lives: notwithstanding, seeing it hath pleased my sovereign lord the Great Turk to pardon your said offences, and to give you your freedom and liberty, behold, here I make deliverance of you to this English gentleman: so he delivered us all that were there. Thus we were set at liberty the 28 day of April, 1585. For the which we are bound to praise Almighty God, during our lives, and as duty bindeth us, to pray for the preservation of our most gracious Queen, for the great care Her Majesty had over us, her poor subjects, in seeking and procuring of our deliverance aforesaid: and also for her honourable Privy Council, and I especial for the prosperity and good estate of the late deceased, the right honourable the Earl of Bedford, whose soul I doubt not, but is already in the heavens in joy, with the Almighty, unto which place he vouchsafe to bring us all, that for our sins suffered most vile and shameful death upon the cross, there to live perpetually world without end, Amen.

XLV

A letter of directions of the English Ambassador to Mr Richard Forster, appointed the first English Consul at Tripolis in Syria.

Cousin Forster, these few words are for your remembrance when it shall please the Almighty to send you safe arrival in Tripolis of Syria. The next, second, or third day, after your coming, give it out that you be crazed and not well disposed, by means of your travel at sea, during which time, you and those there are most wisely to determine in what manner you are to present yourself to the Beglerbey, Cadi, and other officers. They are to give you there also another janissary according as the French hath; whose outward proceedings you are to imitate and follow, in such sort as you be not his inferior, according as those of our nation heretofore with him resident can inform you. Touching your demeanour after your placing, you are wisely to proceed considering both French and Venetian will have an envious eye on you: whom if they perceive wise and well advised, they will fear to offer you an injury. But if they shall perceive any insufficiency in you, they will not omit any occasion to harm you. They are subtile, malicious, and dissembling people, wherefore you must always have their doings for suspected, and warily walk in all your actions. Touching any outlopers of our nation, which may happen to come thither to traffic, you are not to suffer, but to imprison the chief officers, and suffer the rest not to traffic at any time, that they shall not deal in the Grand Signior's dominions. And touching those there for the Company, you are to defend them according to your privilege and such commandments as you have had hence, in the best order you may. Touching your dealings in their affairs of merchandise, you are not to deal otherwise than in secret and council. You are carefully to foresee the charge of the house, that the same may be in all honest measure to the Company's profit and your own health through moderation in diet, and in due time to provide things needful: for he that buyeth every thing when he needeth it, harmeth his own house, and helpeth the retailer. Touching yourself, you are to cause to be employed fifty or three score ducats, videlicet, twenty in soap, and the rest in spices, whereof the most part to be pepper, whereof we spend very much. From our mansion Rapamat, the fifth of September 1583.

XLVI

The voyage of Mr Ralph Fitch merchant of London by the way of Tripoli in Syria, to Ormuz, and so to Goa in the East India, and all the kingdom of Zelabdim Echebar the great Mogul, to the mighty river Ganges, and down to Bengal, to Pegu, to the kingdom of Siam, and from thence to Malacca, Cochin, and all the coast of the East India: begun in the year of our Lord 1583, and ended 1591, wherein the strange rites, manners, and customs of those people, and the exceeding rich trade and commodities of those countries are faithfully set down and diligently described.

In the year of our Lord 1583, I Ralph Fitch of London merchant being desirous to see the countries of the East India, in the company of Mr John Newberie merchant (which had been at Ormuz once before) of William Leedes jeweller, and James Story painter, being chiefly set forth by the right worshipful Sir Edward Osborne knight and Mr Richard Staper citizens and merchants of London, did ship myself in a ship of London called the *Tiger*, wherein we went for Tripoli in Syria: and from thence we took the way for Aleppo, which we went in seven days with the caravan. Being in Aleppo, and finding good company, we went from thence to Birra, which is two days and a half travel with camels.

Birra is a little town, but very plentiful of victuals: and near to the wall of the town runneth the river of Euphrates. Here we bought a boat and agreed with a master and bargemen, for to go to Babylon. These boats be but for one voyage; for the stream doth run so fast downwards that they cannot return. They carry you to a town Felugia, and there you sell the boat for a little money, for that which cost you fifty at Birra you sell there for seven or eight. It is not good that one boat go alone, for if it should chance to break, you should have much ado to save your goods from the Arabians, which be always there abouts robbing: and in the night when your boats be made fast, it is necessary that you keep good watch. A gun is very good, for they do fear it very much.

Babylon is a town not very great but very populous, and of great traffic of strangers, for that it is the way to Persia, Turkey and Arabia: from thence do go caravans for these and other places. Here are great store of victuals, which come from Armenia down the river of Tigris. They are brought upon rafts made of goats' skins blown full of wind and boards laid upon them: and thereupon they lade their goods, which being discharged they open their skins, and carry them back by camels, to serve another time. Babylon in times past did belong to the kingdom of Persia, but now is subject to the Turk.

The Tower of Babel is built on this side the river Tigris, from the town about seven or eight miles, which tower is ruined on all sides, and with the fall thereof hath made as it were a little mountain, so that it hath no shape at all: it was made of bricks dried in the sun, and certain canes and leaves of the palm tree laid between the bricks. It doth stand upon a great plain between the rivers of Euphrates and Tigris.

Two days journey from Babylon at a place called Ait, in a field near unto it, is a strange thing to see: a mouth that doth continually throw forth against the air boiling pitch with a filthy smoke: which pitch doth run abroad into a great field which is always full thereof. The Moors say that it is the mouth of hell. By reason of the great quantity of it, the men of that country do pitch their boats two or three inches thick on the outside, so that no water doth enter into them.

Basra in times past was under the Arabians, but now is subject to the Turk. They be thieves all and have no settled dwelling, but remove from place to place with their camels, goats, and horses, wives and children and all. They have large blue gowns, their wives' ears and noses are ringed very full of rings of copper and silver, and they wear rings of copper about their legs.

Basra stands near the Gulf of Persia, and is a town of great trade of spices and drugs which come from Ormuz. Also there is a great store of wheat, rice, and dates growing thereabout, wherewith they serve Babylon and all the country, Ormuz, and all the parts of India. I went from Basra to Ormuz down the Gulf of Persia in a certain ship made of boards, and sewed together with cayro, which is thread made of the husk of coconuts, and certain canes or straw leaves sewed upon the seams of the boards which is the cause that they leak very much. And so having Persia always on the left hand, and the coast of Arabia on the right hand we passed many islands, and among others the famous island Bahrein from whence come the best pearls which be round and orient.

Ormuz is an island in circuit about five and twenty or thirty miles, and is the driest island in the world: for there is nothing growing on it but only salt; for their water, wood or victuals, and all things necessary come out of Persia, which is about twelve miles from thence. All the islands thereabout be very fruitful, from whence all kind of victuals are sent into Ormuz. The Portuguese have a castle here which standeth near unto the sea, wherein there is a captain for the King of Portugal having under him a convenient number of soldiers, whereof some part remain in the castle, and some in the town. In this town are merchants of all nations, and many Moors and gentiles. Here is very great trade of all sorts of spice, drugs, silk, cloth of silk, fine tapestry of Persia, great store of pearls which come from the isle of Bahrein, and are the best pearls of all others, and many horses of Persia, which serve all India. They have a Moor to their king, which is chosen and governed by the Portuguese. Their women are very strangely attired, wearing on their noses, ears, necks, arms and legs many rings set with jewels, and locks of silver and gold in their ears, and a long bar of gold upon the side of their noses. Their ears with the weight of their jewels be worn so wide, that a man may thrust three of his fingers into them. Here very shortly after our arrival we were put in prison, and had part of our goods taken from us by the captain of the castle, whose name was Don Mathias de Albuquerque; and from hence the eleventh of October he shipped us and sent us for Goa unto the Viceroy, which at that time was Don Francisco de Mascarenhas. The ship wherein we were embarked for Goa belonged to the captain, and carried one hundred twenty and four horses in it. All merchandise carried to Goa in a ship wherein are horses pay no custom in Goa. The horses pay custom, the goods pay nothing; but if you come in a ship which bringeth no horses, you are then to pay eight in the hundred for your goods. The first city of India that we arrived at upon the fifth of November, after we had passed the coast of Sind, is called Diu, which stands in an island in the kingdom of Cambay, and is the strongest town that the Portuguese have in those parts. It is but little, but well stored with merchandise; for here they lade many great ships with divers commodities for the Straits of Mecca, for Ormuz, and other places. Cambay is the chief city of that province, which is great and very populous, and fairly built: but if there happen any famine, the people will sell their children for very little. The last king of Cambay was Sultan Badu, who was killed at the siege of Diu, and shortly after his city was taken by the great Mogul, which is the king of Agra and of Delhi, which are forty days journey from the country of Cambay. Here the women wear upon their arms infinite numbers of rings made of elephants' teeth, wherein they take so much delight, that they had rather be without their meat than without their bracelets. The tenth of November we arrived at Chaul which standeth in the firm land. There be two towns, the one belonging to the Portuguese, and the other to the Moors. That of the Portuguese is nearest to the sea, and commandeth the bay, and is walled round about. A little above that is the town of the Moors which is governed by a Moorish king called Xa-Maluco. Here is great traffic for all sorts of spices and drugs, silk, and cloth of silk, sandals, elephants' teeth, and much china work, and much sugar: the tree is called the palm: which is the most profitable tree in the world: it doth always bear fruit, and doth yield wine, oil, sugar, vinegar, cords, coals, of the leaves are made thatch for the houses, sails for ships, mats to sit or lie on: of the branches they make their houses, and brooms to sweep, of the tree wood for ships. The wine doth issue out of the top of the tree. They cut a branch of a bough and bind it hard, and hang an earthen pot upon it, which they empty every morning and every evening, and still it and put in certain dried raisins, and it becometh very strong wine in short time. They have a very strange order among them, they worship a cow, and esteem much of the cows' dung to paint the walls of their houses. They will kill nothing not so much as a louse: for they hold it a sin to kill anything. They eat no flesh, but live by roots and rice, and milk. And when the husband dies his wife is burned with him, if she be alive: if she will not, her head is shaven, and then is never any account made of her after. In Cambay they will kill nothing, nor have anything killed: in the town they have hospitals to keep lame dogs and cats, and for birds. They will give meat to the ants.

Goa is the most principal city which the Portuguese have in India, wherein the Viceroy remaineth with his court. It standeth in an island, which may be 25 or 30 miles about. It is a fine city, and for an Indian town very fair. The island is very fair, full of orchards and gardens, and many palm trees, and hath some villages. Here be many merchants of all nations. And the fleet which cometh every year from Portugal, and then goes to Cochin where they load their pepper for Portugal. At our coming we were cast into the prison, and examined before the Justice and demanded for letters, and were charged to be spies, but they could prove nothing by us. We continued in prison until the two and twenty of December, and then we were set at liberty, putting in sureties for two thousand ducats not to depart the town; which sureties Father Stevens an English Jesuit which we found there, and another religious man a friend of his procured for us. Whereupon we presently determined rather to seek our liberties, than to be in danger for ever to be slaves in the country, for it was told us that we should have the strappado. Whereupon presently, the fifth day of April 1585 in the morning we ran from thence. And being set over the river, we went two days on foot not without fear, not knowing the way nor having any guide, for we durst trust no-one. One of the first towns which we came unto, is called Belgaum, where there is a great market kept of diamonds, rubies, sapphires, and many other soft stones. From Belgaum we went to Bijapor which is a very great town where the king doth keep his court. They have their idols standing in the woods, which they call pagodas. Some be like a cow, some like a monkey, some like buffaloes, some like peacocks, and some like the devil. Here be very many elephants which they go to war withal. Here they have a good store of gold and silver: their houses are of stone very fair and high. From hence we went for Golconda.[98] Here be the diamonds found of the old water. It is a very fair town, pleasant, with fair houses of brick and timber, it aboundeth with great store of fruit and fresh water. Here the men and the women do go with a cloth bound about their middles without any more apparel. We found it here very hot.

The winter beginneth here about the last of May. In these parts is a port or haven called Masulipatam, which standeth eight days journey from hence towards the gulf of Bengal, whither come many ships out of India, Pegu,[99] and Sumatra, very richly laden with pepper, spices, and other commodities. The country is very good and fruitful. From thence I went to Berhampur. In this place their money is made of a kind of silver round and thick, to the value of twenty pence, which is very good silver. It is marvellous great and a populous country. In their winter which is in June, July, and August, there is no passing in the streets but with horses, the waters be so high. The houses are made of loam and thatched. Here is great store of cotton cloth made, and painted cloths of cotton wool: here groweth a great store of corn and rice. We found marriages both in towns and villages in many places where we passed, of boys of eight or ten years, and girls of five or six years old. They both do ride upon one horse very trimly decked, and are carried through the town with great piping and playing, and so return home and eat of a banquet made of rice and fruits, and there they dance the most part of the night and so make an end of the marriage. They lie not together until they be ten years old. From thence we went to Agra passing many rivers, which by reason of the rain were so swollen, that we waded and swam often times for our lives. Agra is a very great city and populous, built with stone, having fair and large streets, with a fair river running by it, which falleth into the Gulf of Bengal. It hath a fair castle and a strong with a very fair ditch. The king hath in Agra and Fatehpur as they do credibly report

1000 elephants, thirty thousand horses, 1400 tame deer, 800 concubines: such store of ounces, tigers, buffaloes, cocks and hawks, that is very strange to see. He keepeth a great court. They have many fine carts, and many of them carved and gilded with gold, with two wheels which be drawn with two little bulls about the bigness of our great dogs in England, and they will run with any horse and carry two or three men in one of these carts: they are covered with silk or very fine cloth, and be used here as our coaches be in England. Hither is great resort of merchants from Persia and out of India, and very much merchandise of silk and cloth, and of precious stones, both rubies, diamonds and pearls. The king is apparelled in a white cabie [100] made like a shirt tied with strings on the one side, and a little cloth on his head coloured oftentimes with red and yellow. None come into his house but his eunuchs which keep his women. Here in Fatehpur we stayed all three until the 28 of September 1585 and then master John Newberie took his journey toward the city of Lahore, determining from thence to go for Persia and then for Aleppo or Constantinople, whether he could get soonest passage unto, and directed me to go for Bengal and for Pegu, and did promise me, if it pleased God, to meet me in Bengal within two years with a ship out of England. I left William Leedes the jeweller in service with the king Zelabdim Echebar in Fatehpur, who did entertain him very well, and gave him an house and five slaves, a horse and every day six shillings in money. I went from Agra to Bengal, in the company of one hundred and fourscore boats laden with salt, opium, hing,[101] lead, carpets, and divers other commodities down the river Jumna. In these countries they have many strange ceremonies. The Brahmin which are their priests, come to the water and have a string about their necks made with great ceremonies, and lade up water with both their hands, and turn the string first with both their hands within, and then one arm after the other out. They live with rice, butter, milk, and fruits. They pray in the water naked, and dress their meat and eat it naked, and for their penance they lie flat upon the earth, and rise up and turn themselves about 30 or 40 times, and use to heave up their hands to the sun, and to kiss the earth, with their arms and legs outstretched along out, and their right leg always before the left. Every time they lie down, they make a score on the ground with their finger to know when their stint is finished. The Brahmins mark themselves on their foreheads, ears and throats with a kind of yellow powder. And their wives do come by 10 20 and 30 together to the water side singing, and there do wash themselves, and then use their ceremonies, and mark themselves in their foreheads and faces, and carry some with them, and so depart singing. Their daughters be married, at, or before the age of 10 years. The men may have 7 wives. They be a kind of crafty people, worse than the Jews. The mighty river Ganges, cometh out of the northwest, and runneth east into the Gulf of Bengal. In those parts there are many tigers and many partridges and turtle-doves, and many other fowl. Here be many beggars in these countries which go naked, and the people make great account of them. Here I saw one which was a monster among the rest. He would have nothing upon him, his beard was very long, and with the hair of his head he covered his privities. The nails of some of his fingers were two inches long, for he would cut nothing from him, neither would he speak. He was accompanied with eight or ten, and they spake for him. When any man spake to him, he would lay his hand upon his breast and bow himself, but would not speak. He would not speak to the king. Ganges, is here very broad. The country is very fruitful and populous. In this river of Ganges are many islands. His water is very sweet and pleasant, and the country adjoining very fruitful. We went to Benares which is a great town, and great store of cloth is made there of cotton, and the sashes for the Moors. In this place they be the greatest idolaters that ever I saw. To this town came the gentiles on pilgrimage out of far countries. Here alongst the water's side be very many fair houses, and in all of them, or for the most part they have their images standing, which be evil favoured, made of stone and wood, some like lions, leopards, and monkeys, some like men and women, and peacocks, and some like the devil with four arms and four hands.

The people go all naked save a little cloth round about their middle. Their women have their necks, arms and ears decked with rings of silver, copper, tin, and with round hoops made of ivory, adorned with amber stones, and with many agates, and they are marked with a grest spot of red in their foreheads, and a stroke of red up to the crown, and so it runneth three manner of ways. In their winter, which is our May, the men wear quilted gowns of cotton like to our mattresses and quilted caps like to our great grocer's mortars, with a slit to look out at, and so tied down beneath their ears. If a man or woman be sick and like to die, they will lay him before their idols all night, and that shall help him or make an end of him. And if he do not mend that night, his friends will come and sit with him a little and cry, and afterwards will carry him to the water's edge and set him upon a little raft made of reeds, and so let him go down the river. Their chief idols be black and evil favoured, their mouths monstrous, their ears gilded, and full of jewels, their teeth and eyes of gold, silver, and glass, some having one thing in their hands, and some another. You may not come into the houses where they stand, with your shoes on. They have continually lamps burning before them. Here the women be so decked with silver and copper, that it is strange to see, they use no shoes by reason of the rings of silver and copper which they wear on their toes. Here at Patna, I saw a dissembling prophet which sat upon a horse in the market place, and made as though he slept, and many of the people came and touched his feet with their hands, and then kissed their hands. They took him for a great man, but sure he was a lazy lubber. I left him there sleeping. The people of these countries be much given to such prating and dissembling hypocrites.

From Agra down the river Jumna, and down the river Ganges, I was five months coming to Bengal, but it may be sailed in much shorter time.

Hooghly is the place where the Portuguese keep in the country of Bengal which standeth in 23 degrees of northerly latitude: they call it Porto Piqueno. Not far from Porto Piqueno southwestward, standeth an haven which is called Angul, in the country of Orissa. It was a kingdom of itself, and the king was a great friend to strangers. To this haven of Angul come every year many ships out of India, Sumatra, Malacca, and divers other places; and lade from thence great store of rice, and much cloth of cotton wool, much sugar, and long pepper, great store of butter and other victuals for India.

I went the 28 of November 1586 for Pegu in a small ship or foist of one Albert Caravallos, and so passing down Ganges, with a fair wind at northwest: our course was south and by east, which brought us to the bar of Negrais [102] in Pegu: if any contrary wind had come, we had thrown many of our things over-board: for we were so pestered with people and goods, that there was scant place to lie in. Three days after we came to Cosmin, which is a very pretty town and standeth very pleasantly, very well furnished with all things. The people be very tall and well disposed; the women white, round faced, with little eyes: the houses are high built, set upon great high posts, and they go up to them with long ladders for fear of the tigers which be very many. The country is very fruitful of all things. Here are very great figs, oranges, coconuts, and other fruits. The land is very high that we fall withall, but after we be entered the bar, it is very low and full of rivers, for they go all too and fro in boats, which they call paraos, and keep their houses with wife and children in them.

Pegu is a city very great, strong, and very fair, with walls of stone, and great ditches round about it. There are two towns, the old town and the new. In the old town are all the merchants strangers, and very many merchants of the country. All the goods are sold in the old town which is very great, and hath many suburbs round about it, and all the houses are made of canes which they call bamboos, and be covered with straw. In your house you have a warehouse which they call godown, which is made of brick to put your goods in. It is a city very great and populous, and is made square and with very fair walls, and a great ditch round about it full of water, with many crocodiles in it: it hath twenty gates, and they be made of stone, for every square five gates. There are also many turrets for sentinels to watch, made of wood, and gilded with gold very fair. The streets are the fairest that ever I saw, as straight as a line from one gate to the other, and so broad that ten or twelve men may ride afront through them. On both sides of them at every man's door is set a palm tree which is the nut tree: which make a very fair show and a very commodious shadow, so that a man may walk in the shade all day. The houses be made of wood, and covered with tiles. The king's house is in the middle of the city, and is walled and ditched round about: and the buildings within are made of wood very sumptuously gilded, and great workmanship is upon the forefront, which is likewise very costly gilded. And the house wherein his pagoda or idol standeth is covered with tiles of silver, and all the walls are gilded with gold. Within the first gate of the king's house is a great large room, on both sides whereof are houses made for the king's elephants, which be marvellous great and fair, and are brought up to wars and in service of the king. And among the rest he hath four white elephants, which are very strange and rare: for there is none other king which hath them but he: if any other king hath one, he will send unto him for it. When any of these white elephants is brought unto the king, all the merchants in the city are commanded to see them, and to give him a present of half a ducat, which doth come to a great sum: for that there are many merchants in the city. After that you have given your present you may come and see them at your pleasure, although they stand in the king's house. This king in his title is called the king of the white elephants. If any other king hath one, and will not send it him, he will make war with him for it: for he had rather lose a great part of his kingdom, than not to conquer him. They do very great service unto these white elephants; every one of them standeth in an house gilded with gold, and they do feed in vessels of silver and gilt. One of them when he doth go to the river to be washed, as every day they do, goes under a canopy of cloth of gold or of silk carried over him by six or eight men, and eight or ten men go before him playing on drums, shawms, or other instruments: and when he is washed and cometh out of the river, there is a gentleman which doth wash his feet in a silver basin: which is his office given him by the king. There is no such account made of any black elephant, be he never so great. The king hath above five thousand elephants of war. There be many huntsmen, which go into the wilderness with she elephants: for without the she they are not to be taken. And they be taught for that purpose: and every hunter hath five or six of them: and they say that they anoint the she elephants with a certain ointment, which when the wild elephant doth smell, he will not leave her. When they have brought the wild elephant near unto the place, they send word unto the town, and many horsemen and footmen come out and cause the she elephant to enter into a strait way which doth go to the palace, and the she and the he do run in: for it is like a wood: and when they be in, the gate doth shut. Afterward they get out the female: and when the male seeth that he is left alone, he weepeth and crieth, and runneth against the walls, which be made of so strong trees, that some of them do break their teeth with running against them. Then they prick him with sharp canes, and cause him to go into a strait house, and there they put a rope about his middle and about his feet, and let him stand there three or four days without eating or drinking: and then they bring the female to him, with meat and drink, and within few days he becometh tame. The chief force of the king is in these elephants. And when they go into the wars they set a frame of wood upon their backs, bound with great cords, wherein sit four or six men, which fight with guns, bows and arrows, darts and other weapons. And they say that their skins are so thick that a pellet of an arquebus will scarce pierce them, except it be in some tender place. Their weapons be very bad. They have guns, but shoot very badly in them, darts and swords short without points. The king keepeth a very great state. If any man will speak with the king, he is to kneel down, to heave up his hands to his head, and to put his head to the ground three times, when he entereth, in the middle way, and when he cometh near to the king: and then he sitteth down and talketh with the king: if the king like well of him, he sitteth near him within three or four paces: if he think not well of him, he sitteth further off. These people do eat roots, herbs, leaves, dogs, cats, rats, serpents, and snakes; they refuse almost nothing. When the king rideth abroad, he rideth with a great guard, and many noblemen, oftentimes upon an elephant with a fine castle upon him very fairly gilded with gold; and sometimes upon a great frame like a horse litter, which hath a little house upon it covered overhead, but open on the sides, which is all gilded with gold, and set with many rubies and sapphires, whereof he hath an infinite store in his country, and is carried upon sixteen or eighteen men's shoulders. This king hath little force by sea, because he hath but very few ships. He has houses full of gold and silver, and bringeth in often, but spendeth very little, and hath mines of rubies and sapphires. The king hath one wife and above three hundred concubines, by which they say he hath four score or fourscore and ten children. He sitteth in judgement almost every day. They use no speech, but give up their supplications written in the leaves of a tree with the point of an iron bigger than a bodkin. These leaves are an ell long, and about two inches broad; they are also double. He which giveth in his supplication, doth stand in a place a little distance off with a present. If his matter be liked of, the king accepteth of his present, and granteth his request: if his suit be not liked of, he returneth with his present; for the king will not take it.

There are in Pegu eight brokers, whom they call Tareghe, which are bound to sell your goods at the price which they be worth, and you give them for their labour two in the hundred: and they be bound to make your debt good, because you sell your merchandises upon their word. If the broker pay you not at his day, you may take him home, and keep him in your house: which is a great shame for him. And if he pay you not presently, you may take his wife and children and his slaves, and bind them at your door, and set them in the sun; for that is the law of the country. The merchandises which be in Pegu, are gold, silver, rubies, sapphires, spinels, musk, benjamin or frankincense, long pepper, tin, lead, copper, lac whereof they make hard wax, rice, and wine made of rice, and some sugar. The elephants do eat the sugar canes, or else they would make very much. And they consume many canes likewise in making of their idol temples, which are in great number both great and small. They be made round like a sugar loaf, some are as high as a church, very broad beneath, some a quarter of a mile in compass: within they be all earth done about with stone. They consume great quantity of gold; for that they be all gilded aloft: and many of them from the top to the bottom: and every ten or twelve years they must be new gilded, because the rain consumeth off the gold: for they stand open abroad. If they did not consume their gold in these vanities, it would be very plentiful and good cheap in Pegu.

Caplan [103] is the place where they find the rubies and sapphires, and spinels; it stands six days journey from Ava in the kingdom of Pegu. There are many great high hills out of which they dig them. None may go to the pits but only those which dig them.

In Pegu, the men wear bunches or little round balls in their privy members: some of them wear two and some three. They cut the skin and so put them in, one into one side and another into the other side: which they do when they be 25 or 30 years old, and at their pleasure they take one or more of them out as they think good. When they be married the husband is for every child which his wife has, to put in one until he comes to three and then no more: for they say the women do desire them. They were invented because they should not abuse the male sex. For in times past all those countries were so given to that villainy, that they were very scarce of people. The bunches be of divers sorts: the least be as big as a little walnut, and very round: the greatest are as big as a little hen's egg: some are of brass and some of silver: but those of silver be for the king and his noble men. These are gilded and made with great cunning, and ring like a little bell. The king sometimes taketh his out, and giveth them to his noblemen as a great gift: and because he hath used them, they esteem them greatly. They will put one in, and heal up the place in seven or eight days.

These people wear no beards: they pull out the hair on their faces with little pincers made for that purpose. Some of them will let 16 or 20 hairs grow together, some in one place of his face and some in another, and pulleth out all the rest: for he carrieth his pincers always with him to pull the hairs out as soon as they appear. If they see a man with a beard they wonder at him.

The Pegus if they have a suit in law which is so doubtful, that they cannot well determine it, put two long canes into the water where it is very deep: and both the parties go into the water by the poles, and there sit men to judge, and they both do dive under the water, and he which remaineth longest under the water doth win the suit.

The 10 of January I went from Pegu and so came to Malacca the 8 of February, where the Portuguese have a castle which standeth near the sea. And the country fast without the town belongeth to the Malays, which is a kind of proud people. They go naked with a cloth about their middle, and a little roll of cloth about their heads. Hither come many ships from China and from the Malaccas, Banda, Timor, and from many other islands of the Javas, which bring great store of spices and drugs, and diamonds and other jewels. The voyages into many of these islands belong unto the captain of Malacca: so that none may go thither without his license: which yield him great sums of money every year.

When the Portuguese go from Macao in China to Japan, they carry much white silk, gold, musk, and porcelains: and they bring from thence nothing but silver. They have a great carrack which goeth thither every year, and she bringeth from thence every year above six hundred thousand crusadoes: and all this silver of Japan, and two hundred thousand crusadoes more in silver which they bring yearly out of India, they employ to their great advantage in China: and they bring from thence gold, musk, silk, copper, porcelains, and many other things very costly and gilded. When the Portuguese come to Canton in China to traffic, they must remain there but certain days: and when they come in at the gate of the city, they must enter their names in a book, and when they go out at night they must put out their names. They may not lie in the town all night, but must lie in their boats without the town. The Chinese are very suspicious, and do not trust strangers. The order of China is when they mourn, that they wear white thread shoes, and hats of straw. A man may keep as many concubines as he will, but one wife only. All the Chinese, Japanese, and Cochin Chinese do write downwards, and they do write with a fine pencil made of dog's or cat's hair.

The 29 of March 1588, I returned from Malacca to Pegu, where I remained the second time until the 17 of September; and passing many dangers by reason of contrary winds, it pleased God that we arrived in Bengal in November following: where I stayed for want of passage until the third of February 1589, and then I shipped myself for Cochin.[104] In which voyage we endured great extremity for lack of fresh water: for the weather was extremely hot, and we were many merchants and passengers, and we had very many calms, and hot weather. Yet it pleased God that we arrived in Ceylon the sixth of March, where we stayed five days to water, and to furnish ourselves with other necessary provision. This Ceylon is a brave island, very fruitful and fair. When the king talketh with any man, he standeth upon one leg, and setteth the other foot upon his knee with his sword in his hand: it is not their order for the king to sit but to stand. His apparel is a fine painted cloth made of cotton wool about his middle: his hair is long and bound up with a little fine cloth about his head: all the rest of his body is naked. His guard are a thousand men, which stand round about him, and he in the middle; and when he marcheth, many of them go before him, and the rest come after him. Their ears are very large; for the greater they are, the more honourable they are accounted. Some of them are a span long. The wood which they burn is cinnamon wood, and it smelleth very sweet. There is a great store of rubies, sapphires, and spinels in this island: the best kind of all be here; but the king will not suffer the inhabitants to dig for them, lest his enemies should know of them, and make wars against him, and so drive him out of his country for them. They have no horses in all the country. The elephants be not so great as those of Pegu, which be monstrous huge: but they say all other elephants do fear them, and none dare fight with them, though they be very small. Their women have a cloth bound about them from their middle to their knee: and all the rest is bare. All of them be black and but little, both men and women. Their houses are very little, made of the branches of the palm or coconut tree, and covered with the leaves of the same tree.

The eleventh of March we sailed from Ceylon. Thus passing the coast we arrived in Cochin the 22 of March, where we found the weather warm, but scarcity of victuals: for here groweth neither corn nor rice. These people here are Malabars. These have heads very full of hair, and bound up with a string: and there doth appear a bush without the band wherewith it is bound. The men be tall and strong, and good archers with a longbow and a long arrow, which is their best weapon: yet there be some calivers among them, but they handle them badly.

Here groweth the pepper; and it springeth up by a tree or a pole, and is like our ivy berry, but something longer like the wheat ear: and at the first the bunches are green, and as they wax ripe they cut them off and dry them. The leaf is much lesser than the ivy leaf and thinner. All the inhabitants here have very little houses covered with the leaves of the coconut trees. The men be of a reasonable stature; the women little; all black, with a cloth bound about their middle hanging down to their hams: all the rest of their bodies be naked: they have horrible great ears with many rings set with pearls and stones in them. The best cinnamon does come from Ceylon, and is peeled from fine young trees. Here are very many palm or coconut trees, which is their chief food: for it is their meat and drink.

I remained in Cochin until the second of November, which was eight months; for that there was no passage that went away in all that time: if I had come two days sooner I had found a passage presently. From Cochin I went to Goa, where I remained three days. From Goa I departed to Ormuz; where I stayed for a passage to Basra fifty days.

Here I thought good, before I make an end of this my book, to declare some things which India and the country farther eastward do bring forth.

The pepper groweth in many parts of India, especially about Cochin. The shrub is like unto our ivy tree: and if it did not run about some tree or pole, it would fall down and rot. When they first gather it, it is green; and then they lay it in the sun, and it becometh black.

The ginger groweth like unto our garlic, and the root is the ginger: it is to be found in many parts of India.

The cloves do come from the Isles of the Moluccas: their tree is like to our bay tree.

The nutmegs and maces grow together, and come from the Isle of Banda:[105] the tree is like to our walnut tree, but somewhat lesser.

Camphor is a precious thing among the Indians, and is sold

dearer than gold. I think none of it cometh for Christendom. That which is compounded cometh from China: but that which groweth in canes and is the best cometh from the great isle of Borneo.

The long pepper groweth in Bengal, in Pegu, and in the islands of the Javas.

The musk cometh out of Tartary, and is made after this order, by report of the merchants which bring it to Pegu to sell; in Tartary there is a little beast like unto a young roe, which they take in snares, and beat him to death with the blood: after that they cut out the bones, and beat the flesh with the blood very small, and fill the skin with it: and hereof cometh musk.

Of the amber they hold divers opinions: but most men say it cometh out of the sea, and that they find it upon the shore's side.

The rubies, sapphires, and spinels are found in Pegu.

The diamonds are found in divers places, as in Agra, in Delhi, and in the islands of the Javas.

The best pearls come from the island of Bahrein in the Persian sea, the worser from the piscaria near the isle of Ceylon, and from Hainan a great island on the southernmost coast of China.

Spodium and many other kinds of drugs come from Cambay.

From Ormuz I went to Basra, and from Basra to Babylon: and we passed the most part of the way by the strength of men by hauling the boat up the river with a long cord. From Babylon I came by land to Mosul, which standeth near to Nineveh, which is all ruinated and destroyed: and so past the river of Euphrates to Aleppo, where I stayed certain months for company; and then I went to Tripoli; where finding English shipping, I came with a prosperous voyage to London, where by God's assistance I safely arrived the 29 of April 1591, having been eight years out of my native country.

XLVII

The first voyage made to the coasts of America, with two barks, captains Mr Philip Amadas, and Mr Arthur Barlowe, who discovered part of the country now called Virginia, An. 1584. Sent to Sir Walter Raleigh knight, at whose charge the voyage was set forth.

The 27 day of April, 1584 we departed the West of England, with two barks well furnished with men and victuals.

The tenth of May we arrived at the Canaries, and the tenth of June in this present year, we were fallen with the islands of the West Indies, keeping a more southeasterly course than was needful, because we doubted that the current of the Bay of Mexico, had been of greater force than afterwards we found it to be. At which islands we found the air very unwholesome, and our men grew for the most part ill disposed: so that having refreshed ourselves with sweet water, and fresh victuals, we departed the twelfth day of our arrival there.

The second of July, we found shoal water, where we smelt so sweet, and so strong a smell, as if we had been in the midst of some delicate garden by which we were assured that the land could not be far distant: and keeping good watch, and bearing but slack sail, the fourth of the same month we arrived upon the coast, which we supposed to be a continent and firm land, and we sailed along the same a hundred and twenty English miles before we could find any entrance, or river issuing into the sea. The first that appeared unto us, we entered, though not without some difficulty, and cast anchor about three arquebus-shot within the haven's mouth.

We passed from the sea side towards the tops of those hills next adjoining, being but of mean height, and from thence we beheld the sea on both sides. This land we found to be but an island of twenty miles long, and not above six miles broad. Under the bank or hill whereon we stood, we beheld the valleys replenished with goodly cedar trees, and having discharged our arquebus-shot, such a flock of cranes (the most part white) arose under us, with such a cry redoubled by many echoes, as if an army of men had shouted all together.

This island had many goodly woods full of deer, coneys, hares, and fowl: even in the midst of summer in incredible abundance. The woods are the highest and reddest cedars of the world. We remained by the side of this island two whole days before we saw any people of the country: the third day we saw one small boat rowing towards us having in it three persons: this boat came to the island side, four arquebus-shot from our ships, and there two of the people remaining, the third come along the shoreside towards us. Captain Philip Amadas, myself, and others rowed to the land, whose coming this fellow attended, never making any show of fear or doubt. And after he had spoken of many things not understood by us, we brought him with his own good liking, aboard the ships, and gave him a shirt, a hat and some other things, and made him taste of our wine, and our meat, which he liked very well: and after having viewed both barks, he departed, and went to his own boat again: as soon as he was two bow shoot into the water, he fell to fishing, and in less than half an hour, he had laden his boat as deep, as it could swim, with which he came again to the point of the land, and there he divided his fish into two parts, pointing one part to the ship, and the other to the pinnace: which, after he had (as much as he might) requited the former benefits received, departed out of our sight.

The next day there came unto us divers boats, and in one of them the king's brother, accompanied with forty or fifty men, very handsome and goodly people, and in their behaviour as mannerly and civil as any of Europe. The king is called Wingina, the country Wingandacoa, and now by her Majesty Virginia. After he had made a long speech unto us, we presented him with divers things, which he received very joyfully, and thankfully.

The king is greatly obeyed, and his brothers and children reverenced: the king himself in person was at our being there, sore wounded in a fight which he had with the king of the next country, and was shot in two places through the body, and once clean through the thigh, but yet he recovered.

A day or two after this, we fell to trading with them, exchanging some things that we had, for chamois, buff, and deer skins: when we showed him all our packet of merchandise, of all things that he saw, a bright tin dish most pleased him, which he presently took up and clapped it before his breast, and after made a hole in the brim thereof and hung it about his neck, making signs that it would defend him against his enemies' arrows: for those people maintain a deadly and terrible war, with the people and king adjoining. We exchanged our tin dish for twenty skins, worth twenty crowns, or twenty nobles: and a copper kettle for fifty skins worth fifty crowns. They offered us good exchange for our hatchets, and axes, and for knives, and would have given any thing for swords: but we would not depart with any. After two or three days the king's brother came aboard the ships, and drank wine, and ate of our meat and of our bread, and liked exceedingly thereof: and after a few days overpassed, he brought his wife with him to the ships, his daughter and two or three children: his wife was very well favoured, of mean stature, and very bashful: she had on her back a long cloak of leather, with the fur side next to her body, and before her a piece of the same: about her forehead she had a band of white coral, and so had her husband many times: in her ears she had bracelets of pearls hanging down to her middle, (whereof we delivered your worship a little bracelet) and those were of the bigness of good peas. The rest of her women of the better sort had pendants of copper hanging in either ear, and some of the children of the king's brother and other noble men, have five or six in either ear: he himself had upon his head a broad plate of gold, or copper, for being unpolished we knew not what metal it should be, neither would he by any means suffer us to take it off his head, but feeling it, it would bow very easily.

Their boats are made of one tree, either of pine or of pitch trees: a wood not commonly known to our people, nor found growing in England. They have no edge-tools to make them withal: if they have any they are very few, and those it seems they had twenty years since, which, as those two men declared, was out of a wreck which happened upon their coast of some Christian ship, being beaten that way by some storm and outrageous weather, whereof none of the people were saved, but only the ship, or some part of her being cast upon the sand, out of whose sides they drew the nails and the spikes, and with those they made their best instruments. The manner of making their boats is thus: they burn down some great tree, or take such as are wind fallen, and putting gum and resin upon one side thereof, they set fire into it, and when it hath burnt it hollow, they cut out the coal with their shells, and ever where they would burn it deeper or wider they lay on gums, which burn away the timber, and by this means they fashion very fine boats, and such as will transport twenty men. Their oars are like scoops, and many times they set with long poles.

The king's brother had great liking of our armour, a sword, and divers other things which we had: and offered to lay a great box of pearl in gage for them: but we refused it for this time, because we would not make them know, that we esteemed thereof, until we had understood in what places of the country the pearl grew.

He sent us every day a brace or two of fat bucks, coneys, hares, fish the best of the world. He sent us divers kinds of fruits, melons, walnuts, cucumbers, gourds, pease, and divers roots, and fruits very excellent good, and of their country corn, which is very white, fair, and well tasted, and groweth three times in five months: they cast the corn into the ground, breaking a little of the soft turf with a wooden mattock, or pickaxe: ourselves proved the soil, and put some of our pease in the ground, and in ten days they were of fourteen inches high: they have also beans very fair of divers colours and wonderful plenty.

After they had been divers times aboard our ships, myself, with seven more went twenty mile into the river: and the evening following, we came to an island, which they call Roanoke, distant from the harbour by which we entered, seven leagues: and at the north end thereof was a village of nine houses, built of cedar, and fortified round with sharp trees, to keep out their enemies; the wife of Granganimo the king's brother came running out to meet us very cheerfully and friendly. When were come in, having five rooms in her house, she caused us to sit down by a great fire, and after took off our clothes and washed them, and dried them again: some of the women plucked off our stockings and washed them, some washed our feet in warm water.

She brought us into the inner room, where she set on the board standing along the house, some wheat like furmenty, sodden venison, and roasted, fish sodden, boiled, and roasted, melons raw, and sodden, roots of divers kinds, and divers fruits: their drink is commonly water. We found the people most gentle, loving and faithful, void of all guile and treason, and such as live after the manner of the golden age. The people only care how to defend themselves from the cold in their short winter, and to feed themselves with such meat as the soil affordeth. While we were at meat, there came in at the gates two or three men with their bows and arrows from hunting, whom when we espied, we began to look one towards another, and offered to reach our weapons: but as soon as she espied our mistrust, she was very much moved, and caused some of her men to run out, and take away their bows and arrows and break them, and withall beat the poor fellows out of the gate again.

Beyond this island there is the main land, and over against this island falleth into this spacious water, the great river called Occam[106] by the inhabitants.

Into this river falleth another great river, called Cipo, in which there is found great store of mussels in which there are pearls. Towards the southwest, four days' journey, six and twenty years past there was a ship cast away, whereof some of the people were saved, and those were white people, whom the country people preserved.

They had our ships in marvellous admiration, and all things else were so strange unto them, as it appeared that none of them had ever seen the like. When we discharged any piece, they would tremble thereat for very fear, and for the strangeness of the same: for the weapons which themselves use are bows and arrows: the arrows are but of small canes, headed with a sharp shell or tooth of a fish sufficient enough to kill a naked man. Their swords be of wood hardened: likewise they use wooden breastplates for their defence. They have besides a kind of club, in the end whereof they fasten the sharp horns of a stag, or other beast. When they go to wars they carry about with them their idol, of whom they ask counsel.

When we first had sight of this country, some thought the first land we saw to be the continent: but after we entered into the haven, we saw before us another mighty long sea: for there lieth along the coast a tract of islands, and between the islands, two or three entrances: when you are entered between them then there appeareth another great sea, containing in breadth in some places, forty, and in some fifty, in some twenty miles over, before you come unto the continent: and in this enclosed sea there are above a hundred islands of divers bignesses, whereof one is sixteen miles long, at which we were, finding it a most pleasant and fertile ground, replenished with goodly cedars, and divers other sweet woods, full of currants, of flax, and many other notable commodities, which we at that time had no leisure to view.

Thus Sir, we have acquainted you with the particulars of our discovery, made this present voyage, as far forth as the shortness of the time we there continued would afford us to take view of: we arrived safely in the West of England about the midst of September.

We brought home also two of the savages being lusty men, whose names were Wanchese and Manteo.

XLVIII

The prosperous voyage of the worshipful Thomas Candish of Trimley in the County of Suffolk Esquire, into the South Sea, and from thence round about the circumference of the whole earth, begun in the year of our Lord 1586, and finished 1588.

We departed out of Plymouth on Thursday the 21 of July 1586 with 3 sails, to wit, the *Desire* a ship of 120 tons, the *Content* of 60 tons, and the *Hugh Gallant* a bark of 40 tons: in which small fleet were 123 persons and victuals sufficient for the space of two years, Thomas Candish being our general.

The first of August we came in sight of Fuerteventura, one of the isles of the Canaries, about ten of the clock in the morning.

The 25 day we fell with the point on the south side of Sierra Leone.

On Sunday the 28 the general sent some of his company on shore, and there they played and danced all the forenoon among the negroes.

On Monday morning being the 29 day, our general landed with 70 men or thereabout, and went up to their town, where we burnt 2 or 3 houses, and took what spoil we would, which was but little, but all the people fled: and in our retiring aboard in a very little plain at their town's end they shot their arrows at us out of the woods, and hurt three or four of our men; their arrows were poisoned, but yet none of our men miscarried at that time, thanked be God. Their town is marvellous artificially builded with mud walls, and built round, with their yards paled in and kept very clean as well in their streets as in their houses. These negroes use good obedience to their king. There were in their town by estimation about one hundred houses.

The first of September there went many of our men on shore at the watering place, and did wash shirts very quietly all the day: and the second day they went again, and the negroes were in ambush round about the place: the negroes rushed out upon our men so suddenly, that in retiring to our boats, many of them were hurt: among whom one William Pickman a soldier was shot into the thigh, who plucking the arrow out, broke it, and left the head behind: the poison wrought so that night, that he was marvellously swollen, and all his belly and privy parts were as black as ink, and the next morning he died.

The last of October running west southwest about 24 leagues from Cape Frio in Brazil, we fell with a great mountain[107] which had an high round knop on the top of it standing from it like a town, with two little islands from it.

The first of November we went in between the Ilha de São

Sebastião[108] and the mainland, and had our things on shore, and set up a forge, and had our cask on shore: our coopers made hoops, and so we remained there until the 23 day of the same month: in which time we fitted our things, built our pinnace, and filled our fresh water.

The 17 day of December in the afternoon we entered into an harbour, where there is a wonderful great store of seals, and another island of birds which are grey gulls. These seals are of a wonderful great bigness, huge, and monstrous of shape, and for the fore-part of their bodies cannot be compared to any thing better than to a lion: their head, and neck, and fore-parts of their bodies are full of rough hair: their feet are in manner of a fin, and in form like unto a man's hand: they breed and cast every month, giving their young milk, yet continually get they their living in the sea, and live altogether upon fish: their young are marvellous good meat, and being boiled or roasted, are hardly to be known from lamb or mutton. The old ones be of such bigness and force, that it is as much as 4 men are able to do to kill one of them with great cowlstaves: and he must be beaten down with striking on the head of him: for his body is of that bigness that four men could never kill him, but only on the head. For being shot through the body with an arquebus or a musket, yet he will go his way into the sea, and never care for it at the present.

This harbour is a very good place to trim ships in, and bring them on ground, and grave them in: for there ebbeth and floweth much water: therefore we graved and trimmed all our ships there.

The 24 of December being Christmas Eve, a man and a boy went into a very fair green valley at the foot of the mountains, where was a little pit or well which our men had digged and made some 2 or 3 days before to get fresh water: this man and boy came thither to wash their linen: there were a great store of Indians which were come down and found the man and boy in washing. These Indians being divided on each side of the rocks, shot at them with their arrows and hurt them both, but they fled presently, being about fifty or threescore, though our general followed them but with 16 or 20 men. The man's name which was hurt was John Garge, the boy's name was Lutch: the man was shot clean through the knee, the boy into the shoulder: either of them having very sore wounds. Their arrows are made of little canes, and their heads are of a flint stone, set into the cane very artificially: they seldom or never see any Christians: they are as wild as ever was a buck or any other wild beast: for we followed them, and they ran from us as it had been the wildest thing in the world. We took the measure of one of their feet, and it was 18 inches long. Their use is when any of them die, to bring him or them to the cliffs by the sea side, and upon the top of them they bury them, and in their graves are buried with them their bows and arrows, and all their jewels which they have in their lifetime, which are fine shells which they find by the sea side, which they cut and square after an artificial manner: and all is laid under their heads. The grave is made all with great stones of great length and bigness, being set all along full of the dead man's darts which he used when he was living. And they colour both their darts and their graves with a red colour which they use in colouring of themselves.

The 6 day [of January] we put in for the Straits.

The 7 day between the mouth of the Straits and the narrowest place thereof, we took a Spaniard whose name was Hernando, who was there with 23 Spaniards more, which were all that remained of four hundred, which were left there three years before in these Straits of Magellan, all the rest being dead with famine. And the same day we passed through the narrowest of the Straits.

The ninth day we departed from Penguin Island, and ran south southwest to King Philip's city which the Spaniards had built: which town or city had four forts, and every fort had in it one cast piece, which pieces were buried in the ground, the carriages were standing in their places unburied: we digged for them and had them all. They had contrived their city very well, and seated it in the best place of the Straits for wood and water: they had built up their churches by themselves: they had laws very severe among themselves, for they had erected a gibbet, whereon they had done execution upon some of their company. It seemed unto us that their whole living for a great space was altogether upon mussels and limpets: for there was not anything else to be had, except some deer which came out of the mountains down to the fresh rivers to drink. These Spaniards which were there, were only come to fortify the Straits, to the end that no other nation should have passage through into the South Sea saving only their own: but as it appeared, it was not God's will so to have it. For during the time that they were there, which was two years at the least, they could never have anything to grow or in any wise prosper. And on the other side the Indians oftentimes preyed upon them, until their victuals grew so short, that they died like dogs in their houses, and in their clothes, wherein we found them still at our coming, until that in the end the town being wonderfully tainted with the smell and the savour of the dead people, the rest which remained alive were driven to bury such things as they had there in their town either for provision or for furniture, and so to forsake the town, and to go along the sea side, and seek their victuals to preserve them from starving, taking nothing with them, but every man his arquebus (some were not able to carry them for weakness) and so lived for the space of a year and more with roots, leaves, and sometimes a fowl which they might kill with their piece. To conclude, they were determined to have travelled towards the river of Plate, only being left alive 23 persons, whereof two were women, which were the remainder of 4 hundred. In this place we watered and wooded well and quietly.

There was a fresh water river, where our general went up with the ship-boat about three miles, which river hath very good and pleasant ground about it, and it is low and champaign soil, and so we saw none other ground else in all the Straits but that was craggy rocks and monstrous high hills and mountains. In this river are great store of savages which we saw, and had conference with them: they were men eaters, and fed altogether upon raw flesh, and other filthy food: which people had preyed upon some of the Spaniards before spoken of. For they had gotten knives and pieces of rapiers to make darts of. They used all the means they could possibly to have allured us up farther into the river, of purpose to have betrayed us, which being espied by our general, he caused us to shoot at them with our arquebuses, whereby we killed many of them.

During this time, which was a full month, we fed almost altogether upon mussels and limpets, and birds, or such as we could get on shore, seeking every day for them, as the fowls of the air do, where they can find food, in continual rainy weather.

The 24 day of February we entered into the South Sea: the first of March a storm took us. This storm continued 3 or 4 days, and for that time we in the *Hugh Gallant* being separated from the other 2 ships looked every hour to sink, our bark was so leak, and ourselves so weakened with freeing it of water, that we slept not in three days and three nights.

The 15 of March in the morning the *Hugh Gallant* came in between the island of Santa Maria[109] and the main where she met with the admiral and the *Content* which had rid at the island called La Mocha 2 days: at which place some of our men went on shore with the vice-admiral's boat, where the Indians fought with them with their bows and arrows, and were marvellous wary of their calivers. These Indians were enemies to the Spaniards, and belonged to a great place called Arauco,[110] and took us for Spaniards, as afterwards we learned.

This place which is called Arauco is wonderfully rich, and full of gold mines, and yet it could not be subdued at any time by the Spaniards, but they always returned with the greatest loss of men. For these Indians are marvellous desperate and careless of their lives to live at their own liberty and freedom.

We weighed anchor, and ran under the west side of Santa Maria island, where we rode very well in six fathoms of water.

There came down to us certain Indians with two which were the principals of the island to welcome us on shore, thinking we had been Spaniards, for it is subdued by them: who brought us up to a place where the Spaniards had erected a church with crosses and altars in it. And there were about this church 2 or 3 store houses, which were full of wheat and barley ready threshed. The wheat and barley was as fair, as clean, and every way as good as any we have in England. There were also cades full of potato roots, which were very good to eat, ready made up in the store houses for the Spaniards against they should come for their tribute. This island also yieldeth many sorts of fruits, hogs, and hens. These Indians are held in such slavery by them, that they dare not eat a hen or an hog themselves. Thus we fitted ourselves here with corn as much as we would have, and as many hogs as we had salt to powder them withal, and great store of hens, with a number of bags of potato roots, and about 500 dried dog-fishes, and Guinea wheat, which is called maize.

The fifteenth [of April, 1587] we came thwart of a place called Morro Moreno,[111] an excellent harbour: here we went with our general on shore to the number of 30 men: and at our going on shore upon our landing, the Indians of the place came down from the rocks to meet with us, with fresh water and wood on their backs. They are in marvellous awe of the Spaniards, and very simple people, and live marvellously savagely: for they brought us to their bidings about two miles from the harbour, where we saw their women and lodging, which is nothing but the skin of some beast laid upon the ground: and over them instead of houses, is nothing but five or six sticks laid across, which stand upon two forks with sticks on the ground and a few boughs laid on it. Their diet is raw fish, which stinketh most vilely. And when any of them die, they bury their bows and arrows with them, with their canoe and all that they have: for we opened one of their graves, and saw the order of them. Their canoes or boats are marvellous artificially made of two skins like unto bladders, and are blown full at one end with quills: they have two of these bladders blown full, which are sewn together and made fast with a sinew of some wild beast; which when they are in the water swell, so that they are as tight as may be. They go to sea in these boats, and catch very much fish with them, and pay much of it for tribute unto the Spaniards: but they use it marvellous beastly.

The 27 day we took a small bark, which came from Santiago. In this bark was one George a Greek, a reasonable pilot for all the coast of Chile. There were also in the said bark one Fleming and three Spaniards: and they were all sworn and received the sacrament before they came to sea by three or four friars, that if we should chance to meet them, they should throw those letters over board: which (as we were giving them chase with our pinnace) before we could fetch them up, they had accordingly thrown away. Yet our general wrought so with them, that they did confess it: but he was fain to cause them to be tormented with their thumbs in a wrench, and to continue them at several times with extreme pain. Also he made the old Fleming believe that he would hang him; and the rope being about his neck he was pulled up a little from the hatches, and yet he would not confess, choosing rather to die, than he would be perjured. In the end it was confessed by one of the Spaniards, whereupon we burnt the bark, and carried the men with us.

The tenth day [of May] the *Hugh Gallant* in which bark I Francis Pretty was lost company with our admiral.

The 17 of May we met with our admiral again, and all the rest of our fleet. They had taken two ships, the one laden with sugar, molasses, maize, Cordovan skins, montego de porco, many packs of pintados, many Indian coats, and some marmalade, and 1000 hens: and the other ship was laden with wheat meal, and boxes of marmalade. One of these ships which had the chief merchandise in it, was worth twenty thousand pounds, if it had been in England or in any other place of Christendom where we might have sold it. We filled all our ships with as much as we could bestow of these goods: the rest we burnt and the ships also; and set the men and women that were not killed on shore.

The 20 day in the morning we came into the road of Paita, and being at an anchor, our general landed with sixty or seventy men, skirmished with them of the town, and drove them all to flight to the top of the hill which is over the town. We found the quantity of 25 pounds weight in silver in pieces of eight reals, and abundance of household stuff and storehouses full of all kind of wares: but our general would not suffer any man to carry much cloth or apparel away, because they should not cloy themselves with burthens: for they were five men to one of us: and we had an English mile and a half to our ships. Thus we came down in safety to the town, which was very well builded, and marvellous clean kept in every street, with a town-house or guild hall in the midst, and had to the number of two hundred houses at the least in it. We set it on fire to the ground, and goods to the value of five or six thousand pounds: there was also a bark riding in the road which we set on fire and departed.

The 25 day of May we arrived at the island of Puna,[12] where is a very good harbour, where we found a great ship of the burthen of 250 tons riding at an anchor with all her furniture, which was ready to be hauled on ground: for there is a special good place for that purpose. We sunk it, and went on shore where the lord of the island dwelt, who had a sumptuous house marvellous well contrived: and out of every chamber was framed a gallery with a stately prospect into the sea on one side, and into the island on the other side, with a marvellous great hall below, and a very great storehouse at the one end of the hall: the most part of the cables in the South Sea are made upon that island. This great cacique doth make all the Indians upon the island to work and to drudge for him: and he himself is an Indian born, but is married to a marvellous fair woman which is a Spaniard.

This Spanish woman his wife is honoured as a queen in the island, and never goeth on the ground upon her feet: but when her pleasure is to take the air, she is always carried in a shadow like unto an horse-litter upon four men's shoulders, with a veil or canopy over her for the sun or the wind, having her gentlewomen still attending about her with a great troop of the best men of the island with her. But both she and the lord of the island with all the Indians in the town were newly fled out of the island before we could get to an anchor, by reason we were becalmed before we could get in, and were gone over unto the mainland, having carried away with them to the sum of 100,000 crowns.

This island is very pleasant: but there are no mines of gold nor silver in it. There are at least 200 houses in the town about the cacique's palace, and as many in one or two towns more upon the island, which is almost as big as the Isle of Wight. There is planted on the one side of the cacique's house a fair garden, with all herbs growing in it, and at the lower end a well of fresh water, and round about it are trees set, whereon bombazine cotton groweth after this manner: the tops of the trees grow full of cods, out of which the cotton groweth, and in the cotton is a seed of the bigness of a pea, and in every cod there are seven or eight of these seeds: and if the cotton be not gathered when it is ripe, then these seeds fall from it, and spring again.

There are also in this garden fig-trees which bear continually, also pompions, melons, cucumbers, radishes, rosemary and thyme, with many other herbs and fruits. At the other end of the house there is also another orchard, where grow oranges sweet and sour, lemons, pomegranates and limes, with divers other fruits.

There is very good pasture ground in this island; and withal many horses, oxen, bullocks, sheep very fat and fair, great store of goats which be very tame, and are used continually to be milked. They have moreover abundance of pigeons, turkeys, and ducks of a marvellous bigness.

There was also a very large and great church hard by the cacique's house, whither he caused all the Indians in the island to come and hear mass: for he himself was made a Christian when he was married to the Spanish woman before spoken of, and upon his conversion he caused the rest of his subjects to be christened. In this church was an high altar with a crucifix, and five bells hanging in the nether end thereof. We burnt the church and brought the bells away.

The second day of June in the morning, by and by after break of day, every one of the watch being gone abroad to seek to fetch in victuals, some one way, some another, upon the sudden there came down upon us an hundred Spanish soldiers with muskets and an ensign, which were landed on the other side of the island that night, and all the Indians of the island with them, every one with weapons. Thus being taken at advantage we had the worst: for our company was not past sixteen or twenty; whereof they had slain one or two before they were come to the houses: yet we skirmished with them an hour and a half: at the last being sore overcharged with multitudes, we were driven down from the hill to the water's side, and there kept them play a while, until in the end Zachary Saxie, who with his halberd had kept the way of the hill, and slain a couple of them, as he breathed himself being somewhat tired, had an honourable death and a short; for a shot struck him to the heart: who feeling himself mortally wounded cried to God for mercy, and fell down presently dead. But soon after the enemy was driven somewhat to retire from the bank's side to the green: and in the end our boat came and carried as many of our men away as could go in her, which was in hazard of sinking while they hastened into it: and one of our men whose name was Robert Maddocke was shot through the head with his own piece, being a snap-hance, as he was hasting into the boat. But four of us were left behind which the boat could not carry: to wit, myself Francis Pretty, Thomas Andrewes, Stephen Gunner, and Richard Rose: which had our shot ready and retired ourselves unto a cliff, until the boat came again, which was presently after they had carried the rest aboard. There were six and forty of the enemy's slain by us, whereof they had dragged some into bushes, and some into old houses, which we found afterwards. We lost twelve men.

The self same day, we went on shore again with seventy men, and had a fresh skirmish with the enemies, and drove them to retire, being an hundred Spaniards serving with muskets, and two hundred Indians with bows, arrows and darts. This done, we set fire on the town and burnt it to the ground, having in it to the number of three hundred houses: and shortly after made havoc of their fields, orchards and gardens, and burnt four great ships more which were in building on the stocks.

The fifth day of June we departed, and turned up for a place which is called Rio Dolce, where we watered: at which place also we sunk the *Hugh Gallant* for want of men, being a bark of forty tons.

The 27 in the morning by the break of day we came into the road of Aguatulco, where we found a bark of 50 tons, laden with cacaos and anil which they had there landed: and the men were all fled on shore. We landed there, and burnt their town, with the church and custom-house which was very fair and large: in which house were 600 bags of anil to dye cloth; every bag whereof was worth 40 crowns, and 400 bags of cacaos: every bag whereof is worth ten crowns. These cacaos go among them for meat and money. For 150 of them are in value one real of plate in ready payment. They are very like unto an almond, but are nothing so pleasant in taste: they eat them, and make drink of them. This the owner of the ship told us.

[The 8 we came to the road of Chaccalla.] Our general sent up Captain Havers with forty men of us before day, we went unto a place about two leagues up into the country in a most villainous desert path through the woods and wilderness: and in the end we came to a place where we took three householders with their wives and children and some Indians, we bound them all and made them come to the sea side with us.

Our general made their wives to fetch us plantains, lemons, and oranges, pine-apples and other fruits whereof they had abundance, and so let their husbands depart.

The 4 of November the *Desire* and the *Content*, beating up and down upon the headland of California, between seven and 8 of the clock in the morning one of the company which was the trumpeter of the ship going up into the top espied a sail bearing in from the sea with the cape, whereupon he cried out with no small joy, a sail, a sail: we gave them chase some 3 or 4 hours, standing with our best advantage and working for the wind. In the afternoon we got up unto them, giving them the broadside with our great ordnance and a volley of small shot, and presently laid the ship aboard, whereof the King of Spain was owner, called the *Santa Anna* and thought to be 700 tons in burthen. As we were ready on their shipside to enter her, being not past 50 or 60 men at the uttermost, we perceived that the captain had made fights fore and after, and having not one man to be seen, stood close under their fights, with lances, javelins, rapiers, and targets, and an innumerable sort of great stones, which they threw overboard upon our heads being so many of them, they put us off the ship again, with the loss of 2 of our men which were slain, and with the hurting of 4 or 5. We new trimmed our sails, and gave them a fresh encounter with our great ordnance and also with our small shot, raking them through and through, to the killing and maiming of many of their men. Their captain still like a valiant man with his company stood very stoutly not yielding as yet: our general encouraging his men afresh with the whole noise of trumpets gave them the third encounter with our great ordnance. They being thus discomforted and spoiled, and their ship being in hazard of sinking by reason of the great shot, whereof some were under water, within 5 or 6 hours fight set out a flag of truce, desiring our general to save their lives and to take their goods, and that they would presently yield. Our general of his goodness promised them mercy, and willed them to strike their sails, and to hoist out their boat and to come aboard: one of their chief merchants came aboard unto our general: and falling down upon his knees, offered to have kissed our general's feet, and craved mercy. The general of his great humanity, promised their lives and good usage. The said pilot and captain presently certified the general what goods they had within board, to wit, an hundred and 22 thousand pesos of gold: with silks, satins, damasks, with musk and divers other merchandise, and great store of all manner of victuals with the choice of many conserves for to eat, and of sundry sorts of very good wines. On the 6 day of November following we went into an harbour which is called by the Spaniards, Puerto Seguro.

Here the whole company of the Spaniards, to the number of 190 persons were set on shore: where they had a fair river of fresh water, with great store of fresh fish, fowl and wood, and also many hares and coneys upon the mainland. Our general also gave them great store of victuals, of garbanzos, pease, and some wine. Also they had all the sails of their ship to make them tents on shore, with licence to take such store of planks as should be sufficient to make them a bark. Then we fell to hoisting in of our goods, sharing of the treasure, and allotting to every man his portion. In division whereof the eighth of this month, many of the company fell into a mutiny against our general, which nevertheless were after a sort pacified for the time.

Our general discharged the captain, with provision for his defence against the Indians, both of swords, targets, pieces, shot and powder to his great contentment: but before his departure, he took out of this great ship two young lads born in Japan, which could both write and read their own language. He took also with him out of their ship, 3 boys born in the isle of Manila, the one about 15, the other about 13, and the youngest about 9 years old. The third remaineth with the right honourable the Countess of Essex.

He took also from them a Spaniard, which was a very good pilot unto the islands of Ladrones, where the Spaniards do put in to water, sailing between Acapulco and the Philippines: in which isles of Ladrones, they find fresh water, plantains, and potato roots: howbeit the people be very rude and heathens. The 19 day of November about 3 of the clock in the afternoon, our general caused the King's ship to be set on fire, which having to the quantity of 500 tons of goods in her we saw burnt into the water, and then set sail joyfully homewards towards England with fair wind: we left the *Content* astern of us. Thinking she would have overtaken us, we lost her company and never saw her after. We were sailing unto the isles of Ladrones the rest of November, and all December, and so forth until the 3 of January 1588, with a fair wind for the space of 45 days: and we esteemed it to be between 17 and 18 hundred leagues. We were coming up within 2 leagues of the island, where we met with 60 or 70 sails of canoes full of savages, who came off to sea unto us, and brought with them in their boats plantains, cocos, potato roots, and fresh fish, which they had caught at sea, and held them up unto us for to exchange with us; we made fast little pieces of old iron upon small cords and fishing lines, and so veered the iron into their canoes, and they caught hold of them and took off the iron, and in exchange of it they would make fast unto the same line either a potato root, or a bundle of plantains, which we hauled in: and thus our company exchanged with them until they had satisfied themselves with as much as did content them: yet we could not be rid of them. For afterward they were so thick about the ship, that it stemmed and broke 1 or 2 of their canoes: but the men saved themselves being in every canoe 4, 6, or 8 persons all naked and excellent swimmers and divers. They are of a tawny colour and marvellous fat, and bigger ordinarily of stature than the most part of our men in England, wearing their hair long: their canoes were as artificially made as any that ever we had seen: considering they were made and contrived without any edge-tool. They are not above half a yard in breadth and in length some seven or eight yards, and their heads and sterns are both alike: their sail is made of mats of sedges, square or triangle wise: and they sail as well right against the wind, as before the wind: these savages followed us so long, that we could not be rid of them: until in the end our general commanded some half dozen arquebuses to be made ready; and himself struck one of them and the rest shot at them: but they were so nimble, that we could not discern whether they were killed or not, because they would fall back into the sea and prevent us by diving.

The 14 day of January, by the break of day we fell with a headland of the isles of the Philippines. Manila is well planted and inhabited with Spaniards to the number of six or seven hundred persons: which dwell in a town unwalled, which hath 3 or 4 small block houses, part made of wood, and part of stone being indeed of no great strength: they have one or two small galleys belong to the town. It is a very rich place of gold and many other commodities; and they have yearly traffic from Acapulco in Nueva España, and also 20 or 30 ships from China, which bring them many sorts of merchandise. They bring great store of gold with them, which they traffic and exchange for silver, and give weight for weight.

The fifteenth of January we fell with an island called Capul.[113] Our ship was no sooner come to an anchor, but presently there came a canoe rowing aboard us, wherein was one of the chief caciques of the island, who supposing that we were Spaniards, brought us potato roots, and green cocos, in exchange whereof we gave his company pieces of linen to the quantity of a yard for four cocos, and as much linen for a basket of potato roots of a quart in quantity; which roots are very good meat, and excellent sweet either roasted or boiled.

This cacique's skin was carved and cut with sundry and many strokes and devices all over his body. Presently the people of the island came down with their cocos and potato roots, and brought with them hens and hogs. Thus we rode at anchor all day, doing nothing but buying roots, cocos, hens, hogs, refreshing ourselves marvellously well.

The same day at night being the fifteenth of January 1588, Nicholas Roderigo the Portuguese, whom we took out of the great *Santa Anna* at the cape of California, desired to speak with our general in secret: our general understood, and asked him what he had to say. The Portuguese made him this answer. That the Spaniard which was taken out of the great *Santa Anna* for a pilot, had written a letter, and secretly sealed it and locked it up in his chest, meaning to convey it by the inhabitants of this island to Manila, the contents whereof were: that there had been two English ships along the coast of Chile, Peru, [and] Nueva España, and that they had taken many ships and merchandise in them, and burnt divers towns, and spoiled all that ever they could come unto, and that they had taken the King's ship which came from Manila and all his treasure, with all the merchandise that was therein: and had set the people on shore. Therefore he willed them that they should make strong their bulwarks with their two galleys. He further signified, that we were riding at an island called Capul at the end of the island of Manila, being but one ship with small force in it: if they could use any means to surprise us being there at an anchor, they should dispatch it: for our force was but small, and our men but weak, and that the place where we rode was but 50 leagues from them. Our general called for him, and charged him with these things, which at the first he utterly denied: but in the end, the matter being made manifest, the next morning our general willed that he should be hanged.

The people of this island go almost all naked and are tawny of colour. The men wear only a strop about their waists, of some kind of linen of their own weaving, which is made of plantain leaves, and another strop coming from their back under their twists, which covers their privy parts, and is made fast to their girdles at their navels.

Every man and man-child among them hath a nail of tin thrust quite through the head of his privy part, being split in the lower end and rivetted, and on the head of the nail is as it were a crown: which is driven through their privities when they be young, and the place groweth up again, without any great pain to the child: and they take this nail out and in, as occasion serveth: and for the truth whereof we ourselves have taken one of these nails from a son of one of the kings which was of the age of 10 years, who did wear the same in his privy member.

This custom was granted at the request of the women of the country, who finding their men to be given to the foul sin of sodomy, desired some remedy against that mischief. Moreover all the males are circumcised, having the foreskin of their flesh cut away. These people wholly worship the devil, and often times have conference with him, which appeareth unto them in most ugly and monstrous shape.

On the 23 day of January, our general Mr Thomas Candish caused all the principals of this island, and of an hundred islands more, which he had made to pay tribute unto him (in hogs, hens, potatoes and cocos) to appear before him, and made himself and his company known unto them, that they were English men, and enemies to the Spaniards: and thereupon spread his ensign and sounded up the drums, which they much marvelled at: they promised both themselves and all the islands thereabout to aid him, whensoever he should come again to overcome the Spaniards. Also our general gave them, in token that we were enemies to the Spaniards, money back again for all their tribute which they had paid: which they took marvellous friendly, and rowed about our ship to show us pleasure: at the last our general caused a saker to be shot off, whereat they wondered, and with great contentment took their leaves of us.

On the 21 day of February, being Ash Wednesday Captain Havers died of a most fervent and pestilent ague, to the no small grief of our general, who caused two falcons and one saker to be shot off, who after he was shrouded in a sheet and a prayer said, was heaved overboard with great lamentation of us all. After his death myself with divers others in the ship fell marvellously sick, and so continued in very great pain for the space of three weeks or a month by reason of the extreme heat of the climate.

The first day of March having passed through the straits of Java, we came to an anchor under the southwest parts of Java: where we espied certain of the people fishing by the sea side. Our general taking into the ship-boat certain of his company, and a negro which could speak the tongue, made towards those fishers, which having espied our boat ran on shore into the wood for fear: but our general caused his negro to call unto them: presently one of them came out to the shore side and made answer. Our general by the negro enquired of him for fresh water, which they found, and caused the fisher to go to the king and to certify him of a ship that was come to have traffic for victuals, and for diamonds, pearls, or any other rich jewels that he had: for which he should have either gold or other merchandise in exchange.

Two or three canoes came from the town unto us with eggs, hens, fresh fish, oranges, and limes. Our general weighed anchor and stood in nearer for the town: and as we were under sail we met with one of the king's canoes coming toward us. In this canoe was the king's secretary, who had on his head a piece of dyed linen cloth folded up like a Turk's turban: he was all naked saving about his waist, his breast was carved with the broad arrow upon it: he went barefooted: he had an interpreter with him, which was a mestizo, that is, half an Indian and half a Portuguese, who could speak very good Portuguese. This secretary signified unto our general that he had brought him an hog, hens, eggs, fresh fish, sugar-canes and wine: (which wine was as strong as any aquavitae, and as clear as any rock water). Our general used him singularly well, banqueted him most royally with the choice of many and sundry conserves, wines both sweet and other, and caused his musicians to make him music. This done our general told him that he and his company were Englishmen; and that we had been at China and had had traffic there with them, and that we were come thither to discover, and purposed to go to Malacca. The people of Java told our general that there were certain Portuguese in the island which lay there as factors continually to traffic with them, to buy negroes, cloves, pepper, sugar, and many other commodities. This secretary of the king with his interpreter lay one night aboard our ship. In the evening at the setting of the watch, our general commanded every man in the ship to provide his arquebus and his shot, and so with shooting off 40 or 50 small shot and one saker, himself set the watch with them. This was no small marvel unto these heathen people, who had not commonly seen any ship so furnished with men and ordnance.

After the break of day there came to the number of 9 or 10 of the king's canoes so deeply laden with victuals as they could swim with two great live oxen, half a score of wonderful great and fat hogs, a number of hens which were alive, drakes, geese, eggs, plantains, sugar canes, sugar in plates, cocos, sweet oranges and sour, limes, great store of wine and aquavitae, salt to season victuals withal, and almost all manner of victuals else. Among all the rest came two Portuguese of middle stature, and men of marvellous proper personage; they were each of them in a loose jerkin and hose, which came down from the waist to the ankle, because of the use of the country, and partly because it was Lent, and a time for doing of their penance: they had on each of them a very fair and a white lawn shirt, very decently, only their bare legs excepted. These Portuguese were no small joy unto our general and all the rest of our company: for we had not seen any Christian that was our friend of a year and a half before. Our general used and entreated them singularly well, with banquets and music: they told us that they were no less glad to see us, than we to see them, and enquired of the estate of their country, and what was become of Dom Antonio their King, and whether he be living or no: for the Spaniards had always brought them word that he was dead. Then our general satisfied them in every demand: assuring them, that their King was alive, and in England, and had honourable allowance of our Queen, and that there was war between Spain and England, and that we were come under the King of Portugal into the South Sea, and had warred upon the Spaniards there, and had fired, spoiled and sunk all the ships along the coast that we could meet withal, to the number of eighteen or twenty sails. With this report they were sufficiently satisfied.

On the other side they declared unto us the state of the island of Java. First the plentifulness and great choice and store of victuals of all sorts, and of all manner of fruits as before is set down: then the great and rich merchandise which are there to be had. The name of the king of that island was Raja Bolamboam, who was a man had in great majesty and fear among them. The common people may not bargain, sell or exchange any thing with any other nation without special licence from their king: and if any so do, it is present death for him. The king himself is a man of great years, and hath an hundred wives, his son hath fifty. The custom of the country is, that whensoever the king doth die, they take the body so dead and burn it and preserve the ashes, and within five days next after, the wives go together to a place appointed, and the chief of the women, hath a ball in her hand, and throweth it from her, and to the place where the ball resteth, thither they go all, and turn their faces to the eastward, and every one with a dagger in their hand, (which dagger they call a kris, and is as sharp as a razor) stab themselves to the heart, and falling grovelling on their faces so end their days. This thing is as true as it seemeth to any hearer to be strange.

The men of themselves be very politic and subtile, and singularly valiant, and wonderfully at commandment and fear of their king. For example: if their king command them to undertake any exploit, be it never so dangerous or desperate, they dare not nor will refuse it, though they die every man in the execution of the same. For he will cut off the heads of every one of them which return alive without bringing of their purpose to pass: they never fear any death. If any of them feeleth himself hurt with lance or sword, he will willingly run himself upon the weapon quite through his body to procure his death more speedily, and in this desperate sort end his days, or overcome his enemy. Moreover, although the men be tawny of colour and go continually naked, yet their women be fair of complexion and go more apparelled.

They told us further, that if their King, Dom Antonio would come unto them, they would warrant him to have all the Moluccas at commandment, besides, China, and the isles of the Philippines, and that he might be assured to have all the Indians on his side that are in the country. The next day being the 16 of March we set sail towards the Cape of Good Hope, on the southernmost coast of Africa.

The rest of March and all the month of April we spent in traversing that mighty and vast sea, between the isle of Java and the main of Africa, observing the stars, the fowls, which are marks unto the seamen of fair weather, approaching lands or islands, the winds, the tempests, the rains and thunders, with the alterations of the tides and currents.

The 11 of May in the morning one of the company went into the top, and espied land. This cape is very easy to be known. For there are right over it three very high hills standing but a small way one off another, and the highest standeth in the midst and the ground is much lower by the seaside.

This cape of Buena Esperanza is set down and accompted for two thousand leagues from the island of Java in the Portuguese sea charts: but it is not so much almost by an hundred and fifty leagues, as we found by the running of our ship. We were in running of these eighteen hundred and fifty leagues just nine weeks.

The eighth day of June by break of day we fell in sight of the island of Saint Helena.

This island is very high land, and lieth in the main sea standing as it were in the midst of the sea between the main land of Africa, and the main of Brazil and the coast of Guinea.

The same day about two or three of the clock in the afternoon we went on shore, where we found a marvellous fair and pleasant valley, wherein divers handsome buildings and houses were set up, and especially one which was a church, which was tiled and whited on the outside very fair, and made with a porch, and within the church at the upper end was set an altar, whereon stood a very large table set in a frame having in it the picture of Our Saviour Christ upon the Cross and the image of Our Lady praying.

There are two houses adjoining to the church, which serve for kitchens to dress meat in: the coverings of the said houses are made flat, whereon is planted a very fair vine, and through both the said houses runneth a very good and wholesome stream of fresh water.

There is also right over against the said church a fair causeway made up with stones reaching unto a valley by the seaside, in which valley is planted a garden, wherein grow great store of pompions and melons: and upon the said causeway is a frame erected whereon hang two bells wherewith they ring to mass; and hard unto it is a cross set up, which is squared, framed and made very artificially of free stone, whereon is carved in ciphers what time it was builded, which was in the year of Our Lord 1571.

This valley is the fairest and largest low plot in all the island, and it is marvellous sweet and pleasant, and planted in every place either with fruit trees, or with herbs. There are fig trees, which bear fruit continually, and marvellous plentifully: for on every tree you shall have blossoms, green figs, and ripe figs, all at once: and it is so all the year long. There be also great store of lemon trees, orange trees, pomegranate trees, pomecitron trees, date trees, which bear fruit as the fig trees do, and are planted carefully and very artificially with very pleasant walks under and between them, and the said walks be overshadowed with the leaves of the trees: and in every void place is planted parsley, sorrel, basil, fennel, aniseed, mustard seed, radishes, and many special good herbs: and the fresh water brook runneth through divers places of this orchard, and may with very small pains be made to water any one tree in the valley.

This fresh water stream cometh from the tops of the mountains, and falleth from the cliff into the valley the height of a cable, and hath many arms out of it, which refresh the whole island, and almost every tree in it. The island is altogether high mountains and steep valleys, except it be in the tops of some hills, and down below in some of the valleys, where marvellous store of all these kind of fruits before spoken of do grow: there is greater store growing in the tops of the mountains than below in the valleys: but it is wonderful laboursome and also dangerous travelling up unto them and down again by reason of the height and steepness of the hills.

There is also upon this island great store of partridges, which are very tame, not making any great haste to fly away though one come very near them, but only to run away, and get up into the steep cliffs: we killed some of them with a fowling piece. They differ very much from our partridges which are in England both in bigness and also in colour. For they be within a little as big as an hen, and are of an ash colour, and live in coveys twelve, sixteen, and twenty together.

There are likewise no less store of pheasants in the island, which are also marvellous big and fat, surpassing those which are in our country in bigness.

There are in this island thousands of goats, which the Spaniards call cabritos, which are very wild: you shall see one or two hundred of them together: they will climb up the cliffs which are so steep that a man would think it a thing unpossible for any living thing to go here. We took and killed many of them for all their swiftness.

Here are in like manner great store of swine which be very wild and very fat, and of a marvellous bigness: they keep altogether upon the mountains, and will very seldom abide any man to come near them.

We found in the houses at our coming 3 slaves which were negroes, which told us that the East Indian fleet, which were in number 5 sails, the least whereof were in burthen of 8 or 900 tons, all laden with spices and Calicut cloth, with store of treasure and very rich stones and pearls, were gone from the said island of Saint Helena but 20 days before we came thither.

This island hath been found of a long time by the Portuguese, and hath been altogether planted by them, for their refreshing as they come from the East Indies.

The 20 day of June having taken in wood and water and refreshed ourselves with such things as we found there, and made clean our ship, we set sail about 8 of the clock in the night toward England.

The third of September we met with a Flemish hulk which came from Lisbon, and declared unto us the overthrowing of the Spanish fleet, to the singular rejoicing and comfort of us all.

The 9 of September, after a terrible tempest which carried away most part of our sails, by the merciful favour of the Almighty we recovered our long wished port of Plymouth in England, from whence we set forth at the beginning of our voyage.

XLIX

The second voyage attempted by Mr John Davis with others, for the discovery of the Northwest Passage, in Anno 1586.

The 7th day of May, I departed from the port of Dartmouth for the discovery of the Northwest Passage, with a ship of an hundred and twenty tons named the *Mermaid*, a bark of 60 tons named the *Sunshine*, a bark of 35 tons named the *Moonshine*, and a pinnace of ten tons named the *North Star*.

The 15 of June I discovered land in the latitude of 60 degrees, mightily pestered with ice and snow, so that there was no hope of landing.

The 29 of June after many tempestuous storms we again discovered land, in latitude 64. The year before I had been in the same place, and found it very convenient, well stored with float wood, and possessed by a people of tractable conversation. This land is very high and mountainous, having before it on the west side a mighty company of isles full of fair sounds, and harbours. This land was very little troubled with snow, and the sea altogether void of ice.

The ships being within the sounds we sent our boats to search for shoal water, where we might anchor, and as the boat went sounding and searching, the people of the country came in their canoes towards them with many shouts and cries: but after they had seen in the boat some of our company that were the year before here with us, they presently rowed to the boat, and took hold on the oar: they came with the boats to our ships, making signs that they knew all those that the year before had been with them. After I perceived their joy and small fear of us, myself with the merchants and others of the company went ashore, bearing with me twenty knives: I had no sooner landed, but they leapt out of their canoes and came running to me and the rest, and embraced us with many signs of hearty welcome: at this present there were eighteen of them, and to each of them I gave a knife: they offered skins to me for reward, but I made signs that they were not sold, but given them of courtesy.

The people came continually unto us sometime an hundred canoes at a time, sometime forty, fifty, more and less, as occasion served. They brought with them seal skins, stag skins, white hares, seal fish, salmon, small cod, with other fish, and birds such as the country did yield.

The boats that went from me found the tents of the people made with seal skins set up on timber, wherein they found great store of dried caplin, being a little fish no bigger than a pilchard: they found bags of train oil, many little images cut in wood.

They also found ten miles within the snowy mountains a plain champaign country, with earth and grass, such as our moory and waste grounds of England are.

I was desirous to have our men leap with them, which was done, but our men did overleap them: from leaping they went to wrestling, we found them strong and nimble, and to have skill in wrestling, for they cast some of our men that were good wrestlers.

The people are of good stature, well in body proportioned, with small slender hands and feet, with broad visages, and small eyes, wide mouths, the most part unbearded, great lips, and close toothed. They are witches, and have many kinds of enchantments, which they often used, but to small purpose, thanks be to God.

Being among them at shore the fourth of July one of them began to kindle a fire in this manner: he took a piece of a board wherein was a hole half through: into that hole he puts the end of a round stick like unto a bedstaff, wetting the end thereof in train, and in fashion of a turner with a piece of leather, by his violent motion doth very speedily produce fire: which done, with turfs he made a fire, into which with many words and strange gestures, he put divers things, which we supposed to be a sacrifice. I then took one of them, and thrust him into the smoke, and willed one of my company to tread out the fire, and to spurn it into the sea, which was done to show them that we did contemn their sorcery. These people are very simple in all their conversation, but marvellous thievish, especially for iron, which they have in great account. They began through our levity to show their vile nature: they began to cut our cables: they cut away the *Moonlight's* boat from her stern, they cut our cloth where it lay to air, though we did carefully look unto it, they stole our oars, a caliver, a boar spear, a sword: whereupon there was a caliver shot among them, which strange noise did sore amaze them, so that with speed they departed: notwithstanding their simplicity is such, that within ten hours after they came again to us to entreat peace; which being promised, we again fell into a great league. They brought us seal skins, and salmon, but seeing iron, they could in no wise forbear stealing: which when I perceived it, it did but minister unto me an occasion of laughter, to see their simplicity.

They eat all their meat raw, they live most upon fish, they drink salt water, and eat grass and ice with delight: they are never out of the water, but live in the nature of fishes, save only when dead sleep taketh them, and then under a warm rock laying his boat upon the land, he lyeth down to sleep. They pronounce their language very hollow, and deep in the throat: these words we learned from them.

Kesinyoh, Eat some
Madlycoyte, Music
Aginyoh, Go fetch
Yliaoute, I mean no harm
Ponameg, A boat
Asanock, A dart
Sawygmeg, A knife
Uderah, A nose
Aoh, Iron.

The 17th of this month being in the latitude of 63 degrees 8 minutes, we fell upon a most mighty and strange quantity of ice in one entire mass, so big as that we knew not the limits thereof, and being withall so very high in form of a land, with bays and capes and like high cliff land, as that we supposed it to be land and therefore sent our pinnaces off to discover it: but at her return we were certainly informed that it was only ice.

We coasted this mighty mass of ice until the 30th of July, finding it a mighty bar to our purpose. All our shrouds, ropes and sails were so frozen, and compassed with ice, as seemed to me more than strange, since the last year I found this sea free and navigable.

Our men through this extremity began to grow sick and feeble, and withall hopeless of good success: whereupon very orderly, with good discretion they entreated me that in conscience I ought to regard the safety of mine own life with the preservation of theirs, and that I should not through my over-boldness leave their widows and fatherless children to give me bitter curses: yet considering the excellency of the business if it might be attained, the great hope of certainty by the last year's discovery, and that there was yet a third way not put in practice, I thought it would grow to my great disgrace, if this action by my negligence should grow into discredit. Whereupon falling into consideration that the *Mermaid*, albeit a very strong and sufficient ship, yet by reason of her burthen was not so convenient and nimble as a smaller bark, especially in such desperate hazards. I determined to furnish the *Moonlight* with revictualling and sufficient men, and to proceed in this action as God should direct me.

The second of August we harboured ourselves in a very excellent good road, where with all speed we graved the *Moonlight*, and revictualled her: we searched this country with our pinnace while the bark was trimming. We found it very hot, and we were much troubled with a fly which is called mosquito, for they did sting grievously.

The sixth of August, the people came unto us without fear, and did barter with us for skins, as the other people did: they differ not from the other, neither in their canoes nor apparel.

The fifteenth day at three o'clock in the morning we departed from this land to the south, and the eighteenth of August we discovered land northwest from us in the morning, being a very fair promontory, in latitude 65 degrees, having no land on the south. Here we had great hope of a through passage.

This land is nothing in sight but isles, which increaseth our hope. This nineteenth of August at six o'clock in the afternoon, it began to snow, and so continued all night with foul weather, and much wind, so that we were constrained to lie at hull all night five leagues off the shore: in the morning being the twentieth of August, the fog and storm breaking up, we bare in with the land, and at nine o'clock in the morning we anchored in a very fair and safe road and locked for all weathers. At ten of the clock I went on shore to the top of a very high hill, where I perceived that this land was islands.

We coasted this land till the eight and twentieth of August, finding it still to continue towards the south, from the latitude of 67 to 57 degrees. We arrived in a very fair harbour in the latitude of 56 degrees, with very fair woods on both sides: in this place we continued until the first of September, in which time we had two very great storms. I landed and went six miles by guess into the country, and found that the woods were fir, pineapple, alder, yew, withy, and birch: here we saw a black bear: this place yieldeth great store of birds, as pheasant, partridge, barbary hens or the like, wild geese, ducks, blackbirds, jays, thrushes, with other kinds of small birds. Of the partridge and pheasant we killed great store with bow and arrows: in this place at the harbour mouth we found great store of cod.

The sixth of September having a fair northnorthwest wind, we purposed to depart, and sent five of our young men ashore to an island to fetch certain fish: the brutish people of this country lay secretly lurking in the wood, and upon the sudden assaulted our men; which when we perceived, we presently let slip our cables, and under our foresail bear into the shore, and with all expedition discharged a double musket upon them twice, at the noise whereof they fled: notwithstanding to our very great grief, two of our men were slain with their arrows, and two grievously wounded, only one escaped by swimming, with an arrow shot through his arm.

This present evening it pleased God to further increase our sorrows with a mighty tempestuous storm, which lasted unto the tenth of this month very extreme. We unrigged our ship, and purposed to cut down our masts, the cable of our shut anchor broke, so that we only expected to be driven on shore among these cannibals for their prey. Yet in this deep distress the mighty mercy of God, when hope was past, gave us succour, and sent us a fair lee, so as we recovered our anchor again, and new moored our ship.

The eleventh day with a fair westnorthwest wind we departed

with trust in God's mercy, shaping our course for England, and arrived in the West Country in the beginning of October.

Master Davis being arrived, wrote his letter to Mr William Sanderson of London.

Sir, the *Sunshine* came into Dartmouth the fourth of this month: she hath been at Iceland, and from thence to Greenland, from thence to Desolation, where she made trade with the people staying in the country twenty days. They have brought home five hundred seal skins, and a hundred and forty half skins and pieces of skins. I stand in great doubt of the pinnace, God be merciful unto the poor men, and preserve them.

I have now experience of much of the northwest part of the world, and I can assure you upon the peril of my life, that this voyage may be performed without further charge, nay with certain profit to the adventurers, if I may have but your favour in the action. This fourteenth of October 1586.

Yours to command,
JOHN DAVIS.

L

The third voyage made by a ship sent in the year 1586, to the relief of the colony planted in Virginia, at the sole charges of Sir Walter Raleigh.

In 1586 Sir Walter Raleigh at his own charge prepared a ship of an hundred ton, freighted with all manner of things in most plentiful manner, for the supply and relief of his colony then remaining in Virginia: our colony half despaired of the coming of any supply: wherefore every man prepared for himself, determining resolutely to spend the residue of their life time in that country. And for the better performance of this their determination, they sowed, planted, and set such things as were necessary for their relief in so plentiful a manner as might have sufficed them two years without any further labour. Thus trusting to their own harvest, they passed the summer till the tenth of June: at which time their corn which they had sowed was within one fortnight of reaping: but then it happened that Sir Francis Drake in his prosperous return from the sacking of Santo Domingo, Cartagena, and San Agustín, determined in his way homeward to visit his countrymen the English colony then remaining in Virginia. So passing along the coasts of Florida, he fell with the parts where our English colony inhabited: and having espied some of that company, there he anchored and went a'land, where he conferred with them of their state and welfare, and how things had passed with them. They answered him that they lived all; but hitherto in some scarcity: and as yet could hear of no supply out of England: therefore they requested him that he would leave with them some two or three ships, that if in some reasonable time they heard not out of England, they might then return themselves. Which he agreed to. Whilst some were then writing their letters to send into England, and some others making reports of the accidents of their travels each to other, some on land, some on board, a great storm arose, and drove the most of their fleet from their anchors to sea, in which ships at that instant were the chiefest of the English colony: the rest on land perceiving this, hasted to those three sails which were appointed to be left there; and for fear they should be left behind they left all things confusedly, as if they had been chased from thence by a mighty army: and no doubt so they were; for the hand of God came upon them for the cruelty and outrages committed by some of them against the native inhabitants of that country.

Immediately after the departing of our English colony out of this paradise of the world, the ship sent and set forth at the charges of Sir Walter Raleigh and his direction, arrived; who after some time spent in seeking our colony up in the country, and not finding them, returned with all the provision into England.

About fourteen or fifteen days after the departure of the aforesaid ship, Sir Richard Grenville General of Virginia, accompanied with three ships well appointed for the same voyage, arrived there; who not finding the ship according to his expectation, nor hearing any news of our English colony there seated, and left by him anno 1583: after some time spent therein, not hearing any news of them, and finding the places which they inhabited desolate, yet unwilling to lose the possession of this country which Englishmen had so long held: after good deliberation, he determined to leave some men behind to retain possession of the country: whereupon he landed fifteen men in the isle of Roanoke, furnished plentifully with all manner of provision for two years, and so departed for England.

LI

A voyage to the Azores with two pinnaces belonging to Sir Walter Raleigh, written by John Evesham, Gentleman, in the year 1586.

The 10 of June 1586 we departed from Plymouth with two pinnaces, the one named the *Serpent*, the other the *Mary Spark* of Plymouth both belonging to Sir Walter Raleigh Knight. Directing our course towards the Isles of the Azores we took a small bark laden with sumach and other commodities. From thence we sailed to the island of Graciosa, where we descried a sail, and bearing with her we found her to be a Spaniard. For that we would not be known of what nation we were, we displayed a white silk ensign in our main top, which they seeing, made accompt that we had been some of the King of Spain's armadas, lying in wait for English men of war: but when we came within shot of her, we took down our white flag, and spread abroad the Cross of St George, which when they saw, it made them to fly as fast as they might. Our ships were swifter of sail than they, which they fearing did presently cast their ordnance and small shot with many letters and the draft of the Straits of Magellan into the sea, and thereupon immediately we took her. We also took a gentleman of Spain, named Pedro Sarmiento, governor of the Straits of Magellan, which said Pedro we brought into England with us, and presented him to our sovereign lady the Queen.

After this, lying off and about the islands, we descried two other sails, the one a ship, and the other a caravel, to whom we gave chase, which they seeing, with all speed made in under the Isle of Graciosa, to a certain fort there for their succour, where they came to an anchor. Having the wind of us, we could not hurt them with our ships, but we having a small boat, which we called a light horseman,[114] wherein myself was, being a musketeer, and four more with calivers, and four that rowed, came near unto the shore against the wind, which when they saw us come towards them they carried a great part of their merchandise on land, whither also the men of both vessels went and landed, and as soon as we came within musket shot, they began to shoot at us with great ordnance and small shot, and we likewise at them.

In the end we boarded one ship wherein was no man left, so we cut her cables, hoisted her sails, and sent her away with two of our men, and the other seven of us passed more near unto the shore, and boarded the caravel, which did ride within a stone's cast from the shore, and so near the land that the people did cast stones at us. Yet in despite of them all we took her, and one only negro therein; and cutting her cables in the hawse we hoisted her sails and being becalmed under the land, we were constrained to row her out with our boat, the fort still shooting at us, and the people on land with muskets and calivers, to the number of 150 or thereabouts: and we answered them with the small force we had. The shot of my musket being a crossbar-shot happened to strike the gunner of the fort to death, even as he was giving level to one of his great pieces.

Setting our course for England, being off the islands in the height of 41 degrees or there about, one of our men being in the top descried a sail, then 10 sail, then 15 whereupon it was concluded to send home those prizes we had, and so left in both our pinnaces not above 60 men. Thus we returned again to the fleet we had descried, where we found 24 sail of ships, whereof two of them were carracks, the one of 1200 and the other of 1000 tons, and ten galleons, the rest were small ships and caravels, all laden with treasure, spices and sugars, with which 24 ships we with two small pinnaces did fight, and kept company the space of 32 hours, continually fighting with them and they with us, but the two carracks kept still betwixt the fleet and us, that we could not take any one of them, so wanting powder, we were forced to give them over against our wills.

Thus we again set our course for England, and so come to Plymouth within six hours after our prizes, which we sent away 40 hours before us, where we were received with triumphant joy, not only with great ordnance then shot off, but with the willing hearts of all the people of the town, and of the country thereabout. And from thence we brought our prizes to Southampton, where Sir Walter Raleigh being our owner, rewarded us with our shares.

Our prizes were laden with sugars, elephant's teeth, wax, hides, rice, brazil.[115]

LII

A discourse of the West Indies and South Sea written by Lopez Vaz a Portuguese, continued unto the year 1587. Wherein, certain voyages of our Englishmen are truly reported.

Francis Drake an Englishman being on the sea, and having knowledge of the small strength of the town of Nombre de Dios,[116] came into the harbour on a night with four pinnaces, and landed an hundred and fifty men: and leaving one half of his men with a trumpet in a fort which was there, he with the rest entered the town without doing any harm till he came at the market place: and there his company discharging their calivers, and sounding their trumpets (which made a great noise in the town) were answered by their fellows in the fort, who discharged and sounded in like manner. This attempt put the townsmen in such extreme fear, that leaving their houses, they fled into the mountains. But the Spaniards being men for the most part of good discretion, getting to a corner of the market place discovered the Englishmen, and perceiving that they were but a few, discharged their pieces at them; and their fortune was such, that they slew the trumpeter, and shot the captain (whose name was Francis Drake) into the leg: who feeling himself hurt retired toward the fort, where he had left the rest of his men. Now Francis Drake (whom his men carried because of his hurt) when he came to the fort where he left his men and saw them fled, he and the rest of his company were in so great fear, that putting off their hose, they swam and waded all to their pinnaces, and departed forth of the harbour. Thus Captain Drake did no more harm at Nombre de Dios, neither was there in this skirmish any more than one Spaniard slain, and of the Englishmen only their trumpeter, whom they left behind with his trumpet in his hand.

But Captain Drake being discontent with the repulse that the men of Nombre de Dios gave him, went with his pinnaces into the bay of Darien, where having conference with certain negroes which were run away from their masters of Panama and Nombre de Dios, he was informed that at the very same time many mules were coming from Panama to Nombre de Dios laden with gold and silver. Upon this news Francis Drake taking with him an hundred shot, and the negroes stayed in the way till the treasure came by, accompanied and guarded only by those that drove the mules, who mistrusted nothing at all. When Captain Drake met with them, he took away their gold: but the silver he left behind, because he could not carry it over the mountains. And two days after this he went to the house where all the merchants leave their goods, where he slew six or seven of the merchants, but found neither gold nor silver, but great store of merchandise: and so he fired the said house, with all the goods, which were judged to be worth above two hundred thousand ducats. He forthwith recovered his pinnaces: where fortune so favoured his proceedings, that he had not been aboard half an hour, but there came to the sea side above three hundred soldiers, which were sent of purpose to take him: but God suffered him to escape their hands, to be a further plague unto the Spaniards.

Captain Drake carried from the coast of Peru, of silver eight hundred sixty six quintals, at 100 pound weight the quintal, all which sum amounteth to a million and thirty nine thousand and two hundred ducats. He carried away a hundred thousand pesos of gold, that is ten quintals, which last sum amounteth to an hundred and fifty thousand ducats: over and besides the treasure consisting of pearls, precious stones, reals of plate, and other things of great worth.

LIII

A brief relation of the notable service performed by Sir Francis Drake upon the Spanish fleet prepared in the Road of Cadiz: and of his destroying of 100 sails; and thence surprised a mighty carrack called the San Felipe *coming out of the East India, which was the first of that kind that ever was seen in England: performed in the year 1587.*

Her Majesty being informed of a mighty preparation by sea begun in Spain for the invasion of England, by good advice of her grave and prudent Council thought it expedient to prevent the same. Whereupon she caused a fleet of some 30 sails to be rigged and furnished with all things necessary. Over that fleet she appointed general Sir Francis Drake (of whose manifold former good services she had sufficient proof) to whom she caused 4 ships of her Navy Royal to be delivered, to wit, the *Bonaventure* wherein himself went as general; the *Lion*, the *Dreadnought*, and the *Rainbow*, unto which 4 ships two of her pinnaces were appointed as handmaids. There were also added to this fleet certain tall ships of the City of London, of whose especial good service the general made particular mention in his private letters directed to Her Majesty. This fleet set sail from the sound of Plymouth in the month of April towards the coast of Spain.

We met with two ships of Middelburg which came from Cadiz; by which we understood that there was great store of warlike provision at Cadiz and thereabout ready to come for Lisbon.

Upon this information our General with all speed possible, bending himself thither, upon the 19 of April entered with his fleet into the harbour of Cadiz: where at our first entering we were assailed over against the town by six galleys, which notwithstanding in short time retired under their fortress.

At our first coming in we sunk with our shot a ship of Ragusa of 1000 tons, furnished with 40 pieces of brass and very richly laden. There came two galleys more from St Maryport, and two from Puerto Real,[117] which shot freely at us, but altogether in vain: for they went away with the blows well beaten for their pains.

Before night we had taken 30 of the said ships, and became masters of the road, in despite of the galleys, which were glad to retire them under the fort. Five of them were great ships of Biscay, whereof 4 we fired, as they were taking in the King's provision of victuals for the furnishing of his fleet at Lisbon: the fifth being a ship about 1000 tons in burthen, laden with iron spikes, nails, iron hoops, horse-shoes, and other like necessaries bound for the West Indies we fired in like manner. Also we took a ship of 250 tons laden with wines for the King's provision, which we carried out to sea with us, and there discharged the said wines for our own store, and afterward set her on fire. Moreover we took 3 flyboats of 300 tons apiece laden with biscuit, whereof one was half unladen by us in the harbour and there fired, and the other two we took in our company to the sea. The whole number of ships and craft (as we suppose) then burnt, sunk, and brought away with us, amounted to about 10,000 tons of shipping.

We found little ease during our abode there, by reason of their continual shooting from the galleys, the fortresses and from the shore: where continually at places convenient they planted new ordnance to offend us with: besides the inconvenience which we suffered from their ships, which, when they could defend no longer they set on fire to come among us. By the assistance of the Almighty, and the invincible courage and industry of our general, this strange and happy enterprise was achieved in one day and two nights, to the great astonishment of the King of Spain.

Having performed this notable service, we came out of the road of Cadiz on the Friday morning the 21 of the said month of April, with very small loss not worth the mentioning.

After our departure ten of the galleys that were in the road came out, as it were in disdain of us, to make some pastime with their ordnance.

We now have had experience of galley fight: wherein I can assure you, that only these 4 of Her Majesty's ships will make no account of 20 galleys, if they may be alone, and not busied to guard others. There were never galleys that had better place and fitter opportunity for their advantage to fight with ships: but they were still forced to retire, we riding in a narrow gut, the place yielding no better, and driven to maintain the same, until we had discharged and fired the ships, which could not conveniently be done but upon the flood, at which time they might drive clear off us. Thus being victualled with bread and wine at the enemy's cost for divers months (besides the provisions that we brought from home) our general shaped our course toward Cape Sagres,[118] and in the way thither we took at several times of ships, barks, and caravels well near a hundred, laden with hoops, galley oars, pipe staves, and other provisions of the King of Spain, for the furnishing of his forces intended against England, all which we burned, having dealt favourably with the men and sent them on shore.

Our general shaped his course towards the Isles of the Azores, and passing towards the Isle of Saint Michael, within 20 or 30 leagues thereof, it was his good fortune to meet with a Portuguese carrack called *San Felipe*. This carrack without any great resistance he took, bestowing the people thereof in certain vessels well furnished with victuals, and sending them courteously home into their country: and this was the first carrack that ever was taken coming forth of the East Indies; which the Portuguese took for an evil sign, because the ship bare their king's own name.

The riches of this prize seemed so great unto the whole company (as in truth it was) that they assured themselves every man to have a sufficient reward for his travail: and thereupon they all resolved to return home for England: which they happily did, and arrived in Plymouth the same summer with their whole fleet and this rich booty, to their own profit and due commendation, and to the great admiration of the whole kingdom.

And here by the way it is to be noted, that the taking of this carrack wrought two extraordinary effects in England: first, that it taught others, that carracks were no such bugs but that they might be taken, and secondly in acquainting the English nation more generally with the particularities of the exceeding riches and wealth of the East Indies: whereby themselves and their neighbours of Holland have been encouraged, being men as skilful in navigation and of no less courage than the Portuguese to share with them in the East Indies: where their strength is nothing so great as heretofore hath been supposed.

LIV

The miraculous victory achieved by the English fleet, upon the Spanish huge Armada sent in the year 1588, for the invasion of England. Recorded by Emanuel van Meteren in his History of the Low Countries.

The Spanish King having for above twenty years together, waged war against the Netherlanders, after deliberation with his councillors thereabout, thought it most convenient to assault them once again by sea. Unto the which expedition it stood him now in hand to join great puissance, as having the English people his professed enemies; whose island is so situated, that it may either greatly help or hinder all such as sail into those parts. For which cause he thought good first of all to invade England, being persuaded by divers well experienced Spaniards and Dutchmen, and by many English fugitives, that the conquest of that island was less difficult than the conquest of Holland. Moreover the Spaniards were of opinion, that it would be far more behoveful for their King to conquer England and the Low Countries all at once, than to be constrained continually to maintain a warlike navy to defend his East and West Indies fleets, from the English Drake, and from such like valiant enemies.

For the same purpose the King Catholic had given commandment long before in Italy and Spain, that a great quantity of timber should be felled for the building of ships; and had besides made great preparation: as namely in founding of brazen ordnance, in storing up of corn and victuals, in training of men to use warlike weapons, in levying and mustering of soldiers: in so much that about the beginning of the year 1588 he had finished such a mighty navy, and brought it into Lisbon haven, as never the like had before that time sailed upon the ocean sea.

A particular description of this navy was put in print and published by the Spaniards: scarce was there any family of account, or any one principal man throughout all Spain, that had not a brother, son or kinsman in that fleet: who all of them were in good hope to purchase unto themselves in that navy invincible, endless glory and renown, and to possess themselves of great seigneuries and riches in England, and in the Low Countries. All the ships appertaining to this navy amounted unto the sum of 150. The number of mariners were above 8000 of slaves 2088 of soldiers 20,000 (besides noblemen and gentlemen voluntaries) of great cast pieces 2650. The foresaid ships were of an huge and incredible capacity. The whole fleet was large enough to contain the burden of 60 thousand tons.

The galleons were 64 in number, being of an huge bigness, and very stately built, of marvellous force also, and so high, that they resembled great castles, most fit to defend themselves and to withstand any assault, but far inferior unto the English and Dutch ships, which can with great dexterity wield and turn themselves. The upper work of the said galleons was of thickness and strength sufficient to bear off musket shot. The lower work and the timbers thereof were out of measure strong, being framed of planks and ribs four or five foot in thickness, insomuch that no bullets could pierce them, but such as were discharged hard at hand: which afterward proved true, for a great number of bullets were found to stick fast within the massy substance of those thick planks.

The galliasses were rowed with great oars, there being in each one of them 300 slaves for the same purpose, and were able to do great service with the force of their ordnance. All these together with the residue aforenamed were furnished and beautified with trumpets, streamers, banners, warlike ensigns, and other such like ornaments.

Their pieces of brazen ordnance were 1600 and of iron 1000.

The bullets thereto belonging were 120 thousand.

Item of gunpowder 5600 quintals. Of match 1200 quintals.

Moreover they had great store of cannons, double cannons culverins and field pieces for land services.

They had in like sort great store of mules and horses, and whatsoever else was requisite for a land-army. They were so well stored of biscuit, that for the space of half a year, they might allow each person in the whole fleet half a quintal every month; whereof the whole sum amounteth unto an hundred thousand quintals.

Likewise of wine they had 147 thousand pipes, sufficient also for half a year's expedition. Of bacon 6500 quintals. Of cheese three thousand quintals. To be short, they brought all things expedient either for a fleet by sea, or for an army by land.

There were in the said navy five tercios of Spaniards, (which tercios the Frenchmen call regiments) under the command of five governors termed by the Spaniards, masters of the field, and amongst the rest there were many old and expert soldiers chosen out of the garrisons of Sicily, Naples and Tercera. Besides the which companies there were many bands also of Castilians and Portuguese. It was not lawful for any man, under grievous penalty, to carry any women or harlots in the fleet: for which cause the women hired certain ships, wherein they sailed after the Navy: some of which being driven by tempest arrived upon the coast of France.

The general of this mighty navy, was Don Alonso Perez de Guzman duke of Medina Sidonia, Lord of San Lucar, and Knight of the Golden Fleece. Martin Alorcon was appointed vicar general of the Inquisition, being accompanied with more than a hundred monks, to wit Jesuits, Capuchins, and friars mendicant.

While the Spaniards were furnishing this their navy, the Duke of Parma, at the direction of King Philip, made great preparation in the Low Countries, to give aid and assistance unto the Spaniards; he had assembled above a hundred small ships called hoys being well stored with victuals. In the river of Waten he caused 70 ships with flat bottoms to be built, every one of which should serve to carry 30 horses. He had provided 200 other vessels at Nieuwpoort, but not so great. And at Dunkirk he procured 28 ships of war.

Near unto Nieuwpoort he had assembled an army of 30 bands or ensigns of Italians, of ten bands of Walloons, eight of Scots, and eight of Burgundians, all which together amount unto 56 bands, every band containing a hundred persons. Pope Sixtus Quintus for the setting forth of the foresaid expedition, as they used to do against Turks and infidels, published a Cruzado, with most ample indulgences which were printed in great numbers. Some there be which affirm that the Pope had bestowed the realm of England with the title of *Defensor fidei*, upon the King of Spain, giving him charge to invade it upon this condition, that he should enjoy the conquered realm, as a vassal and tributary, in that regard, unto the see of Rome. To this purpose the said Pope proferred a million of gold, the one half thereof to be paid in ready money, and the other half when the realm of England or any famous port thereof were subdued. And for the greater furtherance of the whole business, he despatched one Allen an Englishman (whom he had made Cardinal) into the Low Countries, unto whom he committed the administration of all matters ecclesiastical throughout England.

But that all matters might be performed with greater secrecy, and that the whole expedition might seem rather to be intended against the Low Countries, than against England, there was a solemn meeting appointed in Flanders for a treaty of peace between Her Majesty and the Spanish King.

Against which treaty the United Provinces making open protestation, used all means possible to hinder it, alleging that it was more requisite to consult how the enemy now pressing upon them might be repelled from off their frontiers. Howbeit some there were in England affirming that peace might easily and upon reasonable conditions be obtained of the Spaniard. Whereupon it came to pass, that England and the United Provinces prepared indeed some defence to withstand that dreadful expedition and huge Armada, but nothing in comparison of the great danger which was to be feared.

At length the French King about the end of May signified unto Her Majesty in plain terms that she should stand upon her guard, because he was now most certainly informed, that there was an invasion imminent upon her realm. The greatest and the strongest ships of the whole navy she sent unto Plymouth under Lord Charles Howard, Lord High Admiral of England etc. Under whom the renowned knight Sir Francis Drake was appointed vice-admiral. The number of these ships was about an hundred. The lesser ships being 30 or 40 in number, were commanded to lie between Dover and Calais.

On land likewise throughout the whole realm, soldiers were mustered and trained in all places, and were committed unto the most resolute and faithful captains. And whereas it was commonly given out that the Spaniard meant to invade by the river of Thames, there was at Tilbury in Essex over-against Gravesend, a mighty army encamped, and on both sides of the river fortifications were erected. Unto the said army came in proper person the Queen's most royal Majesty. Also there were other such armies levied in England.

The principal catholic recusants (lest they should stir up any tumult in the time of the Spanish invasion) were sent to convenient places, as the Isle of Ely and Wisbech, where they were kept from endangering the state of the common wealth, and of Her Sacred Majesty who of her most gracious clemency gave express commandment, that they should be entreated with all humanity and friendship.

The provinces of Holland and Zeeland giving credit unto their intelligence out of Spain, made preparation to defend themselves: but because the Spanish ships were described unto them to be so huge, they relied partly upon the shallow and dangerous seas all along their coasts. Wherefore they stood most in doubt of the Duke of Parma his small and flat-bottomed ships. Howbeit they had all their ships of war to the number of 90 and above, in a readiness: the greater part whereof were of a small burthen, as being more meet to sail upon their rivers and shallow seas: and with these ships they besieged all the havens in Flanders, and fortified all their sea-towns with strong garrisons.

Against the Spanish fleet's arrival, they had provided 25 or 30 good ships, committing the government of them unto Admiral Lonck, whom they commanded to join himself unto the Lord Henry Seymour, lying between Dover and Calais.

The Spanish Armada set sail out of the haven of Lisbon upon the 19 of May, An. Dom. 1588 directing their course for the Bay of La Coruña, where they took in soldiers and warlike provision, this port being in Spain the nearest unto England. As they were sailing along, there arose such a mighty tempest, that the whole fleet was dispersed, so that when the Duke was returned unto his company, he could not escry above 80 ships in all, whereunto the residue by little and little joined themselves, except eight which had their masts blown overboard.

The navy receiving daily commandment from the King to hasten their journey, hoisted up sails the 11 day of July, so holding their course till the 19 of the same month, they came then unto the mouth of the English Channel. The Spanish fleet was escried by an English pinnace, Captain Thomas Fleming.

The Lord High Admiral of England being thus on the 19 of July about four of the clock in the afternoon, informed by the pinnace of Captain Fleming aforesaid, of the Spaniards' approach, with all speed and diligence possible he warped his ships, and caused his mariners and soldiers to come on board. The very next day was the Spanish fleet escried by the English, which with a southwest wind came sailing along, and passed by Plymouth. They were enjoined by their commission to anchor near unto, or about Calais, whither the Duke of Parma with his ships and all his warlike provision was to resort, and while the English and Spanish great ships were in the midst of their conflict, to pass by, and to land his soldiers upon the Downs.

The Spanish captives reported that they were determined first to have entered the river of Thames, and thereupon to have passed with small ships up to London, supposing that they might easily win that rich and flourishing city being but meanly fortified and inhabited with citizens not accustomed to the wars, who durst not withstand their first encounter, hoping moreover to find many rebels against Her Majesty and popish Catholics, or some favourers of the Scottish Queen (which was not long before most justly beheaded) who might be instruments of sedition.

Thus they passed by Plymouth, which the English ships pursuing and getting the wind of them, gave them the chase and the encounter, and so both fleets frankly exchanged their bullets.

The day following which was the 21 of July, the English ships approached within musket shot of the Spanish: at what time the Lord Charles Howard most hotly and valiantly discharged his ordnance upon the Spanish vice-admiral. The Spaniards then well perceiving the nimbleness of the English ships in discharging upon the enemy on all sides, gathered themselves close into the form of an half moon and slackened their sails, lest they should outgo any of their company. And while they were proceeding on in this manner, one of their great galliasses was so furiously battered with shot, that the whole navy was fain to come up rounder together for the safeguard thereof: whereby it came to pass that the principal galleon of Seville falling foul of another ship had her foremast broken, and by that means was not able to keep way with the Spanish fleet, neither would the said fleet stay to succour it, but left the distressed galleon behind. Sir Francis Drake was giving of chase unto five great hulks which had separated themselves from the Spanish fleet: but finding them to be Easterlings, he dismissed them. The Lord Admiral all that night following the Spanish lantern instead of the English, found himself in the morning to be in the midst of his enemy's fleet, but when he perceived it, he cleanly conveyed himself out of that great danger.

The day following, the two and twenty of July, was set on fire one of their greatest ships, being Admiral of the squadron of Guipuzcoa, being the ship of Michael de Oquendo vice admiral of the whole fleet, which contained great store of gunpowder and other warlike provision. The upper part only of this ship was burnt, and all the persons contained (except a very few) were consumed with fire. And thereupon it was taken by the English with a number of miserable burnt and scorched Spaniards.

Upon Tuesday which was the three and twenty of July, the navy being come over against Portland, the wind began to turn northerly, insomuch that the Spaniards had a fortunate and fit gale to invade the English. But the Englishmen having lesser and nimbler ships, recovered again the vantage of the wind from the Spaniards whereat the Spaniards seemed to be more incensed to fight than before. But when the English fleet had continually and without intermission from morning to night, beaten and battered them with all their shot both great and small: the Spaniards uniting themselves, gathered their whole fleet close together into a roundel, so that it was apparent that they meant not as yet to invade others, but only to defend themselves and to make haste unto the place prescribed unto them, which was near unto Dunkirk, that they might join forces with the Duke of Parma, who was determined to have proceeded secretly with his small ships under the shadow and protection of the great ones, and so had intended circumspectly to perform the whole expedition.

This was the most furious and bloody skirmish of all, in which the Lord Admiral of England continued fighting amidst his enemy's fleet.

The English navy in the meanwhile increased, whereunto out of all the havens of the realm resorted ships and men: for they all with one accord came flocking thither as unto a set field, where immortal fame was to be attained, and where faithful service to be performed unto their prince and country. The number of the English ships amounted unto a hundred: which when they were come before Dover, were increased to an hundred and thirty, being notwithstanding of no proportionable bigness to encounter with the Spaniards, except two or three and twenty of the Queen's greater ships, the mariners and soldiers whereof were esteemed to be twelve thousand.

The four and twenty of July when as the sea was calm, and no wind stirring, the fight was only between the four great galliasses and the English ships, which being rowed with oars, had great advantage of the said English ships. They were now constrained to send their men on land for a new supply of gun-powder, whereof they were in great scarcity, by reason they had so frankly spent the greater part in the former conflicts.

The same day, a council being assembled, it was decreed that the English fleet should be divided into four squadrons: the principal whereof was committed unto the Lord Admiral: the second, to Sir Francis Drake: the third, to Captain Hawkins: the fourth, to Captain Frobisher.

The five and twenty of July when the Spaniards were come over against the Isle of Wight, the Lord Admiral with great valour and dreadful thundering of shot, encountered with the Spanish Admiral being in the very midst of all his fleet. Which when the Spaniard perceived, he came forth and entered a terrible combat with the English: for they bestowed on each other the broad sides, and mutually discharged all their ordnance, being within one hundred, or an hundred and twenty yards of one another.

At length the Spaniards hoisted up their sails, and again gathered themselves up close into the form of a roundel. In the meanwhile Captain Frobisher had engaged himself into a most dangerous conflict. Whereupon the Lord Admiral coming to succour him, found that he had valiantly and discreetly behaved himself, and that he had wisely and in good time given over the fight, because that after so great a battery he had sustained no damage.

For which cause the day following, being the six and twenty of July, the Lord Admiral rewarded him with the order of knighthood, together with Mr John Hawkins and others. The same day the Lord Admiral received intelligence from Newhaven[119] in France, by certain of his pinnaces, that all things were quiet in France, and that there was no preparation of sending aid unto the Spaniards: there was a false rumour spread all about, that the Spaniards had conquered England.

The seven and twenty of July, the Spaniards about the sun setting were come over-against Dover, and rode at anchor within the sight of Calais, intending to hold on for Dunkirk, expecting there to join with the Duke of Parma his forces, without which they were able to do little or nothing.

Likewise the English fleet following up hard upon them, anchored just by them within culverin shot.

The Duke of Parma being advertised of the Spanish fleet's arrival upon the coast of England, made all the haste he could. Travelling to Dunkirk he heard the thundering ordnance of either fleet: and the same evening being come to Diksmuide, he was given to understand the hard success of the Spanish fleet.

Upon Tuesday which was the thirtieth of July, about high noon, he came to Dunkirk, when as all the Spanish fleet was now passed by: neither durst any of the ships in the mean space come forth to assist the said Spanish fleet for fear of five and thirty warlike ships of Holland and Zeeland, which there kept watch and ward.

The foresaid five and thirty ships were furnished with most cunning mariners and old expert soldiers, amongst the which were twelve hundred musketeers, whom the States had chosen out of all their garrisons, and whom they knew to have been heretofore experienced in sea-fights.

This navy was given especially in charge not to suffer any ship to come out of the haven, nor to permit any small vessels of the Spanish fleet to enter thereinto, for the greater ships were not to be feared by reason of the shallow sea in that place. Only the English fugitives being seven hundred in number under the conduct of Sir William Stanley, came in fit time to have been embarked, because they hoped to give the first assault against England.

It seemeth that the Duke of Parma and the Spaniards grounded upon a vain and presumptuous expectation, that all the ships of England and of the Low Countries would at the first sight of the Spanish and Dunkirk navy have betaken themselves to flight. Wherefore their intent and purpose was, that the Duke of Parma in his small and flat-bottomed ships, should as it were under the shadow of the Spanish fleet, convey over all his troops, and invade England; or while the English fleet were busied in fight against the Spanish, should enter upon any part of the coast, which he thought to be convenient. Which invasion (as the captives afterward confessed) the Duke of Parma thought first to have attempted by the river of Thames; upon the banks whereof having at his first arrival landed twenty or thirty thousand of his principal soldiers, he supposed that he might easily have won the city of London.

Whenas therefore the Spanish fleet rode at anchor before Calais, the Lord Admiral of England took forthwith eight of his worst and basest ships which came next to hand, and disburdening them of all things which seemed to be of any value, filled them with gun-powder, pitch, brimstone, and with other combustible and fiery matter; and charging all their ordnance with powder, bullets, and stones, he sent the said ships upon the 28 of July being Sunday, about two of the clock after midnight, with the wind and tide against the Spanish fleet: which being forsaken of the pilots and set on fire, were directly carried upon the King of Spain's navy: which fire in the dead of the night put the Spaniards into such a perplexity and horror that cutting their cables whereon their anchors were fastened, and hoisting up their sails, they betook themselves very confusedly unto the main sea.

In this sudden confusion, the principal and greatest of the four galliasses falling foul of another ship, lost her rudder: for which cause when she could not be guided any longer, she was by the force of the tide cast into a certain shoal upon the shore of Calais, where she was immediately assaulted by divers English pinnaces.

This huge and monstrous galliasse, wherein were contained three hundred slaves to lug at the oars, and four hundred soldiers, was in the space of three hours rifled in the same place; and there were found amongst divers other commodities 50,000 ducats of the Spanish king's treasure. At length the slaves were released out of their fetters.

Albeit there were many excellent and warlike ships in the English fleet, yet scarce were there 22 or 23 among them all which matched 90 of the Spanish ships in bigness, or could conveniently assault them. Wherefore the English ships using their prerogative of nimble steerage, whereby they could turn and wield themselves with the wind whichever way they listed, came often times very near upon the Spaniards, and charged them so sore, that now and then they were but a pike's length asunder: and so continually giving them one broad side after another, they discharged all their shot both great and small upon them, spending one whole day from morning till night in that violent kind of conflict, until such time as powder and bullets failed them.

The Spaniards that day sustained great loss and damage having many of their ships shot through and through, and they discharged likewise great store of ordnance against the English; who indeed sustained some hindrance, but not comparable to the Spaniard's loss; for they lost not any one ship or person of account. Albeit Sir Francis Drake's ship was pierced with shot above forty times, and his very cabin was twice shot through, and about the conclusion of the fight, the bed of a certain gentleman lying weary thereupon, was taken quite from under him with the force of a bullet. Likewise, as the Earl of Northumberland and Sir Charles Blunt were at dinner upon a time, the bullet of a demi-culverin broke through the midst of their cabin, touched their feet, and struck down two of the standers by, with many such accidents befalling the English ships, which it were tedious to rehearse. Whereupon it is most apparent, that God miraculously preserved the English nation.

The same night two Portuguese galleons of the burthen of seven or eight hundred tons apiece, to wit the *Saint Philip* and the *Saint Matthew*, were forsaken of the Spanish fleet, for they were so torn with shot, that the water entered into them on all sides.

The 29 of July the Spanish fleet being encountered by the English lying close together under their fighting sails, with a southwest wind sailed past Dunkirk, the English ships still following the chase. The Lord Admiral of England despatched the Lord Henry Seymour with his squadron of small ships unto the coast of Flanders, where, with the help of the Dutch ships, he might stop the Prince of Parma his passage, if perhaps he should attempt to issue forth with his army. And he himself in the mean space pursued the Spanish fleet until the second of August, because he thought they had set sail for Scotland. And albeit he followed them very near, yet did he not assault them any more, for want of powder and bullets. But upon the fourth of August, the wind arising, when as the Spaniards had spread all their sails, betaking themselves wholly to flight, and leaving Scotland on the left hand, trended toward Norway. The English seeing that they were now proceeded unto the latitude of 57 degrees, and being unwilling to participate that danger whereinto the Spaniards plunged themselves, and because they wanted things necessary, and especially powder and shot, returned back for England; leaving behind them certain pinnaces only, which they enjoined to follow the Spaniards aloof, and to observe their course. And so it came to pass that the fourth of August, with great danger and industry, the English arrived at Harwich: for they had been tossed up and down with a mighty tempest for the space of two or three days together, which it is likely did great hurt unto the Spanish fleet, being so maimed and battered. The English now going on shore, provided themselves forthwith of victuals, gunpowder, and other things expedient, that they might be ready at all assays to entertain the Spanish fleet, if it chanced any more to return. But being afterward more certainly informed of the Spaniards' course, they thought it best to leave them unto those boisterous and uncouth northern seas.

The Spaniards seeing now that they wanted four or five thousand of their people and having divers maimed and sick persons, and likewise having lost 10 or 12 of their principal ships, they consulted among themselves, what they were best to do, being now escaped out of the hands of the English. They thought it good at length, so soon as the wind should serve them, to fetch a compass about Scotland and Ireland, and so to return for Spain.

They well understood, that commandment was given throughout all Scotland, that they should not have any succour or assistance there. Neither yet could they in Norway supply their wants. Fearing also lest their fresh water should fail them, they cast all their horses and mules overboard: and so touching nowhere upon the coast of Scotland, but being carried with a fresh gale between the Orkneys and Fair Isle, they proceeded far north, even unto 61 degrees of latitude, being distant from any land at the least 40 leagues. Here the Duke of Medina general of the fleet commanded all his followers to shape their course for Biscay: and he himself with twenty or five and twenty of his ships which were best provided of fresh water and other necessaries, holding on his course over the main ocean, returned safely home.

There arrived at Newhaven in Normandy, being by the tempest enforced to do so, one of the four great galliasses, where they found the ships with the Spanish women which followed the fleet at their setting forth. Two ships also were cast away upon the coast of Norway, one of them being of a great burthen; howbeit all the persons in the said great ship were saved: insomuch that of 134 ships, which set sail out of Portugal, there returned home 53 only small and great: namely of the four galliasses but one, and but one of the four galleys. Of the 91 great galleons and hulks there were missing 58, and 33 returned. Of 30,000 persons which went in this expedition, there perished (according to the number and proportion of the ships) the greater and better part; and many of them which came home, by reason of the toils and inconveniences which they sustained in this voyage, died not long after their arrival.

Likewise upon the Scottish western isles of Lewis, and Islay, and about Cape Kintyre upon the main land, there were cast away certain Spanish ships, out of which were saved divers captains and gentlemen, and almost four hundred soldiers, who for the most part, after their shipwreck, were brought into Edinburgh in Scotland, and being miserably needy and naked, were there clothed at the liberality of the King and the merchants, and afterward were secretly shipped for Spain; but the Scottish fleet wherein they passed touching at Yarmouth on the coast of Norfolk, were there stayed for a time until the Council's pleasure was known; who in regard of their manifold miseries, though they were enemies, winked at their passage.

Upon the Irish coast many of their noblemen and gentlemen were drowned; and divers slain by the barbarous and wild Irish. To conclude, there was no famous nor worthy family in all Spain, which in this expedition lost not a son, a brother, or a kinsman.

For the perpetual memory of this matter, the Zeelanders caused new coin of silver and brass to be stamped: which on the one side contained the arms of Zeeland, with this inscription: GLORY TO GOD ONLY: and on the other side, the pictures of certain great ships, with these words: THE SPANISH FLEET: and in the circumference about the ships: IT CAME, WENT, AND WAS. Anno 1588.

A while after the Spanish fleet was departed, there was in England, by the commandment of her Majesty, and in the United Provinces, by the direction of the States, a solemn festival day publicly appointed, wherein all persons were enjoined to resort unto the Church, and there to render thanks and praises unto God. The solemnity was observed upon the 29 of November; which day was wholly spent in fasting, prayer, and giving of thanks.

The Queen's Majesty herself, rode into London in triumph, in regard of her own and her subjects' glorious deliverance. For being attended upon very solemnly by all the principal estates and officers of her realm, she was carried through her said City of London in a triumphant chariot, and in robes of triumph, from her palace unto the cathedral church of Saint Paul, out of the which the ensigns and colours of the vanquished Spaniards hung displayed.

Thus the magnificent, huge, and mighty fleet of the Spaniards (which themselves termed in all places invincible) such as sailed not upon the ocean sea many hundred years before, in the year 1588 vanished into smoke.

LV

A petition made by certain of the company of the Delight *of Bristol unto the master of the said ship Robert Burnet, in the Straits of Magellan the 12th of February 1589.*

We have thought good to show unto you (being our master)[120] our whole minds and griefs in writing: that whereas our captain Matthew Hawlse, and Walter Street do begin to take into the captain's cabin this 12 of February both bread and butter, (such as was put in for the provision of the ship and company) only to feed themselves, and a few others, which are of their mess: meaning thereby rather to starve us, than to keep us strong and in health: and likewise he hath taken into his cabin, swords, calivers, and muskets: we therefore not well knowing their intents, we may conjecture, that your death, which God forbid, by them hath been determined: being our master, and having charge of the ship, consider: first that by God's visitation we have lost 16 men, so much the rather because they were not allotted such necessary provision, as was in the ship to be had. Also to consider the great loss of 15 of our men with our boat at Penguin Island within the Straits of Magellan: and of 7 good and serviceable men besides near Port Famine: and of three anchors, and our carpenter. Over and besides all these calamities to consider how you have (without all reason and conscience) been overthwarted, disgraced, and out-countenanced by your mate Street, and Matthew Hawlse: also what danger you now are subject unto, your death having been so often conspired, and what danger we should be in, if it were (which God forbid) effected. Furthermore, to weigh with yourself the great want of many necessaries in our ship: namely that we have but 6 sailors, (besides yourself and your mate Street, whom we dare not trust,) also that we have but one anchor, likewise the lack of our boat and a carpenter, of ropes, of pitch, bolts, and planks, and the want of a skilful surgeon. But five months victuals of bread, meal, groats, and pease, and also but three months victuals of beef, penguins and pork, three hogsheads of wine, ten gallons of aquavitae (whereof the sick men could not get any to relieve them) four hogsheads of cider and 18 flitches of bacon etc. the [ship's] company hath but three flitches. Also Captain Hawlse and Street have taken and seized upon 17 pots of butter, with certain cheese, and an hogshead of bread possessed to their own private uses: and have not only immoderately spent the company's provision in butter, cheese, aquavitae etc. but have also consumed those sweet meats, which were laid up in the ship only for the relief of sick persons (themselves being healthy and sound, and withholding the said meats from others in their sickness) and even at this time also (by reason of the small store of our provision, we being enforced to come to a shorter allowance) Captain Hawlse and your mate Street, do find themselves aggrieved at the very same allowance, wherewith other men are well contented. These things being well weighed, you ought likewise to consider the long time that we have lain here in these Straits of Magellan, having been seven or eight times, ten leagues beyond Cape Froward, we have had but a small gale of wind with us: neither could we come to an anchor, the water being so deep: and (you know) the place is so dangerous, that we were once embayed, and could scarce get out again: and likewise, what fogs and mists are here already? Much more here will be, the winter and dark nights being at hand, and we having not so much as a boat to seek out any road to ride in, saving a small weak boat made of men's chests, in which it is not convenient to go on shore in a foreign country, where we must go with force: and having but one anchor left us, there is but little hope of life in us, as you may sufficiently judge, if we should lose either the said anchor or our boat. We having lain here these six weeks and upward, the wind hath continued in the north-west directly against our course, so that we can no way hope to get through the Straits into the South Sea this year, and if we could, yet our provision is not sufficient, having spent so much thereof, in this our lingering abode. Nay we have scarcely victuals enough to carry us home into England, if they be not used sparingly. Therefore we do again most humbly desire you to consider, as you tender your own safety and the safety of us which remain alive, that we may (by God's help) return back into England, rather than die here among wild and savage people: for if we make any longer abode in this place, it will be (without all doubt) to the utter decay and loss, both of ourselves, and of the ship: and in returning back, it may please God, that we may find our fifteen men, and our boat at Penguin Island (although this be contrary to the minds of Matthew Hawlse, and your mate Street) we do not despair in God's mercy, but that in our return homeward, He will send us purchase sufficient, if we would join ourselves together in prayer, and love one another.

Lastly, we do most humbly beseech you to consider, that (after the loss of so many men, as we were taking in of water by Port Famine, our boat-swain, the hooper, and William Magoths being on shore) Matthew Hawlse did halloo to have them in all haste come on-board: saying therewithall these words: he that will come in this voyage, must not make any reckoning to leave two or three men on shore behind him, whereas we had so lately lost all the aforesaid men, having then but six sailors left us on-board. Also Matthew Hawlse did carry a pistol for the space of two days secretly under his gown, intending therewithall to have murdered Andrew Stoning, and William Combe, for William Martin reported unto two of his friends, Richard Hungate, and Emanuel Dornel, that he kneeled upon his knees one whole hour before Matthew Hawlse in his own cabin, desiring him, for God's cause, not to kill either of them, especially because Stoning and Martin came both out of one town.

And thus we end, desiring God to send us well into our native country. In witness whereof we have suscribed our names.

THOMAS BROWN, gunner,
JOHN MORRIS, [etc.]

LVI

An excellent treatise of the kingdom of China, printed in Latin at Macao a city of the Portuguese in China, An. Dom. 1590.

This kingdom of China is situate most easterly: albeit certain islands, as Japan, stand more easterly than China itself. As touching the limits and bounds of this kingdom, we may appoint the first towards the west to be a certain isle commonly called Hainan, which standeth in 19 degrees of northerly latitude. The farthest Chinese inhabitants that way do behold the North Pole elevated, at least 50 degrees: whereupon a man may easily conjecture how large the latitude of this kingdom is, whenas it containeth more than 540 leagues in direct extension towards the north. Certain it is, that according to the map wherein the people of China describe the form of their kingdom, the latitude thereof does not much exceed the longitude.

Almost no lord or potentate in China hath authority to levy unto himself any peculiar revenues, or to collect any rents within the precincts of his seignories, all such power belonging only to the King: whereas in Europe the contrary is most commonly seen. The Kings of China, by reason of the manifold and cruel wars moved by the Tartars, were constrained to defix their princely seat and habitation in that extreme province of the north. Whereupon it cometh to pass, that those northern confines of the kingdom do abound in many more fortresses, martial engines, and garrisons of soldiers.

The number of the greater cities throughout the whole kingdom is more than 150, and there is a greater multitude of inferior cities. Of walled towns not endued with the privileges of cities there are more than 1120: the villages and garrisons can scarce be numbered: it is not easy to find any place void of inhabitants in all that land.

Scarcely in any other realm are so many found that live unto decrepit and extreme old age. Amongst them they have no phlebotomy or letting of blood: but all their cures are achieved by fasting, decoctions of herbs, and light or gentle potions. In fruitfulness of soil this kingdom certes does excel, far surpassing all other kingdoms of the east. This kingdom is most large and full of navigable rivers, so that commodities may easily be conveyed out of one province into another. This region affordeth especially sundry kinds of metals, of which the chief is gold, of which so many pesos are brought from China to India that I have heard say that in one and the same ship, this present year, 2000 such pieces consisting of massy gold, as the Portuguese commonly call golden loaves, were brought unto us for merchandise: and one of these loaves is worth almost 100 ducats. Neither are these golden loaves only bought by the Portuguese, but also great plenty of gold-twine and leaves of gold: for the Chinese can very cunningly beat and extenuate gold into plates and leaves. There is also great store of silver. What should I speak of their iron, copper, lead, tin, and other metals, and also of their quick-silver? Of all which in the realm of China there is great abundance. But now let us proceed unto the silk, whereof there is great plenty in China: the women do employ a great part of their time in preserving of silkworms, and in combing and weaving of silk. Every year the King and Queen with great solemnity come forth into a public place, the one of them touching a plough, and the other a mulberry tree, with the leaves whereof silkworms are nourished: by this ceremony encouraging both men and women in their vocation, whereas otherwise, all the whole year throughout, no man besides the principal magistrates, may once attain to the sight of the King. Of this silk there is such abundance, that three ships for the most part coming out of India to the port of Macao, are laden especially with this freight, and carried even unto Portugal, also sundry stuffs woven thereof, for the Chinese do greatly excel in the art of weaving.

The kingdom of China aboundeth with most costly spices and odours, and especially with cinnamon, with camphire also, and musk. Musk deriveth his name from a beast (which beast resembleth a beaver) from the parts whereof bruised and putrified proceedeth a most delicate and fragrant smell. But who would believe that there is so much cotton-wool in China; whereof such variety of clothes are made like unto linen?

Let us now entreat of that earthen or pliable matter commonly called porcelain, which is pure white, and is to be esteemed the best stuff of that kind in the whole world: whereof vessels of all kinds are very curiously framed. I say, it is the best earthen matter in all the world, for three qualities; namely, the cleanness, the beauty, and the strength thereof.

This nation is endued with excellent wit and dexterity for the attaining of all arts, and, being very constant in their own customs, they lightly regard the customs or fashions of other people. They use one and the same kind of vesture, yet so, that there is some distinction between the apparel of the magistrate, and of the common subject. In old time one language was common to all the provinces: notwithstanding by reason of variety of pronunciation, it is very much altered, and is divided into sundry idioms according to the divers provinces: howbeit, among the magistrates, and in public assemblies of judgement, there is one language used throughout the whole realm.

There are very many painters, using either the pencil or the needle (of which the last sort are called embroiderers) and others also that curiously work gold-twine either upon cloth either of linen or of cotton: whose operations of all kinds are diligently conveyed by the Portuguese into India. Their industry doth no less appear in founding of guns and in making of gun-powder, whereof are made many rare and artificial fireworks. To these may be added the art of printing, albeit their letters be in manner infinite and most difficult, the portraitures whereof they cut in wood or in brass, and with marvellous facility they daily publish huge multitudes of books.

Let us now come unto that art, which the Chinese do most of all profess, and which we may, not unfitly, call literature or learning. In all cities and towns, yea, and in petty villages also, there are certain school-masters hired for stipends to instruct children. And in each city or walled town there is a public house called the school, and unto that all they do resort from all private and petty-schools that are minded to obtain the first degree; where they do amplify a sentence or theme propounded unto them by some magistrate.

All those books are fraught with precepts, wherein such grave and pithy sentences are set down, that, in men void of the light of the Gospel, more can not be desired. Among the five virtues, which the Chinese principally regard, urbanity or courtesy is one; the rest are piety, a thankful remembrance of benefits, true dealing in contracts or bargains, and wisdom in achieving of matters: with the praises and commendations of which virtues the Chinese books are full fraught. The kingdom of China hath hitherto been destitute of true religion, being distracted into sundry opinions, and following manifold sects. The first is of them that profess the doctrine of one Confucius a notable philosopher. This man (as it is reported in the history of his life) was one of most upright and incorrupt manners, whereof he wrote sundry treatises very pithily and largely. The sum of the doctrine is, that men should follow the light of nature as their guide, and that they should diligently endeavour to attain unto the virtues by me before mentioned: and lastly, that they should employ their labour about the orderly government of their families and of the Commonwealth.

LVII

The voyage of Richard Rainolds and Thomas Dassel to the rivers of Senegal and Gambia adjoining upon Guinea, 1591.

By virtue of her Majesty's most gracious charter given in the year 1588, certain English merchants are granted to trade, in and from the river of Senegal to and in the river of Gambia, on the Western coast of Africa. The chiefest places of traffic on that coast are these:

1. Senegal river: hides, gum, elephant's teeth, a few grains, ostrich feathers, ambergris, and some gold.

2. Beseguiache.

3. Rufisque.

4. Palmerin; small hides, and a few elephant's teeth now and then.

5. Porto d'Ally, a town 5 leagues from Palmerin: small hides, teeth, ambergris, and a little gold: and many Portuguese are there.

6. Joal, a town 6 leagues from Palmerin: hides, wax, elephant's teeth, rice, and some gold: and many Spaniards and Portuguese are there.

7. Gambia river: rice, wax, hides, elephant's teeth, and gold.

The Frenchmen of Dieppe and Newhaven have traded thither above thirty years. Where in all places generally they were well beloved. And very often the negroes come into France and return again, which is a further increasing of mutual love and amity. The Frenchmen never use to go into the river of Gambia: which is a river of secret trade and riches concealed by the Portuguese. For long since one Frenchman entered the river with a small bark which was betrayed, surprised, and taken by two galleys of the Portuguese.

In our second voyage there were by vile treacherous means of the Portuguese and the king of the negroes about forty Englishmen cruelly slain and captured, and most or all of their goods confiscated.

LVIII

A memorable fight made against certain Spanish ships and galleys in the West Indies, by three ships of the honourable Sir George Carey, Knight, then marshall of Her Majesty's household, now Lord Hunsdon, Lord Chamberlain.

The 13 of June 1591 being Sunday, at five of the clock in the morning we descried six sail of the King of Spain his ships. Four of them were of 700 tons apiece, and the other two of 600 apiece and the other two were small ships. We met with them off the Cape de Corrientes, which standeth on the island of Cuba. The sight of the foresaid ships made us joyful, hoping that they should make our voyage. But as soon as they descried us, they made false fires one to another and gathered their fleet together, lying all close by a wind to the southwards.

We therefore having made our prayers to Almighty God, prepared ourselves for the fight. We in the *Content* bare up with their viceadmiral, and (ranging along by his broad side aweather of him) gave him a volley of muskets and of our great ordnance: then coming up with another small ship ahead of the former, we hailed her in such sort, that she paid room. Thus being in fight with the little ship, we saw a great smoke come from our admiral, and the *Hopewell* and *Swallow* forsaking him with all the sails they could make. Thus we were forced to stand to the northwards, the *Hopewell* and the *Swallow* not coming in all this while to aid us, as they might easily have done.

All this time we were forced to the northwards with two of their great ships and one of their small. Both their great ships came up fair by us, shot at us, and on the sudden furled their spritsails and mainsails, thinking that we could not escape them. Then falling to prayer, we shipped our oars that we might row to shore, and anchor in shallow water where their great ships could not come nigh us.

Then one of their small ships being manned from one of their great, and having a boat to row themselves in, shipped her oars likewise and rowed after us, thinking with their small shot to have put us from our oars, until the great ships might come up with us: but by the time she was within musket shot, the Lord of His mercy did send us a fair gale of wind at the northwest off the shore.

The other great ship got under our lee, and the small ship on our weather quarter, purposing to make us pay room with the great ship, by force of her small and great shot. Then we by a fortunate shot which our gunner made, pierced her betwixt wind and water. Afterward (commending ourselves to Almighty God in prayer, and giving Him thanks for the wind which He had sent us for our deliverance) we looked forth and descried two sails more to the offing: these we thought to have been the *Hopewell*, and the *Swallow* that had stood in to aid us: but it proved far otherwise, for they were two of the King's galleys.

We shipped our oars, and rowed off the shore: and our watch was no sooner set, but we espied one galley under our lee hard by us, budging up with us. Then (because it was evening) one of the great ships discharged six great shot at us, to the end the galleys should know that we were the ship they looked for. Then the galley came up, and (hailing us of whence our ship was) a Portuguese which we had with us made them answer that we were of the fleet of Tierra Firma, and of Seville: with that they bid us amain English dogs, and came upon our quarter starboard: and giving us five cast pieces out of her prow, they sought to lay us aboard: but we so galled them with our muskets, that we put them from our quarter.

When they began to approach, we heaved into them a ball of fire, and by that means put them off. We went to prayer, and sang the first part of the 25 Psalm, praising God for our safe deliverance. This being done, we might see two galleys and a frigate all three of them bending themselves together to encounter us: hereupon we armed ourselves, and resolved (for the honour of God, Her

Majesty, and our country) to fight it out till the last man. Then shaking a pike of fire in defiance of the enemy, and waving them amain, we bade them come aboard. Every moment expecting the assault, we heard them parley to this effect, that they determined to keep us company till the morning, and then to make an end with us: then giving us another shot from one of the galleys, they fell astern. Thus our fight continued with the ships and with the galleys, from seven of the clock in the morning till eleven at night. Howbeit God (which never faileth them that put their trust in Him) sent us a gale of wind about two of the clock in the morning at eastnortheast, which was for the preventing of their cruelty, and the saving of our lives.

The next day we saw all our adversaries to lee-ward of us, and they espying us, chased us till ten of the clock, and then seeing they could nor prevail, gave us over. We bare up to the southwest, in hope to find our consorts, but we had no sight of them at that time, or afterward.

The bark called *The Content* had but one minion, one falcon, one saker and two port-bases. She continued fight (from seven in the morning till sunset) with three armadas of 600 and 700 tons apiece, and one small ship of 100 tons, not being above musket shot from any of them. And before the sun was set, there came up to her two of the King's galleys. The armadas shot their great ordnance continually at her, not so few as 500 times. And the sides, hull, and masts of the *Content* were sowed thick with musket bullets. Only 13 men continued this fight, the rest being in the hold.

LIX

The last voyage of Thomas Candish, intended for the South Sea, the Philippines, and the coast of China: written by Mr John Jane, a man of good observation.

The 26 of August 1591, we departed from Plymouth with 3 tall ships, and two barks, the *Galleon* wherein Mr Candish went himself, the *Roebuck* whereof Mr Cocke was captain, the *Desire* whereof was captain Mr John Davis (with whom and for whose sake I went this voyage).

The 29 of November we fell upon the coast of Brazil: at which time we took a small bark bound for the River of Plate with sugar, haberdash wares, and negroes. The 15 at evening we anchored at the bar of Santos; and the next morning about nine of the clock we came to Santos, where being discovered, we were enforced to land with 24 gentlemen, our long boat being far astern, by which expedition we took all the people of the town at mass both men and women, whom we kept all that day in the church as prisoners. Master Candish desired to take this town to supply his great wants. But such was the negligence of our governor Master Cocke, that the Indians were suffered to carry out of the town whatsoever they would in open view, and no man did control them: in three days the town that was able to furnish such another fleet with all kinds of necessaries, was left unto us nakedly bare, without people and provisions.

Eight or ten days after Master Candish himself came thither, where he remained until the 22 of January, seeking by entreaty to have that, whereof we were once possessed. But in conclusion we departed out of the town through extreme want of victual, not being able any longer to live there, and were glad to receive a few canisters or baskets of cassava meal; so that in every condition we went worse furnished from the town, than when we came into it. The 22 of January we departed from Santos, and burnt São Vicente to the ground. The 24 we set sail, shaping our course for the Straits of Magellan.

The seventh of February we had a very great storm, and the eighth our fleet was separated by the fury of the tempest. Then our captain concluded to go for Port Desire, which is in the southerly latitude of 48 degrees; hoping that the general would come thither, because that in his first voyage he had found great relief there. For our captain could never get any direction what course to take in any such extremities, though many times he had entreated for it, as often I have heard him with grief report. In sailing to this port by good chance we met with the *Roebuck*, and so we both arrived at Port Desire the sixth of March.

The 18 the *Galleon* came into the road, and Master Candish came into the harbour in a boat which he had made at sea; for his long boat and light horseman were lost at sea, as also a pinnace which he had built at Santos: and being aboard the *Desire* he told our captain of all his extremities, and spoke most hardly of his company, and of divers gentlemen that were with him. We all sorrowed to hear such hard speeches of our good friends; but having spoken with the gentlemen of the *Galleon* we found them faithful, honest, and resolute in proceeding, although it pleased our general otherwise to conceive of them.

The eighth of April 1592, we fell with the Straits of Magellan, enduring many furious storms between Port Desire and the Strait. The 18 we doubled Cape Froward. The 21 we were enforced by the fury of the weather to put into a small cove with our ships, 4 leagues from the said cape, upon the south shore, where we remained until the 15 of May. In the which time we endured extreme storms, with perpetual snow, where many of our men died with cursed famine, and miserable cold, not having wherewith to cover their bodies, nor to fill their bellies, but living by mussels, water, and weeds of the sea, with a small relief of the ship's store in meal sometimes. And all the sick men in the *Galleon* were most uncharitably put ashore into the woods in the snow, rain, and cold, when men of good health could scarcely endure it, where they ended their lives in the highest degree of misery, Master Candish all this while being aboard the *Desire*. In these great extremities of snow and cold, doubting what the end would be, he asked our captain's opinion, because he was a man that had good experience of the northwest parts, in his 3 several discoveries that way, employed by the merchants of London. Our captain told him, that this snow was a matter of no long continuance. Notwithstanding he called together all the company, and told them, that he purposed not to stay in the Straits, but to depart upon some other voyage, or else to return again for Brazil. But his resolution was to go for the Cape of Buena Esperanza. The company answered that if it pleased him, they did desire to stay God's favour for a wind, and to endure all hardness whatsoever, rather than to give over the voyage, considering they had been here but a small time, and because they were within forty leagues of the South Sea, it grieved them now to return. Then our captain, after Master Candish was come aboard the *Desire* from talking with the company, told him, that if it pleased him to consider the slenderness of his provisions, with the weakness of his men, it was no course for him to proceed in that new enterprise: for if the rest of your ships (said he) be furnished answerable to this, it is impossible to perform your determination: for we have no more sails than masts, no victuals, no ground-tackling, no cordage more than is over head, and among seventy and five persons, there is but the master alone that can order the ship, and but fourteen sailors. The rest are gentlemen, serving men, and artificers. Therefore it will be a desperate case to take so hard an enterprise in hand. These persuasions did our captain not only use to Master Candish, but also to Master Cocke. In fine upon a petition delivered in writing by the chief of the whole company, the general determined to depart out of the Straits of Magellan, and to return again for Santos in Brazil.

So the 15 of May we set sail, the general then being in the *Galleon*. The eighteenth we were free of the Straits, but at Cape Froward it was our hard hap to have our boat sunk at our stern in the night.

The sixth and twentieth day of May we came to Port Desire, where not finding our general, as we hoped, being most slenderly victualled, without sails, boat, oars, nails, cordage, and all other necessaries for our relief, we were stricken into a deadly sorrow. But referring all to the providence and fatherly protection of the Almighty, we entered the harbour, and by God's favour found a place of quiet road, which before we knew not. Having moored our ship with the pinnace's boat, we landed upon the south shore, where we found a standing pool of fresh water, which by estimation might hold some ten tons, whereby we were greatly comforted. Because at our first being in this harbour we were at this place and found no water, we persuaded ourselves that God had sent it for our relief. Also there were such extraordinary low ebbs as we had never seen, whereby we got mussels in great plenty. Likewise God sent about our ships great abundance of smelts, so that with hooks made of pins every man caught as many as he could eat: by which means we preserved our ship's victuals, and spent not any during the time of our abode here.

Our captain and master falling into consideration of our estate, found our wants so great, as that in a month we could not fit our ship to set sail. For we must needs set up a smith's forge, to make bolts, spikes, and nails. Whereupon they concluded it to be their best course to take the pinnace, and to furnish her with the best of the company, leaving the ship and the rest of the company until the general's return; for he had vowed to our captain, that he would return again for the Straits.

The general having in our ship two most pestilent fellows, when they heard of this determination they utterly misliked it, and in secret dealt with the company of both ships, vehemently persuading them, that our captain and master would leave them in the country to be devoured of the cannibals, and that they were merciless and without charity: whereupon the whole company joined in secret with them in a night to murder our captain and master, with myself, and all those which they thought were their friends. There were marks taken in his cabin how to kill him with muskets through the ship's side, and bullets made of silver for the execution, if their other purposes should fail. All agreed hereunto, except it were the boatswain, who when he knew the matter, and the slender ground thereof, revealed it unto our master, and so to the captain. Then the matter being called in question, those two most murderous fellows were found out, whose names were Charles Parker and Edward Smith.

The captain being thus hardly beset in peril of famine, and in danger of murdering, was constrained by courteous means to pacify this fury: showing, that to do the general service, unto whom he had vowed faith in this action, was the cause why he purposed to go unto him in the pinnace, considering, that the pinnace was so necessary a thing for him, as that he could not be without her, because he was fearful of the shore in so great ships. Whereupon all cried out with cursing and swearing, that the pinnace should not go unless the ship went. Then the captain desired them to show themselves Christians, and not so blasphemously to behave themselves. By which gentle speeches the matter was pacified, and the captain and master at the request of the company were content to forgive this great treachery of Parker and Smith, who after many admonitions concluded in these words: the Lord judge between you and me: which after came to a most sharp revenge even by the punishment of the Almighty. Thus by a general consent it was concluded not to depart, but there to stay for the general's return. Then our captain and master, made a motion to the company, that they would lay down under their hands the losing of the general, with the extremities wherein we then stood: as followeth.

The testimonial of the company of the DESIRE *touching their losing of their general.*

The 26 of August 1591 we whose names be here under written, with divers others departed from Plymouth under Mr Thomas Candish our general. The 16 of December we took the town of Santos, hoping there to revictual ourselves, but it fell not out to our contentment. The 24 of January we set sail from Santos, shaping our course for the Straits of Magellan. In which time we had great store of snow, with some gusty weather. We were enforced for the preserving of our victuals, to live the most part upon mussels, our provision was so slender; so that many of our men died in this hard extremity. Then our general returned for Brazil there to winter, and to procure victuals against the next year. So we departed the straits the 15 of May. The whole fleet following the admiral, our ship coming under his lee shot ahead of him. This night we were severed, by what occasion we protest we know not, whether we lost them or they us. We had a violent storm, with the wind at northwest, and we were enforced to hull, not being able to bear sail, and this night we perished our main trestle-trees,[121] so that we could no more use our main top-sail, lying most dangerously in the sea. The pinnace likewise received a great leak, so that we were enforced to seek the next shore for our relief. And because famine was like to be the best end, we desired to go for Port Desire, hoping with seals and penguins to relieve ourselves. The 26 our fore-shrouds broke, so that if we had not been near the shore, it had been impossible for us to get out of the sea. And now being here moored in Port Desire, our shrouds are all rotten, not having a running rope whereto we may trust, sails all worn, our top-sails not able to abide any stress of weather, neither have we any pitch, tar, or nails, nor any store for the supplying of these wants; and we live only upon seals and mussels, having but five hogsheads of pork within board, and meal three ounces for a man a day, with water for to drink. And forasmuch as it hath pleased God to separate our fleet, and to bring us into such hard extremities, yet because the wonderful works of God in His exceeding great favour towards us His creatures are far beyond the scope of man's capacity, therefore by him we hope to have deliverance in this our deep distress. Also forasmuch as those upon whom God will bestow the favour of life, with return home to their country, may not only themselves remain blameless, but also manifest the truth of our actions, we have thought good in Christian charity to lay down under our hands the truth of all our proceedings even till the time of this our distress.

Given in Port Desire the 2 of June 1592. Beseeching the Almighty God of His mercy to deliver us from this misery.

John Davis captain.
Randolph Cotton.
John Pery.
William Maber gunner.
Charles Parker.
Roland Miller.
Edward Smith. [etc.]

After they had delivered this relation unto our captain under their hands, then we began to travail for our lives, and we built up a smith's forge, and made a coal pit, and burned coals, and there we made nails, bolts, and spikes, others made ropes of a piece of our cable, and the rest gathered mussels, and took smelts for the whole company. Three leagues from this harbour there is an isle with four small isles about it, where there are great abundance of seals, and at the time of the year the penguins come thither in great plenty to breed. We concluded with the pinnace, that she should sometimes go thither to fetch seals for us; upon which condition we would share our victuals with her man for man; whereunto the whole company agreed. So we parted our poor store, and she laboured to fetch us seals to eat, wherewith we lived when smelts and mussels failed: for in the neap streams we could get no mussels. Thus in most miserable calamity we remained until the sixth of August, still keeping watch upon the hills to look for our general, and so great was our vexation and anguish of soul, as I think never flesh and blood endured more. Our captain and master were fully persuaded, that the general might perhaps go directly for the Straits, and not come to this harbour: whereunto the company most willingly consented, as also the captain and master of the pinnace; so that upon this determination we made all possible speed to depart.

The sixth of August we salted twenty hogsheads of seals, which was as much as our salt could possibly do, and so we departed for the Straits the poorest wretches that ever were created. We had full confidence to meet with our general. The ninth we had a sore storm, so that we were constrained to hull, for our sails were not to endure any force. The 21 we doubled Cape Froward. The 22 we anchored in Savage Cove, so named, because we found many savages there: notwithstanding the extreme cold of this place, yet do all these wild people go naked, and live in the woods, painted and disguised, and fly from you like wild deer. They are very strong, and threw stones at us of three or four pound weight an incredible distance. The 25 we anchored in a good cove, within fourteen leagues of the South Sea: in this place we purposed to stay for the general, for the strait in this place is scarce three miles broad, so that he could not pass but we must see him. After we had stayed here a fortnight in the deep of winter, our victuals consuming, (for our seals stank most vilely, and our men died pitifully through cold and famine, for the greatest part of them had not clothes to defend the extremity of the winter's cold) being in this heavy distress, our captain and master thought it best to depart from the Straits into the South Sea, and to go for the isle of Santa Maria, where we might have relief, and be in a temperate climate, and there stay for the general, for of necessity he must come by that isle. So we departed and came in sight of the South Sea. We were forced back again, again we put forth, and being 8 or 10 leagues free of the land, the wind rising furiously at westnorthwest, we were enforced again into the Straits only for want of sails; for we never durst bear sail in any stress of weather, they were so weak: we endured most furious weather, so that one of our two cables broke, whereby we were hopeless of life. Yet it pleased God to calm the storm, and we unrived our sheets, tacks, halyards, and other ropes, and moored our ship to the trees close by the rocks. We laboured to recover our anchor again, but could not by any means, it lay so deep in the water, and as we think clean covered with ooze. Now had we but one anchor which had but one whole fluke, a cable spliced in two places, and a piece of an old cable. In the midst of these our troubles it pleased God that the wind came fair the first of October; whereupon with all expedition we loosed our moorings, and weighed our anchor, and so towed off into the channel; for we had mended our boat in Port Desire, and had five oars of the pinnace. When we had weighed our anchor, we found our cable broken, only one strand held: in the channel, we rived our ropes, and again rigged our ship, no man's hand was idle, but all laboured even for the last gasp of life. Here our company was divided; some desired to go again for Port Desire, and there to be set on shore, and some stood with the captain and master to proceed. Whereupon the captain said to the master: you see the wonderful extremity of our estate, and the great doubts among our company of the truth of your reports, as touching relief to be had in the South Sea: some say in secret, as I am informed, that we undertake these desperate attempts through blind affection that we bear to the general. And because I see in reason, that the limits of our time are now drawing to an end, I do in Christian charity entreat you all, first to forgive me in whatsoever I have been grievous unto you; secondly that you will rather pray for our general, than use hard speeches of him; lastly, let us forgive one another and be reconciled as children in love and charity, and not think upon the vanities of this life: so shall we in leaving this life live with our glorious Redeemer, or abiding in this life, find favour with God. And now (good master) forasmuch as you have been in this voyage once before, satisfy the company of such truths, as are to you best known. Then the master began: Captain your request is very reasonable, and now if you think good to return, I will not gainsay it: but this I assure you, if life may be preserved by any means, it is in proceeding. For at the isle of Santa Maria I do assure you of wheat, pork, and roots enough. Also I will bring you to an isle, where pelicans be in great abundance, besides all our possibility of intercepting some ships upon the coast of Chile and Peru. But if we return there is nothing but death to be hoped for: therefore do as you like, I am ready, but my desire is to proceed. There was a general consent of proceeding; and so the second of October we out into the South Sea, and were free of all land. This night the wind began to blow very much, and still increased in fury, so that we were in great doubt what course to take: to put into the Straits we durst not for lack of ground-tackle: to bear sails we doubted, the tempest was so furious, and our sails so bad. We stood under our courses in view of the lee-shore, still expecting our ruinous end.

The storm growing beyond all reason furious, the pinnace being in the wind of us, struck suddenly ahull, so that we thought she had received some grievous sea, or sprung a leak. This night we lost the pinnace, and never saw her again.

Our foresail was split, and all torn: then our master took the mizzen, and brought it to the fore-mast, to make our ship work, and with our sprit sail we mended our foresail, the storm continuing without all reason in fury, with hail, snow, rain, and wind such and so mighty, as that in nature it could not possibly be more, the seas such and so lofty, with continual breach, that many times we were doubtful whether our ship did sink or swim.

The tenth of October being by the account of our captain and master very near the shore, the weather dark, the storm furious, and most of our men having given over to travail, we yielded ourselves to death, without further hope of succour. Our captain sitting in the gallery very pensive, I came and brought him some Rosa Solis[122] to comfort him; for he was so cold, that he was scarce able to move a joint. After he had drunk, and was comforted in heart, he began for the ease of his conscience to make a large repetition of his forepassed time, and with many grievous sighs he concluded in these words: Oh most glorious God, with whose power the mightiest things among men are matters of no moment, I most humbly beseech Thee, that the intolerable burden of my sins may through the blood of Jesus Christ be taken from me: and end our days with speed, or show us some merciful sign of Thy love and our preservation. Having thus ended, he desired me not to make known to any of the company his intolerable grief and anguish of mind, because they should not thereby be dismayed. And so suddenly, before I went from him the sun shined clear; so that he and the master both observed the true elevation of the Pole, whereby they knew by what course to recover the Straits. The next day being the 11 of October, we saw the cape on the south shore. This cape being within two leagues to leeward off us, our master greatly doubted that we could not double the same: whereupon the captain told him: you see there is no remedy, either we must double it, or before noon we must die: therefore loose your sails, and let us put it to God's mercy. The master being a man of good spirit resolutely made quick despatch and set sails. Our sails had not been half an hour aboard, but the footrope of our fore-sail broke, so that nothing held but the eyelet holes. The seas continually broke over the ship's poop, and flew into the sails with such violence, that we still expected the tearing of our sails, or oversetting of the ship, and withal to our utter discomfort, we perceived that we fell still more and more to leeward, so that we could not double the cape: we were now come within half a mile of the cape, and so near the shore, that the counter-surf of the sea would rebound against the ship's side, so that we were much dismayed with the horror of our present end. Being thus at the very pinch of death, the wind and the seas raging beyond measure, our master veered some of the main sheet; and whether it was by that occasion, or by some current, or by the wonderful power of God, as we verily think it was, the ship quickened her way, and shot past that rock, where we thought she would have shored. Then between the cape and the point there was a little bay; so that we were somewhat further from the shore: and when we were come so far as the cape, we yielded to death: yet our good God the father of all mercies delivered us, and we doubled the cape about the length of our ship, or very little more. We were shot in between the high lands, without any inch of sail, we spooned before the sea, three men not being able to guide the helm, and in six hours we were put five and twenty leagues within the Straits, where we found a sea answerable to the ocean.

In this time we freed our ship from water, and after we had rested a little, our men were not able to move; their sinews were stiff, and their flesh dead, and many of them (which is most lamentable to be reported) were so eaten with lice, as that in their flesh, did lie clusters of lice as big as peas, yea and some as big as beans. Being in this misery we were constrained to put into a cove for the refreshing our men. Our master knowing the shore and every cove very perfectly, put in with the shore and moored to the trees, as before time we had done, laying our anchor to the seaward. Here we continued until the twentieth of October; but not being able any longer to stay through extremity of famine, the one and twentieth we put off into the channel, the weather being reasonable calm: but before night it blew most extremely at westnorthwest. The storm growing outrageous, our men could scarcely stand by their labour; and the Straits being full of turning reaches we were constrained by discretion of the captain and master in their accounts to guide the ship in the hell-dark night, when we could not see any shore, the channel being in some places scarce three miles broad. But our captain, as we first passed through the Straits drew such an exquisite plat of the same, as I am assured it cannot in any sort be bettered: which plat he and the master so often perused, and so carefully regarded, as that in memory they had every turning and creek, and in the deep dark night without any doubting they conveyed the ship through that crooked channel: so that I conclude, the world hath not any so skilful pilots for that place, as they are: for otherwise we could never have passed in such sort as we did.

In a mighty fret of weather the seven and twentieth day of October we were free of the Straits, and the thirtieth of October we came to Penguin Isle being three leagues from Port Desire, the place which we purposed to seek for our relief.

When we were come to this isle we sent our boat on shore, which returned laden with birds and eggs; and our men said that the penguins were so thick upon the isle, that ships might be laden with them; for they could not go without treading upon the birds, whereat we greatly rejoiced. Then the captain appointed Charles Parker and Edward Smith, with twenty others to go on shore, and to stay upon the isle, for the killing and drying of those penguins, and promised after the ship was in harbour to send the rest, not only for expedition, but also to save the small store of victuals in the ship. But Parker, Smith, and the rest of their faction suspected, that this was a device of the captain to leave his men on shore, that by these means there might be victuals for the rest to recover their country: and when they remembered, that this was the place where they would have slain their captain and master, surely (thought they) for revenge hereof will they leave us on shore. Which when our captain understood, he used these speeches unto them: I understand that you are doubtful of your security through the perverseness of your own guilty consciences: it is an extreme grief unto me, that you should judge me blood-thirsty, in whom you have seen nothing but kind conversation: if you have found otherwise, speak boldly, and accuse me of the wrongs that I have done; if not, why do you then measure me by your own uncharitable consciences? Be void of these suspicions, for, God I call to witness, revenge is no part of my thought. The last of October we entered the harbour. Our master at our last being here having taken careful notice of every creek in the river, in a very convenient place, upon sandy ooze, ran the ship on ground, laying our anchor to seaward, and with our running ropes moored her to stakes upon the shore, which he had fastened for that purpose; where the ship remained till our departure.

The third of November our boat with water, wood, and as many as she could carry, went for the Isle of Penguins: but being deep, she durst not proceed, but returned again the same night. Then Parker, Smith, Townsend, Purpet, with five others, desired that they might go by land, and that the boat might fetch them when they were against the isle, it being scarce a mile from the shore. The captain caused them to carry weapons, calivers, swords and targets: so the sixth of November they departed by land, and the boat by sea; but from that day to this day we never heard of our men. Only the captain and master with six others being left in the ship, there came a great multitude of savages to the ship, throwing dust in the air, leaping and running like brute beasts, having vizards on their faces like dog's faces, or else their faces are dog's faces indeed. We greatly feared lest they would set our ship on fire, for they would suddenly make fire, whereat we much marvelled: they came to windward of our ship, and set the bushes on fire, so that we were in a very stinking smoke: but as soon as they came within our shot, we shot at them, and striking one of them in the thigh they all presently fled, so that we never heard nor saw more of them. Hereby we judged that these cannibals had slain our 9 men. When we considered what they were that thus were slain, and found that they were the principal men that would have murdered our captain and master with the rest of their friends, we saw the just judgement of God. They sent the boat to the Isle of Penguins; whereby we understood that the penguins dried to our heart's content, and that the multitude of them was infinite. This penguin hath the shape of a bird, but hath no wings, only two stumps in the place of wings, by which he swimmeth under water with as great swiftness as any fish. They live upon smelts, whereof there is great abundance upon this coast: in eating they be neither fish nor flesh: they lay great eggs, and the bird is of a reasonable bigness, very near twice so big as a duck. All the time that we were in this place, we fared passing well with eggs, penguins, young seals, young gulls, besides other birds, such as I know not: of all which we had great abundance. In this place we found a herb called scurvy[123] grass, which we fried with eggs, using train oil instead of butter. This herb did so purge the blood, that it took away all kinds of swellings, of which many died, and restored us to perfect health of body, so that we were in as good case as when we came first out of England. We stayed in this harbour until the 22 of December, in which time the captain, the master, and myself had made some salt, by laying salt water upon the rocks in holes, which in 6 days would be kerned.

The 22 of December we departed with our ship for the isle, at night we departed with 14,000 dried penguins, and shaped our course for Brazil. Now our captain rated our victuals, and brought us to such allowance, as that our victuals might last six months; for our hope was, that within six months we might recover our country, though our sails were very bad. So the allowance was two ounces and a half of meal for a man a day, and to have so twice a week, so that 5 ounces did serve for a week. Three days a week we had oil, three spoonfuls for a man a day; and 2 days in a week pease, a pint between 4 men a day, and every day 5 penguins for 4 men, and 6 quarts of water for 4 men a day. This was our allowance; wherewith (we praise God) we lived, though weakly, and very feeble. The 30 of January we arrived at the isle of Placencia in Brazil, the first place that outward bound we were at. The last of January at sun-rising they suddenly landed, hoping to take the Portuguese in their houses, and by that means to recover some cassava meal, or other victuals for our relief: but when we came to the houses they were all razed, and burnt to the ground, so that we thought no man had remained on the island.

Then the captain went to the gardens, and brought from thence fruits and roots for the company, and came aboard the ship, and brought her into a fine creek which he had found out, where we might moor her by the trees, and where there was water, and hoops to trim our casks. Our case being very desperate, we presently laboured for dispatch away; some cut hoops, which the coopers made, others laboured upon the sails and ship, every man travailing for his life, and still a guard was kept on shore to defend those that laboured, every man having his weapon likewise by him. The 3 of February our men with 23 shot went again to the gardens, being 3 miles from us upon the north shore, and fetched cassava roots out of the ground, to relieve our company instead of bread; for we spent not of our meal while we stayed here. The 5 of February being Monday, our captain and master hasted the company to their labour; so some went with the coopers to gather hoops, and the rest laboured aboard. This night many of our men in the ship dreamed of murder and slaughter: in the morning they reported their dreams, one saying to another; this night I dreamt, that thou wert slain; another answered, and I dreamt, that thou wert slain: and this was general through the ship. The captain hearing this, who likewise had dreamed very strangely himself, gave very strait charge, that those which went on shore should take weapons with them, and saw them himself delivered into the boat, and sent some of purpose to guard the labourers. All the forenoon they laboured in quietness, and when it was ten of the clock, the heat being extreme, they came to a rock near the wood's side (for all this country is nothing but thick woods) and there they boiled cassava roots, and dined: after dinner some slept, some washed themselves in the sea, all being stripped to their shirts, and no man keeping watch, no match [124] lighted, not a piece charged. Suddenly as they were thus sleeping and sporting, having gotten themselves into a corner out of sight of the ship, there came a multitude of Indians and Portuguese upon them, and slew them sleeping: only two escaped, one very sore hurt, the other not touched, by whom we understood of this miserable massacre: with all speed we manned our boat, and landed to succour our men; but we found them slain, and laid naked on a rank one by another, with their faces upward, and a cross set by them: and withal we saw two very great pinnaces come from the river very full of men. Of 76 persons which departed in our ship out of England, we were now left but 27, having lost 13 in this place, with their chief furniture, as muskets, calivers, powder, and shot. Our cask was all in decay, so that we could not take in more water than was in our ship, for want of cask, and that which we had was marvellous ill conditioned: and being there moored by trees, for want of cables and anchors, we still expected the cutting of our moorings, to be beaten from our decks with our own furniture, and to be assailed. To depart with 8 tons of water in such bad cask was to starve at sea, and in staying our case was ruinous. These were hard choices; but being thus perplexed, we made choice rather to fall into the hands of the Lord, than into the hands of men: for His exceeding mercies we had tasted, and of the others' cruelty we were not ignorant. So concluding to depart, the 6 of February we were off in the channel, with our ordnance and small shot in a readiness, for any assault that should come, and having a small gale of wind, we recovered the sea in most deep distress. Considering what they were that were lost, we found that all those that conspired the murdering of our captain and master were now slain by savages, the gunner only excepted. Some desired to go to Bahía [125], and to submit themselves to the Portuguese, rather than to die for thirst: but the captain with fair persuasions altered their purpose. In this distress it pleased God to send us rain in such plenty, as that we were well watered, and in good comfort to return. But after we came near unto the sun, our dried penguins began to corrupt, and there bred in them a most loathsome and ugly worm of an inch long. This worm did so mightily increase, and devour our victuals, that there was in reason no hope how we should avoid famine, but be devoured of these wicked creatures: there was nothing that they did not devour, only iron excepted: our clothes, boots, shoes, hats, shirts, stockings: and for the ship they did so eat the timbers, as that we greatly feared they would undo us, by gnawing through the ship's side. Great was the care and diligence of our captain, master, and company to consume these vermin, but the more we laboured to kill them, the more they increased; so that at the last we could not sleep for them, but they would eat our flesh, and bite like mosquitoes. In this woeful case, after we had passed the equinoctial toward the north, our men began to fall sick of such a monstrous disease, as I think the like was never heard of: for in their ankles it began to swell; from thence in two days it would be in their breasts, so that they could not draw their breath, and then fell into their cods; and their cods and yards did swell most grievously, and most dreadfully to behold, so that they could neither stand, lie, nor go. Divers grew raging mad, and some died in most loathsome and furious pain. It were incredible to write our misery as it was: there was no man in perfect health, but the captain and one boy. To be short, all our men died except 16, of which there were but 5 able to move. The captain was in good health, the master indifferent, Captain Cotton and myself swollen and short winded, yet better then the rest that were sick, and one boy in health: upon us 5 only the labour of the ship did stand. The captain and master, as occasion served, would take in, and heave out the top-sails, the master only attended on the sprit-sail, and all of us at the capstan without sheets and tackles. In fine our misery and weakness was so great, that we could not take in, nor heave out a sail: so our top-sail and sprit-sails were torn all in pieces by the weather. The master and captain taking their turns at the helm, were mightily distressed and monstrously grieved with the woeful lamentation of our sick men. Thus as lost wanderers upon the sea, the 11 of June 1593, it pleased God that we arrived at Bear-haven [126] in Ireland, and there ran the ship on shore: where the Irishmen helped us to take in our sails, and to moor our ship for floating: which slender pains of theirs cost the captain some ten pounds before he could have the ship in safety.

Thus without victuals, sails, men, or any furniture God only guided us into Ireland, where the captain left the master and three or four of the company to keep the ship; and within 5 days after he and certain others had passage in an English fisher-boat to Padstow in Cornwall. In this manner our small remnants by God's only mercy were preserved, and restored to our country, to whom be all honour and glory world without end.

LX

The truth of the fight about the Isles of Azores, the last of August 1591, betwixt the Revenge, *one of her Majesty's ships, and an Armada of the king of Spain; penned by the honourable Sir Walter Raleigh knight.*

The Lord Thomas Howard with six of her Majesty's ships, six victuallers of London, the Bark *Raleigh*, and two or three pinnaces riding at anchor near unto Flores, one of the westerly islands of the Azores, the last of August in the afternoon, had intelligence by one Captain Middleton of the approach of the Spanish Armada. He had no sooner delivered the news but the fleet was in sight: many of our ship's companies were on shore in the island; some providing ballast for their ships; others filling of water and refreshing themselves from the land with such things as they could either for money, or by force recover. By reason whereof our ships being all pestered and rummaging every thing out of order, very light for want of ballast, and that which was most to our disadvantage, the one half part of the men of every ship sick, and utterly unserviceable: for in the *Revenge* there were ninety diseased: in the *Bonaventure*, not so many in health as could handle her main sail. The Spanish fleet having shrouded their approach by reason of the island; were now so soon at hand, as our ships had scarce time to weigh their anchors, but some of them were driven to let slip their cables and set sail. Sir Richard Grenville was the last that weighed, to recover the men that were upon the island, which otherwise had been lost. The Lord Thomas with the rest very hardly recovered the wind, which Sir Richard Grenville not being able to do, was persuaded by the master and others to cut his main sail, and cast about, and to trust to the sailing of the ship; for the squadron of Seville were on his weather bow. But Sir Richard utterly refused to turn from the enemy, alleging that he would rather choose to die, than to dishonour himself, his country, and Her Majesty's ship, persuading his company that he would pass through the two squadrons, in despite of them, and enforce those of Seville to give him way. Which he performed upon divers of the foremost, who, as the mariners term it, sprang their luff, and fell under the lee of the *Revenge*. But the other course had been better. Notwithstanding out of the greatness of his mind, he could not be persuaded. In the meanwhile as he attended those which were nearest him, the great *San Felipe* being in the wind of him, and coming towards him, becalmed his sails in such sort, as the ship could neither make way, nor feel the helm: so huge and high charged was the Spanish ship, being of a thousand and five hundred tons. Who after laid the *Revenge* aboard. When he was thus bereft of his sails, the ships that were under his lee luffing up, also laid him aboard: of which the next was the Admiral of the Biscaines, a very mighty and puissant ship. The said *Felipe* carried three tiers of ordnance on a side, and eleven pieces in every tier. She shot eight forth right out of her chase, besides those of her stern ports.

After the *Revenge* was entangled with this *Felipe*, four others boarded her; two on her larboard and two on her starboard. The fight thus beginning at three of the clock in the afternoon, continued very terrible all that evening. But the great *San Felipe* having received the lower tier of the *Revenge*, discharged with crossbar-shot, shifted herself with all diligence from her sides, utterly misliking her first entertainment. Some say that the ship foundered, but we cannot report it for truth, unless we were assured. The Spanish ships were filled with companies of soldiers. In ours there were none at all besides the mariners, but the servants of the commanders and some few voluntary gentlemen only. After many interchanged volleys of great ordnance and small shot, the Spaniards deliberated to enter the *Revenge*, and made divers attempts, hoping to force her by the multitudes of their armed soldiers and musketeers, but were still repulsed again and again, and at all times beaten back into their own ships, or into the seas. In the beginning of the fight, the *George Noble* of London having received some shot through her fell under the lee of the

Revenge, and asked Sir Richard what he would command him, being but one of the victuallers and of small force: Sir Richard bid him save himself, and leave him to his fortune. After the fight had thus, without intermission, continued while the day lasted and some hours of the night, many of our men were slain and hurt, and one of the great galleons of the Armada, and the Admiral of the Hulks both sunk, and in many of the Spanish ships great slaughter was made. Some write that Sir Richard was very dangerously hurt almost in the beginning of the fight, and lay speechless for a time ere he recovered. But two of the *Revenge*'s own company, brought home in a ship from the islands, examined by some of the lords, affirmed that he was never so wounded as that he forsook the upper deck, till an hour before midnight; and then being shot into the body with a musket as he was a dressing, was again shot in the head, and withall his surgeon wounded to death.

But to return to the fight, the Spanish ships which attempted to board the *Revenge*, as they were wounded and beaten off, so always others came in their places, she having never less than two mighty galleons by her sides, and aboard her. But as the day increased, so our men decreased: and as the light grew more and more, by so much more grew our discomforts. For none appeared in sight but enemies, saving one small ship called the *Pilgrim*, commanded by Jacob Whiddon, who hovered all night to see the success: but in the morning was hunted like a hare amongst ravenous hounds, but escaped.

All the powder of the *Revenge* to the last barrel was now spent, all her pikes broken, forty of her best men slain, and the most part of the rest hurt. In the beginning of the fight she had but one hundred free from sickness and four score and ten sick, laid in hold upon the ballast. A small troop to man such a ship, and a weak garrison to resist so mighty an army. By those hundred all was sustained, the volleys, boardings, and enterings of fifteen ships of war, besides those which beat her at large. On the contrary the Spanish were always supplied with soldiers brought from every squadron: all manner of arms and powder at will. Unto ours there remained no comfort at all, no hope, no supply either of ships, men, or weapons; the masts all beaten overboard, all her tackle cut asunder, her upper work altogether razed, and in effect evened she was with the water, but the very foundation or bottom of a ship, nothing being left over head either for flight or defence. Sir Richard finding himself in this distress, and unable any longer to make resistance, having endured in this fifteen hours fight, by estimation eight hundred shot of great artillery, besides many assaults and entries; and that himself and the ship must needs be possessed by the enemy, who were now all cast in a ring round about him (The *Revenge* not able to move one way or other, but as she was moved with the waves and billow of the sea) commanded the master gunner, whom he knew to be a most resolute man, to split and sink the ship; that thereby nothing might remain of glory or victory to the Spaniards: seeing in so many hours' fight, and with so great a navy they were not able to take her, having had fifteen hours time, above ten thousand men, and fifty and three sail of men of war to perform it withall: and persuaded the company, or as many as he could induce, to yield themselves unto God, and to the mercy of none else; but as they had, like valiant resolute men, repulsed so many enemies, they should not now shorten the honour of their nation, by prolonging their own lives for a few hours, or a few days. The master gunner readily condescended and divers others; but the captain and the master were of another opinion, and besought Sir Richard to have care of them: alleging that the Spaniard would be as ready to entertain a composition, as they were willing to offer the same: and that there being divers sufficient and valiant men yet living, and whose wounds were not mortal, they might do their country and prince acceptable service hereafter. And whereas Sir Richard had alleged that the Spaniards should never glory to have taken one ship of Her Majesty, seeing they had so long and so notably defended themselves; they answered that the ship had six foot water in the hold, three shot under water, which were so weakly stopped, as with the first working of the sea, she must needs sink, and was besides so crushed and bruised, as she could never be removed out of the place.

And as the matter was thus in dispute, and Sir Richard refusing to hearken to any of those reasons: the master of the *Revenge* (while the captain won unto him the greater party) was conveyed aboard the General Don Alfonso Baçan. Who (finding none over hasty to enter the *Revenge* again, doubting lest Sir Richard would have blown them up and himself, and perceiving by the report of the master of the *Revenge* his dangerous disposition) yielded that all their lives should be saved, the company sent for England, and the better sort to pay such reasonable ransom as their estate would bear, and in the mean season to be free from galley or imprisonment. To this he so much the rather condescended for fear of further loss and mischief to themselves, as also for the desire he had to recover Sir Richard Grenville; whom for his notable valour he seemed greatly to honour and admire.

When this answer was returned, and that safety of life was promised, the common sort being now at the end of their peril, the most drew back from Sir Richard and the master gunner, being no hard matter to dissuade men from death to life. The master gunner finding himself and Sir Richard thus prevented and mastered by the greater number, would have slain himself with a sword, had he not been by force withheld and locked into his cabin. Then the General sent many boats upon the *Revenge*, and divers of our men fearing Sir Richard's disposition, stole away aboard the General and other ships. Sir Richard thus overmatched, was sent unto by Alfonso Baçan to remove out of the *Revenge*, the ship being marvellous unsavoury, filled with blood and bodies of dead, and wounded men like a slaughter house. Sir Richard answered that he might do with his body what he list, for he esteemed it not, and as he was carried out of the ship he swooned, and reviving again desired the company to pray for him. The General used Sir Richard with all humanity, and left nothing unattempted that tended to his recovery, highly commending his valour and worthiness, and greatly bewailing the danger wherein he was, being unto them a rare spectacle, and a resolution seldom approved, to see one ship turn toward so many enemies, to endure the charge and boarding ot so many.

There were slain and drowned in this fight, well near one thousand of the enemy's.

The Admiral of the Hulks and the *Ascension* of Seville were both sunk by the side of the *Revenge;* one other recovered the road of Saint Michael, and sunk also there; a fourth ran herself with the shore to save her men. Sir Richard died as it is said, the second or third day aboard the General, and was by them greatly bewailed. What became of his body, whether it were buried in the sea or on the land we know not: the comfort that remaineth to his friends is, that he hath ended his life honourably in respect of the reputation won to his nation and country, and of the same to his posterity, and that being dead, he hath not outlived his own honour.

LXI

A voyage to the mainland of Malacca, in the year 1591, performed by Mr James Lancaster, and written from the mouth of Edmund Barker of Ipswich, his lieutenant by Mr Richard Hakluyt.

Our fleet of three tall ships departed from Plymouth the 10 of April 1591, and arrived at the Canary Islands from whence we departed the 29 of April. The second of May we were in the height of Cabo Blanco. The fifth we passed the Tropic of Cancer. The eighth we were in the height of Cape Verde. All this time we went with a fair wind at northeast, when we came within 8 degrees of the equinoctial line, we met with a contrary wind. Here we lay off and on in the sea until the sixth of June. While we lay thus we took a Portuguese caravel laden by merchants of Lisbon for Brazil, in which we had some 60 tuns of wine, 1200 jars of oil, about 100 jars of olives, certain barrels of capers, with divers other necessaries fit for our voyage: which wine, oil, olives and capers were better to us than gold. We had two men died before we passed the line, and divers sick, which took their sickness in those hot climates: for they be wonderful unwholesome from 8 degrees of northerly latitude unto the line: we had nothing but tornados, with such thunder, lighting, and rain, that we could not keep our men dry three hours together, which was an occasion of the infection among them, and their eating of salt victuals, with lack of clothes to shift them. The 28 of July we had sight of Cape of Buona Esperanza: until the 31 we lay off and on with the wind contrary to double the Cape. But our men being weak and sick in all our ships, we thought good to seek some place to refresh them. With which consent we bare up with the land to the northward of the Cape, and going along the shore, we spied a goodly bay with an island lying to seawards of it, into which we did bear, and found it very commodious for our ships to ride in. This bay is called Agoada de Saldanha, lying 15 leagues northward on the hither side of the Cape. The first of August being Sunday we came to an anchor in the bay, sending our men on land, and there came unto them certain black savages very brutish which would not stay, but retired from them. For the space of 15 or 20 days we could find no relief but only fowls which we killed with our pieces, which were cranes and geese: there was no fish but mussels and other shell-fish, which we gathered on the rocks. After 15 or 20 days being here, our admiral went with his pinnace unto the island which lies off this bay, where he found great store of penguins and seals, whereof he brought good plenty with him. After we had been here some time, we got here a negro, whom we compelled to march into the country with us, making signs to bring us some cattle. Within 8 days after, he with 30 or 40 other negroes, brought us down some 40 bullocks and oxen, with as many sheep. We bought an ox for two knives, and a sheep for a knife, and some we bought for less value than a knife. The sheep are very big and very good meat, they have no wool on their backs but hair, and have great tails like the sheep in Syria. There be divers sorts of wild beasts, as the antelope, the red and fallow deer, with other great beasts unknown unto us. Here also great store of over-grown monkeys. It was thought good rather to proceed with two ships well manned, than with three evil-manned: for here we had of sound and whole men but 198, of which there went in the *Penelope* with the admiral 101, and in the *Edward* with Captain

Lancaster 97. We left behind 50 men with the *Royal Merchant,* [sic] whereof there were many pretty well recovered. The disease that had consumed our men hath been the scurvy, which in my own judgement proceedeth of their evil diet at home.

Six days after our sending back for England of the *Merchant Royal,* our admiral Captain Raimond in the *Penelope,* and Mr James Lancaster in the *Edward Bonaventure,* set forward to double the Cape of Buona Esperanza, which they did very speedily. The 14 of September we were encountered with a mighty storm and extreme gusts of wind, wherein we lost our general's company, and could never hear of him nor his ship any more, though we did our best endeavour to seek him up and down for a long while. Four days after this, in the morning toward ten of the clock we had a terrible clap of thunder, which slew four of our men outright, their necks being wrung in sunder without speaking any word, and of 94 men there was not one untouched whereof some were stricken blind, others were bruised in their legs and arms, and others in their breasts, so that they voided blood two days after, others were drawn out at length as though they had been racked. But (God be thanked) they all recovered saving only the four which were slain outright. Also with the same thunder our main mast was torn very grievously from the head to the deck, and some of the spikes that were ten inches into the timber, were melted with the extreme heat thereof. We came to an island an hundred leagues to the northeast of Mozambique called Comoro,[127] which we found exceeding full of people, which are Moors of tawny colour and good stature, but they be very treacherous and diligently to be taken heed of. Here we desired to store ourselves with water, whereof we stood in great need, and sent sixteen of our men well armed on shore in our boat: whom the people suffered quietly to land and water, and divers of them with their king came aboard our ship in a gown of crimson satin pinked after the Moorish fashion down to the knee, whom we entertained in the best manner. And though we thought ourselves furnished, yet our master William Mace of Radcliffe pretending that it might be long before we should find any good watering place, would needs go himself on shore with thirty men, much against the will of our captain, and he and 16 of his men, together with one boat which was all that we had, and 16 others that were a washing overagainst our ship, were betrayed of the perfidious Moors, and in our sight for the most part slain, we being not able for want of a boat to yield them any succour. From hence with heavy hearts we shaped our course for Zanzibar the 7 of November. Six days before we departed hence, the Cape merchant of the factory wrote a letter unto our captain in the way of friendship, as he pretended, requesting a jar of wine, and a jar of oil, and two or three pounds of gunpowder, which letter he sent by a negro his man, and Moor in a canoe: we sent him his demands by the Moor but took the negro along with us because we understood he had been in the East Indies and knew somewhat of the country. By this negro we were advertised of a small bark of some thirty tons (which the Moors call a junk) which was come from Goa thither laden with pepper for the factory and service of that kingdom. Thus having trimmed our ship as we lay in this road, in the end we set forward for the coast of the East India, in the 15 of February aforesaid, intending if we could to reach Cape Comorin, which is the headland or promontory of the main of Malabar, and there to have lain off and on for such ships as should have passed from Ceylon, São Tomé, Bengal, Pegu, Malacca, the Moluccas, the coast of China, and the Isle of Japan, which ships are of exceeding wealth and riches. But in our course we were very much deceived by the currents that set into the Gulf of the Red Sea. Nevertheless it pleased God to bring the wind more westerly, and so in the month of May 1592 we happily doubled the Cape Comorin without sight of the coast of India. From hence we directed our course for the islands of Nicobar. We ran in six days with a very large wind though the weather were foul with extreme rain and gusts of winds. These islands were missed through our master's default for want of due observation of the South star. Now the winter coming upon us with much contagious weather, we directed our course from hence with the islands of Pulo Pinaou,[128] some five leagues from the main between Malacca and Pegu. Here we continued until the end of August. Our refreshing in this place was very small, only of oysters growing on rocks, great whelks, and some few fish which we took with our hooks. Here we landed our sick men on these uninhabited islands for their health, nevertheless 26 of them died. The winter passed and having watered our ship and fitted her to go to sea, we had left us but 33 men and one boy, of which not past 22 were sound for labour and help, and of them not past a third part sailors: thence we made sail to seek some place of refreshing and went over to the main of Malacca. The next day we came to anchor in a bay in six fathoms water some two leagues from the shore. Then Master James Lancaster our captain, and Mr Edmund Barker his lieutenant, and other of the company manning the boat, went on shore to see what inhabitants might be found. And coming on land we found the tracking of some barefooted people which were departed thence not long before: for we saw their fire still burning, but people we saw none. The next day about two of the clock in the afternoon we spied a canoe which came near unto us, but would not come aboard us, having in it some sixteen naked Indians, with whom nevertheless going afterward on land, we had friendly conference and promise of victuals. The next day in the morning we espied three ships, being all of burthen 60 or 70 tons, one of which we made to strike with our very boat. The night following all the men except twelve, which we took into our ship, being most of them born in Pegu, fled away in their boat, leaving their ship and goods with us. The next day we weighed anchor and went to the leeward of an island hard by, and took in her lading being pepper.

Our sick men being somewhat refreshed and lusty, with such relief as we had found in this ship, we weighed anchor, determining to run in the Straits of Malacca. And when we there arrived, we lay to and again for such shipping as should come that way. Thus having spent some five days, upon a Sunday we espied a sail which was a Portuguese ship that came from Negapatam a town on the main of India over against the northeast part of the Isle of Ceylon; and that night we took her being of 250 tons: she was loaded with rice for Malacca. We continued here until the sixth of October, at which time we met with the ship of seven hundred tons which came for Goa: the people which were to the number of about three hundred men, women and children, got ashore with two great boats and quite abandoned the ship. At our coming aboard we found in her sixteen pieces of brass, and three hundred butts of Canary wine, and wine which is made of the palm trees, and raisin wine which is also very strong: as also all kinds of haberdasher wares, as hats, red caps knit of Spanish wool, worsted stockings knit, shoes, velvets, taffetas, chamlets, and silks, abundance of suckets, rice, Venice glasses, certain papers full of false and counterfeit stones which an Italian brought from Venice to deceive the rude Indians withall, abundance of playing cards, two or three packs of French paper. Whatsoever became of the treasure which is usually brought in royals of plate in this galleon we could not find it. After that the mariners had disorderly pilled this rich ship, the captain because they would not follow his commandment to unlade those excellent wines into the *Edward*, abandoned her and let her drive at sea. We departed from thence to a bay in the kingdom of Junsalaom,[129] which is between Malacca and Pegu eight degrees to the northward to seek for a pitch to trim our ship. We sent commodities to their king to barter for ambergris, and for the horns of abath,[130] whereof the king only hath the traffic in his hands. Now this abath is a beast which hath one horn only in her forehead, and is thought to be the female unicorn, and is highly esteemed of all the Moors in those parts as a most sovereign remedy against poison. We had only two or three of these horns which are the colour of a brown grey, and some reasonable quantity of ambergris. Our men took occasion to come home, our captain lying at that time very sick more like to die than to live. Having set our foresail, and in hand to set all our other sails, our men made answer that they would take their direct course for England and would stay there no longer. Now seeing that they could not be persuaded by any means possible, the captain constrained to give his consent to return, leaving all hope of so great possibilities. This the eight of December 1592 we set sail for the Cape of Buona Esperanza. In our passage we had exceeding great store of bonitos and albacores, which are a greater kind of fish: of which our captain, being now recovered of his sickness, took with an hook as many in two or three hours as would serve forty persons a whole day. In February 1593 we fell with the easternmost land of Africa. We spent a month or five weeks before we could double the Cape. After, we directed our course for the island of St Helena, and arrived there the third day of April, where we stayed to our great comfort nineteen days. After our arrival at Saint Helena, in an house by the chapel I found an Englishman one John Segar of Bury in Suffolk, who was left there eighteen months before by Abraham Kendall, who left him there to refresh him on the island, being otherwise like to have perished on shipboard: and at our coming we found him as fresh in colour and in as good plight of body to our seeming as might be, but crazed in mind and half out of his wits, as afterwards we perceived: for whether he were put in fright of us, not knowing at first what we were, whether friends or foes, or of sudden joy when he understood we were his old consorts and countrymen, he became idle-headed, and for eight days' space neither night nor day took any natural rest, and so at length died for lack of sleep. We found in this place great store of very wholesome and excellent good green figs, oranges, and lemons very fair, abundance of hogs and goats, and great plenty of partridges, guineacocks, and other wild fowl. Our mariners contrary to the will of the captain, would straight home. The captain because he was desirous to go for Pernambuco in Brazil, granted their request. And about the 12 of April 1593 we departed from St Helena, and directed our course for the place aforesaid. The next day our captain calling upon the sailors to finish a foresail which they had in hand, some of them answered that unless they might go directly home, they would lay their hands to nothing; whereupon he was constrained to follow their humour. And from thence-forth we directed our course for our country. We spent some six weeks, with many calm and contrary winds: which loss of time and expense of our victuals, whereof we had very small store, made us doubt to keep our course: and some of our men growing into a mutiny threatened to break up other men's chests, to the overthrow of our victuals and all ourselves, for every man had his share of his victuals before in his own custody, that they might be sure what to trust to. Our captain seeking to prevent this mischief, being advertised by one of our company which had been at the isle of Trinidad, that there we should be sure to have refreshing. hereupon directed his course to that island. There arrived a French ship of Cannes in which was captain one Monsieur de Barbaterre, of whom we bought some two butts of wine and bread, and other victuals. Then we watered and fitted our ship and stopped a great leak. And having thus made ready our ship to go to sea, we determined to go directly for Newfoundland. But before we departed, there arose a storm the wind being northerly, which put us from an anchor and forced us to the southward of Santo Domingo. We directed our course westward along the island of Santo Domingo, and doubled Cape Tiberon, and passed through the old channel between Santo Domingo and Cuba for the cape of Florida. Thus running to the isle of Bermuda, finding the winds there very variable, contrary to our expectation and all men's writings, we lay there a day or two the wind being northerly, and increasing continually more and more, it grew to be a storm and a great freight of wind: which continued with us some 24 hours, with such extremity, as it carried not only our sails away being furled, but also made much water in our ship, so that we had six foot water in hold. The extremity of the storm was such that with the labouring of the ship we lost our foremast, and our ship grew as full of water as before. The storm once ceased, and the wind contrary to go our course, we fell to consultation which might be our best way to save our lives. Our victuals now being utterly spent, and having eaten hides 6 or 7 days, we thought it best to bear back again for Dominica, and the islands adjoining, knowing that there we might have some relief, whereupon we turned back for the said islands. But before we could get thither the wind scanted upon us, which did greatly endanger us for lack of fresh water and victuals: so that we were constrained to bear up to the westward, to certain other islands called the Nieblas[131] or Cloudy Islands towards the island of St Juan de Puerto Rico, where at our arrival we found land crabs and fresh water, and tortoises, which come most on land about the full of the moon. Here having refreshed ourselves some 17 or 18 days, and having gotten some small store of victuals into our ship, we resolved to return again for Mona: upon which our determination five of our men left us, remaining still on the Isles of Nieblas for all persuasions that we could use to the contrary, which afterward came home in an English ship. From these isles we departed and arrived at Mona [132] about the twentieth of November 1593, and there coming to an anchor toward two or three of the clock in the morning, the captain, and Edmund Barker his lieutenant with some few others went on land to the houses of the old Indian and his three sons, thinking to have gotten some food, our victuals being all spent. We spent two or three days in seeking provision to carry aboard to relieve the whole company. But in the night about twelve of the clock our ship did drive away with five men and a boy only in it, our carpenter secretly cut their own cable, leaving nineteen of us on land without boat or anything, to our great discomfort. In the midst of these miseries reposing our trust in the goodness of God, which many times before had succoured us in our greatest extremities, we took our leaves of one another, dividing ourselves into several companies. The greatest relief that we six which were with the captain could find for the space of nine and twenty days was the stalks of the purslain boiled in water, and now and then a pompion. After the end of nine and twenty days we espied a French ship, which afterwards we understood to be of Dieppe, called the *Louisa*, whose captain was one Monsieur Felix, unto whom we made a fire, at sight whereof he took in his topsails, bare in with the land, and showed us his flag, whereby we judged him French: so coming along to the western end of the island, there he anchored, we making down with all speed unto him. Our captain Master James Lancaster, this night went aboard the Frenchman, who gave him good entertainment, and the next day fetched eleven more of us aboard entreating us all very courteously. We departed thence for the northside of Saint Domingo, where we remained until April following 1594, and spent some two months in traffic with the inhabitants for hides. On Sunday the seventh of April 1594 they set homeward, and thence arrived safely in Dieppe within two and forty days after, on the 19 of May, where after we had stayed two days to refresh ourselves, and given humble thanks unto God, and unto our friendly neighbours, we took passage for Rye and landed there on Friday the 24 of May

1594, having spent in this voyage three years, six weeks and two days.

We understood in the East Indies by certain Portuguese that they have lately discovered the coast of China to the latitude of nine and fifty degrees, finding the sea still open to the northward: giving great hope of the North East or North West Passage. Witness Master James Lancaster.

LXII

The casting away of the Toby *near Cape Espartel, without the strait of Gibraltar on the coast of Barbary, 1593.*

The *Toby* of London a ship of 250 tons manned with fifty men, the owner whereof was the worshipful Mr Richard Staper, being bound for Livorno, Zante and Patras in Morea, being laden with merchandise to the value of 11 or 12 thousand sterling pounds, set sail from Blackwall the 16 day of August 1593.

We had sight of Mount Chiego,[133] which is the first high-land which we descry on the Spanish coast at the entrance of the Strait of Gibraltar, where we had very foul weather and the wind scant two days together. Here we lay off to the sea. The master, whose name was George Goodlay, being a young man, and one which never took charge before for those parts, was very proud, neither would take any counsel of any of his company, but did as he thought best himself. In the end of the two days of foul weather the wind being fair, thinking that he was farther off the land than he was, bare sail all that night, and an hour and a half before day had ran our ship upon the ground on the coast of Barbary four leagues to the south of Cape Espartel. Whereupon being all not a little astonished, the master said unto us, I pray you forgive me; for this is my fault and no man's else. The company asked him whether they should cut off the main mast: no said the master we will hoist out our boat. But one of our men coming speedily up, said, sirs, the ship is full of water, well said the master, then cut the main-mast over board: which we did with all speed. But the after part suddenly split a sunder in such sort that no man was able to stand upon it, but all fled upon the foremast up into the shrouds thereof; and hung there for a time: but seeing nothing but present death approach (being so suddenly taken that we could not make a raft which we had determined) we committed ourselves unto the Lord and began with doleful tune and heavy hearts to sing the 12 Psalm. Help Lord for good and godly men &c. Howbeit before we had finished four verses the waves of the sea had stopped the breaths of most of our men. For the foremast with the weight of our men and the force of the sea fell down into the water, and upon the fall thereof there were 38 drowned, and only 12 by God's providence partly by swimming and other [by] means of chests got on shore, which was about a quarter of a mile from the wreck of the ship. None of the officers were saved but the carpenter.

We twelve which the Lord had delivered from extreme danger of the sea at our first coming on shore all fell down on our knees, praying the Lord most humbly for his merciful goodness. Our prayers being done, we consulted together what course to take, seeing that we were fallen into a desert place. We travelled all that day until night, sometimes one way and sometimes another, and could find no kind of inhabitants; only we saw where wild beasts had been, and places where there had been houses, which after we perceived to have been burnt by the Portuguese. So at night falling into certain groves of olive trees, we climbed up and sat in them to avoid the danger of lions and other wild beasts.

The next day we travelled until three of the clock in the afternoon without any food, but water and wild date roots: then going over a mountain, we had sight of Cape Espartel; whereby we knew somewhat better which way to travel. We came to a hedgerow made with great long canes; we spied and looked over it, and beheld a number of men to the number of five thousand in skirmish together with small shot and other weapons. And after consultation what we were best to do, we concluded to yield ourselves unto them, being destitute of all means of resistance. So rising up we marched towards them, who espying us, forthwith some hundred of them with their javelins in their hands came running towards us as though they would have run us through: howbeit they only struck us flatling with their weapons, and said that we were Spaniards: and we told them that we were Englishmen; which they would not believe yet. By and by the conflict being ended, and night approaching, the captain of the Moors, a man of some 56 years old, came himself unto us, and by his interpreter which spake Italian, asked what we were, and from whence we came. One Thomas Henmer of our company which could speak Italian, declared unto him that we were merchants, and how by great misfortune our ship, merchandise and the greatest part of our company were pitifully cast away upon their coast. But he void of humanity and all manhood, for all this, caused his men to strip us out of our apparel even to our shirts to see what money and jewels we had about us: which when they had found to the value of some 200 pounds in gold and pearls they gave us some of our apparel again, and bread and water only to comfort us. The next morning they carried us down to the shore where our ship was cast away, which was some sixteen miles from that place. In which journey they used us like their slaves, making us (being extreme weak) to carry their stuff, and offering to beat us if we went not so fast as they. We asked them why they used us so, and they replied, that we were their captives: we said we were their friends, and that there was never Englishman captive to the King of Morocco. So we came down to the ship, and lay there with them seven days, while they had gotten all the goods they could, and then they parted it amongst them. After the end of these seven days the captain appointed twenty of his men well armed, to bring us up into the country.

We were delivered by those soldiers unto the Cadi,[134] which examined us what we were: and we told him. He gave us a good answer, and sent us to the Jews' house, where we lay seven days. In the mean while that we lay here, there were brought thither twenty Spaniards and twenty Frenchmen, which Spaniards were taken in a conflict on land, but the Frenchmen were by foul weather cast on land within the Straits about Cape de Gato, and so made captives. Thus at the seven days' end we twelve Englishmen, the twenty French, and the twenty Spaniards were all conducted towards Morocco[135] with nine hundred soldiers, horsemen and footmen, and in two days journey we came to the river of Fez, where we lodged all night, being provided of tents. The next day we went to a town called Sallee, and lay without the town in tents. From thence we travelled almost an hundred miles without finding any town, but every night we came to fresh water, which was partly running water and sometimes rain water. So we came at last within three miles of the city of Morocco, where we pitched our tents: and there we met with a carrier which did travel in the country for the English merchants: and by him we sent word unto them of our estate: and they returned the next day unto us a Moor, which brought us victuals, being at that instant very feeble and hungry, and withall sent us a letter with pen, ink, and paper, willing us to write unto them what ship it was that was cast away, and how many and what men there were alive.

We were carried in as captives and with ropes about our necks as well English as the French and Spaniards. And so we were carried before the king: and when we came before him he did commit us all to ward, where we lay 15 days in close prison: and in the end we were cleared by the English merchants to their great charges: for our deliverance cost them 700 ounces, every ounce in that country containing two shillings. When we came out of prison, we continued eight weeks with the English merchants. At the end of which time being well apparelled by the bounty of our merchants we were conveyed down by the space of eight days' journey to Santa Cruz,[136] where we took shipping. Two of our number died in the country of the bloody-flux: the one at our first imprisonment at Morocco, whose name was George Hancock, and the other at Santa Cruz, whose name was Robert Swancon, whose death was hastened by eating of roots and other unnatural things to slake their raging hunger in our travel, and by our hard and cold lodging in the open fields without tents. Thus of fifty persons through the rashness of an unskilful master ten only survived of us, and after a thousand miseries returned home poor, sick, and feeble into our country.

Richard Johnson	Thomas Henmore
William Williams carpenter	John Silvester
John Durham	Thomas Whiting
Abraham Rouse	William Church
John Matthewes	John Fox

LXIII

The sinking of the stout and war-like carrack called Las Cinque Llagas. *Written by the discreet and valiant captain, Mr Nicholas Downton.*

In the latter end of the year 1593 the right honourable Earl of Cumberland at his own charges and his friends' prepared three ships, the *Royal Exchange*, the *Mayflower* and the *Sampson*.

The 13 of June we met with a mighty carrack of the East Indies called *Las Cinque Llagas* or The Five Wounds. The *Mayflower* was in fight with her beforenight. I, in the *Sampson*, fetched her up in the evening, and commanded to give her the broadside, as we term it. At the very first shot she discharged at us, I was shot in a little above the belly, whereby I was made unserviceable for a good while after. Yet by means of an honest truehearted man which I had with me, one Captain Grant, nothing was neglected: until midnight when the admiral came up, the *Mayflower* and the *Sampson* never left by turns to ply her with their great ordnance: until we attempted to board her. The admiral laid her aboard in the midship: the *Mayflower* coming up in the quarter. The *Sampson* went aboard on the bow. The *Exchange* at the first coming had her captain Mr Cave shot into both the legs, the one whereof he never recovered, and in his absence he had not any that would undertake to lead out his company to enter upon the enemy. My friend Captain Grant did lead my men on the carrack's side, which being not manfully backed by the *Exchange*'s men, his forces being small, made the enemy bolder than he would have been, whereby I had six men presently slain and many more hurt. I say not but some of the *Exchange*'s men did very well, and many more (no doubt) would have done the like, if there had been any principal man to have brought all the company to the fight, and not to have run into corners. The Portuguese, peradventure encouraged by our slack working, played the men, and had barricades made, where they might stand without any danger of our shot. They plied us also very much with fire. This unusual casting of fire did much dismay many of our men and made them draw back as they did. We fired a mat on her beak-head, which more and more kindled, and ran from thence to the mat on the bowsprit, and thence to the top-sail yard, which fire made the Portuguese abaft in the ship to stagger, and to make a show of parley.

We desired to be off from her, but had little hope to obtain our desire; nevertheless we plied water very much to keep our ship well. By God's providence only, by the burning asunder of our spritsail-yard with ropes and sail, and the ropes about the spritsail-yard of the carrack, whereby we were fast intangled, we fell apart, with burning of some of our sails which we had then on board. The *Exchange* also being farther from the fire, afterward was more easily cleared, and fell off from abaft. The Portuguese leaped overboard in great numbers. Then I sent Captain Grant with the boat, with leave to use his own discretion in saving of them. So he brought me aboard two gentlemen, the one an old man called Nuno Velio Pereira, which was governor of Mozambique and Sofala in the year 1582, and since that time has been likewise a governor in a place of importance in the East Indies.

The rest which were taken up by the other ship boats, we set all on shore in the isle of Flores, except some two or three negroes. The people we saved told us that they would not yield, because this carrack was for the king, and that the captain of her was in favour with the king, and at his return into the Indies should have been Viceroy there. This ship was more like a ship of war than otherwise: moreover, she had the ordnance of a carrack that was cast away at Mozambique.

The last of June after long traversing of the seas we had sight of another mighty carrack, which after some few shot bestowed upon her we summoned to yield; but they standing stoutly to their defence utterly refused the same. Seeing no good could be done without boarding her I consulted what course we should take. By reason that we which were the chief captains were partly slain and partly wounded in the former conflict, and because of the murmuring of some disordered and cowardly companions, our valiant and resolute determinations were crossed: and to conclude a long discourse in a few words, the carrack escaped our hands. Being disappointed of our expectations, and victuals growing short, we returned for England, where I arrived at Portsmouth the 28 of August.

LXIV

The prosperous voyage of Mr James Lancaster, begun with three ships and a galley-frigate from London in October 1594, and intended for Pernambuco, the port town of Olinda in Brazil. In which voyage (besides the taking of nine and twenty ships and frigates) he surprised the said port-town, being strongly fortified and manned: and held possession thereof thirty days together (notwithstanding many bold assaults of the enemy both by land and water). Here he found the freight of a rich East Indian carrack; which together with great abundance of sugars, Brazil-wood, and cotton he brought from thence.

In September 1594 the worshipful Mr John Wats, alderman, Mr Paul Banning, alderman, and others of worship in the City of London, victualled three good ships; to wit, the *Consent*, the *Salomon*, and the *Virgin*: and appointed for commanders in this voyage, Mr James Lancaster of London, gentleman, admiral of the fleet, Mr Edmund Barker, and Mr John Audely, having in their ships to the number of 275 men and boys.

We departed from Blackwall in October. Not fifty leagues from our own coast, we lost the *Salomon*, and the *Virgin*, by a storm of contrary wind that fell upon us: yet being alone, in hope to meet them about the Canaries, we kept on our course, but could hear no tidings of our consorts; which greatly grieved us.

Thence we went, bearing for the Isle of Tenerife, where in the morning early we had sight of a sail, which being becalmed under the shore, was towing with their boat ahead. We manned our boat, appointing our men well for fight. The Spaniards seeing our boat come, entered theirs, and leaving the ship, sought to save themselves by flight: but our men pursued them so fast, that they boarded them, and brought them with their ship to our general. This ship was laden with 80 tons of Canary-wine, which came not unto us before it was welcome. We kept and manned it. The very next morning we had sight of one other; to whom in like manner we sent our boat: but their gunner made a shot at her, and struck off a proper young man's arm; yet we enforced her to yield, and found 40 tons of wine in her. The Spaniards were all set on shore upon Tenerife, making a quick return of their long voyage intended into the West Indies.

We met again with the *Virgin*, whose men told us for very truth, that the *Salomon* was returned for England: enforced so to do, by spending her mast. Which when our men understood, they were all in a maze, not knowing what to do, now we had lost one half of our strength: some of them came to the captain, giving him counsel to bear up for the West Indies. The captain hearing this new novelty, as not unacquainted with the variable pretences of mariners, made them this answer: Sirs, I made known to you all at my coming out of England that I meant to go for Pernambuco, and although at the present we want one of our ships, yet (God willing) I mean to go forward. I hope you will all be contented herewith: for to go any other course than I have determined, (by God's help) I will not be drawn unto. They rested all satisfied: and at our coming to Cabo Blanco (God be praised) we met with the *Salomon* with no small joy to us all; and there she had taken of Spaniards and Portuguese 24 sail of ships and caravels, fisher men, and had taken out of them such necessaries as she had need of. Of these ships our captain took four along with him, with another that he had taken himself, meaning to employ them as occasion should serve. At this place he understood of one of the pilots of those ships, that one of the carracks that came out of the East Indies, was castaway in the road of Pernambuco, and that all her goods were laid upon the Arrecife which is the lower town. Of these news we were all glad, and rejoiced much; seeing such a booty before us.

We were all joyful, and had great hope of the blessing of God in our intended voyage, making frolick for joy of our meeting one with the other (praising God for all) we plied for Maio[137]: where coming to anchor, our general went ashore to view the place where we might in best safety set our galley frigate together; which frame we brought from England of purpose to land men in the country of Brazil. Here we discharged our great prize of wine, and set her on fire. The choice being made for the place to build the galley-frigate, ashore it was brought, where the carpenters applied their work, keeping good watch: yet one negligent fellow, who had no knowledge of the country, straying from his company, was by the Portuguese taken, and very kindly used, and brought again unto us: for which good the general rewarded them well with gifts. While we were thus busily employed about the foresaid galley, we descried at sea four sails, which we had good hope would have proved Indiamen: but they proved Captain Venner with his fleet, who, seeing us at anchor, anchored also: being acquainted with our general's determination for landing, consorted with us, we to take three parts, and he the fourth, of all that should be taken, whereby our strength was increased, to all our comforts. Three weeks or thereabouts we stayed in this place before the galley was finished; which done, putting men into her, and fitting her with oars, having fourteen banks on a side, a mast and sail, the commandment of her was committed unto Mr Wats, an honest skilful mariner.

From thence we put again to sea, and plied still to our desired port of Pernambuco, and about midnight we came before the harbour; where some plied up and down, holding that the best policy, to forbear the entering till day might give them light, the harbour being hard, and therefore the more perilous. Our ships being in safety well arrived, God was praised: and the general in his boat went from ship to ship, willing them to make ready such men as they could spare, with muskets, pikes, bills, bows, arrows, and what weapons they had to follow him. Himself, with 80 men from his own ship, embarked himself in the galley, which carried in her prow a good saker and two murdering pieces.[138]

Our admiral spent all the night in giving directions to every ship to have their men ready shipped in their boats, for he intended to enter the harbour at the break of day, and to leave his ships without, till he had gotten the fort and the town: for he would not adventure the ships in, till the harbour was gotten. At the entrance of the harbour rode three great Holland ships, which our admiral doubted would impeach his going in; and therefore he gave order to the men of these five small ships, which were not above 60 tons apiece, if the Hollanders did offer any resistance, to run aboard them, and to set their own ships on fire, and scape in their boats. But when the morning was come, we were fallen above half a mile down to the northward, below the harbour, which was a great inconvenience unto us: so that before we could get up again, the ebb was come upon us, and thereby we were forced to hover before the harbour till two of the clock in the afternoon, in the sight of all the town. In the meantime, our ships rode before the fort without the harbour, about a demi-culverin shot off: in the which time passed many shot between the fort and the ships: but no great harm was done on either part. All this while our admiral kept the men ready hovering in the galley and the boats. The Hollanders that rode in the mouth of the harbour, seeing our resolution, laid out hawsers, and wound themselves out of the way of us. Our admiral was very joyful, and gave great encouragement to all his men: for, to pass these three great Hollanders, he held it the greatest danger of all. About 12 of the clock the governor of the town sent a Portuguese aboard the admiral's ship, to know what he would have, and wherefore he came. He returned him this answer: that he wanted the carrack's goods, and for them he came, and them he would have. In this process of time, the towns-men which saw so much shipping, and perceived us to be enemies, gathered themselves together, three or four ensigns of men, esteemed to the number of some six hundred at the least. These came to the fort or platform lying over against the entry of the harbour, and there attended our landing: but before our admiral set forward with his boats, he gave express order to all that had charge of governing the boats or galley, to run them with such violence against the shore, that they should be all be cast away without recovery, and not one man to stay in them, whereby our men might have no manner of retreat to trust unto, but only to God and their weapons.

Now was the time come of the flood, being about two of the clock in the afternoon, when our admiral set forward, and entered the harbour with the small galley, and all the rest of the boats following him, the Hollanders that rode in the mouth of the harbour, nothing impeached him: but now the fort began to play with their ordnance upon the galley and the boats; and one of their shot took away a great piece of our ensign out of the galley. But our sail being set, it was no time for us to make any stay, but with all the force we could we ran the galley upon the shore right under the fort, with such violence, that we broke her back; and she sank. The boats coming after did the like. At our arrival, those in the fort had loaded all their ordnance, being seven pieces of brass, to discharge them upon us at our landing; which indeed they did: for our admiral leaping into the water, all the rest following him, off came these pieces of ordnance: but Almighty God be praised, they in the fort, with fear to see us land in their faces, had piked their ordnance so steep downwards, that they shot all their shot in the sand, although it was not above a quoit's cast at the most. Our admiral seeing this, cried out, encouraging his men, upon them, upon them: all (by God's help) is ours: and therewith ran to the fort with all violence. Those four ensigns of men that were set to defend our landing, seeing this resolution, began to go back, and retire into certain bushes that were by the same fort; and being followed, fled: and so abandoned the fort, and left it with their ordnance to us. This day of our arrival was their Good Friday, when by custom they usually whip themselves: but God sent us now for a general scourge to them all, whereby that labour among them might be well spared. The fort being taken with all their ordnance, the admiral waved to the ships, willing them to weigh and come in; which they did with all speed, leaving certain men in keeping the fort, and placed the ordnance toward the high town, from whence he suspected the greatest danger; and putting his men in order, marched toward the low town: in which town lay all their merchandise. Approaching to the town, he entered the same, the people embarking themselves in caravels and boats, with all the expedition they could. The base town, of above an hundred houses, being thus taken, we found in it great store of merchandises: as Brazil wood, sugars, calico-cloth, pepper, cinnamon, cloves, mace, nutmegs, with divers other good things. The admiral went up and down the town, and placed at the south end of the same Captain Venner and his company, himself and his company in the midst of the town, and Captain Barker and Captain Addy at the other end, giving great charge, that no man upon pain of great punishment and loss of his shares, should break up or enter into any warehouse, without order and direction from the admiral. And this commandment was as well kept as ever any was kept, where so great spoil and booty was found: for common mariners and soldiers are much given to pillaging and spoiling, making greater account of the same than of their shares.

We kept a very sure watch this first night, and the morning being come, our admiral and Captain Venner, with the rest of the captains, went about the town, and gave order for the fortifying of it with all expedition: so that within two days it was surrounded with posts and planks, all that part of the town next the mainland, at least nine foot high. This town is environed on the one part by the sea, and on the backside by a river that runneth behind it; so that to come to it by land, you must enter it by a small narrow passage not above forty paces over at an high water. At this passage we built a fort, and planted in it five pieces of ordnance. Having the town in possession, our admiral sent for the Hollanders by his surgeon, which had been brought up in that country, a man sober and discreet. At his first coming aboard of them, they seemed to stand upon their own guard and defence, for they were three great and strong ships: they at last willed him to come into the greatest of their ships, which was above 450 tons. Then he declared to them that they should be as sure from any show of violence or injury, as if they were in their own houses, and if they should think so good, his admiral would freight them to England, if they would be content with freight reasonable, he would not force them, unless it were to their own benefit and good liking. Although this people were somewhat stubborn at the first, as that nation is in these causes, yet being satisfied with good words and good dealing they came a land, and after conference had with the admiral, they were so satisfied, that they went through with a freight: they served us as truly and faithfully as their own people did, both at watch and ward, by sea and all other services. Within two days after our coming in, about midnight, a great number of Portuguese and Indians with them, came down upon us with a very great cry and noise; but God be thanked, we were ready for them: for our admiral supposing some such assault, had provided all our muskets with hail shot,[139] which did so gall both the Indians and the Portuguese, that they made them presently retreat. We lost in this conflict but only one man, but had divers hurt. What was lost of their part, we could not tell, for they had before day, after our retreat, carried away all their dead. Within three or four days after our coming in appeared before the harbour 3 ships, and 2 pinnaces, the pinnaces being somewhat near, descried our flags, and one of them came in, which was a French pinnace, declaring all the rest to be French bottoms; which our admiral willed should come in: and so they did. These were Frenchmen of war, and came thither for purchase. The captains came a land, and were welcomed: amongst whom was one, that the year before had taken in our admiral at the island of Mona, where his ship was cast away, coming out of the East Indies. This captain desired of our admiral to bestow upon him his ship's lading of Pernambuco-wood,[140] which he granted him, and also his pinnace, and more, gave him a caravel of about 50 tons, and bid him lade her with wood also; which with other benefits he gratefully received. To the other two captains he granted their ladings of wood, the one captain being of Dieppe, the other of Rochelle. The coming in of these ships did much to strengthen us; for our admiral appointed both these French and the Flemings to keep watch upon the river by night with their boats. This was for fear of fire ships which our admiral had great care unto, and caused our ships to ride by cables and hawsers, at all advantages to shun them, if by that means they should attempt to put us out of the harbour. Thinking ourselves sufficiently fortified, we began to unlade our ships, which came as full laden in as they went forth, but not with so good merchandise. Our men were divided into halves, and the one half wrought one day, and the other half the other day; always those that wrought not kept the watch and none stepped far off or wandered from his colours, and those that wrought had all their weapons in good order set and placed by them, so that at an instant every one knew where to go.

In this meantime, the Portuguese with the country people were not idle, for seeing us so busy, about six nights after our coming in, they privily in the night cast up a trench in the sands about a saker shot from our ships, minding there to plant ordnance, which would have offended our ships greatly. The admiral hearing this, about 3 of the clock in the afternoon marshalled our men, and he and all the rest of the captains marched towards them. The Portuguese and Indians perceiving our coming, began to withdraw themselves within the trench, meaning (as it should appear) to fight it out there: we made no stand, but presently approached the trenches with our muskets and pikes, afore their trenches were thoroughly finished. And the Portuguese left the place and left unto us 4 good pieces of brass ordnance, with powder and shot and divers other necessaries, and among the rest 5 small carts of that country, which to us were more worth than all the rest we took, for the lading of our goods from the town to the water's side: we went forward dividing our merchandise with Captain Venner, and daily loading them aboard, every ship's company according as their turns fell out. The next morning came in a ship with some 60 negroes, 10 Portuguese women, and 40 Portuguese: the women and the negroes we turned out of the town, but the Portuguese our admiral kept to draw the carts when they were laden, which to us was a very great ease. For the country is very hot and ill for our nation to take any great travail in.

In this town there is no fresh water to be had, and therefore we were every 5 or 6 days compelled to pass over the river into the mainland to get fresh water, which after the first or second time the Portuguese kept so that we were driven to water of force, and several times some of our men were hurt, and two or three slain, and with this danger we were forced to get our water.

Within some 20 days after our coming in, they had prepared 5 caravels and filled them with such things as would best take fire and burn: these they brought within a mile or little more of our ships, and there set them on fire, for nearer they could not well come because of our watch of boats, for the admiral had always 6 boats that kept watch above half a mile from the ships for fear of such exploits as these. But these fired caravels had the tide with them, and also the little wind that blew was in their favour: which caused them to come down the stream the faster: which our boats perceiving made to them with as much expedition as conveniently they could, but the tide and wind both serving them, they approached toward the ships with great expedition. Our men in the town began to be in some fear of them. Such as were aboard, were somewhat amazed to see 5 so great fires to be coming down among their ships, but they prepared for to clear them of it, as well as they could, being provided aforehand and judging that some such stratagems would be there used, the river being very fit therefore. But (God be thanked) who was our best defence in this voyage: our company in the boats so played the men when they saw the fires come near our ships, that casting grapnels with iron chains on them, as every boat had one for that purpose, some they towed aground, and some they brought to anchor, where they rode till all their force was burnt out, and so we were delivered by God's help from this fearful danger. Within some 6 nights after this, which might be about the 26 day after our abode there, about 11 of the clock at night, came driving down other three great rafts burning with the hugest fires that I have seen. These were exceeding dangerous, for when our men approached then thinking to clap their grapnels upon them, as they had done upon the caravels the night before, they were prevented: for there stuck out of the rafts many poles, that they could not come to throw their grapnels into them: and yet they had this inconvenience worse than all the rest. There stuck out among the poles certain hollow trunks filled with such provision of fire-works that they ceased not to spout out such sparkles, that our boats having powder in them for our men's use, durst not for fear of firing themselves with their own powder come near those sparkles of the rafts, but seeing them to drive nearer and nearer our ships, they wet certain cloths and laid upon their flasks and bandoliers and so ventured upon them, and with their grapnels took hold of them, and so towed them on ground, where they stuck fast and were not burnt out the next day in the morning. Diverse logs and timbers came driving along by our ships, and burning, but with our boats we easily defended them. And thus (God be praised) we escaped the second fires. Therefore let all men riding in rivers in their enemy's country be sure to look to be provided beforehand, for against fire there is no resistance without preparation. Also it is a practice in these hot countries, where there be such expert swimmers, to cut the cables of ships: and one night it was practised to cut the admiral's cable, and yet the boat rode by the cable with two men in her to watch all the night, and the buoy only was cut, but not the cable: but after that night, seeing then our good watch, they never after attempted it.

Our ships through the diligent labour of our men, began to be wholly laden, and all the best merchandise conveyed aboard, so that our admiral meant to depart that night, which was the 31 day after our entrance, or else on the next day at the farthest, and so warning was given to all men to make themselves ready.

God be thanked of His goodness toward us who sent us a fair wind to go forth withal, so that by 11 of the clock in the night we were all forth in safety. The enemies perceiving our departing, planted a piece or two of ordnance, and shot at us in the night, but did us no harm. We were at our coming forth 15 sails, that is, 3 sails of Hollanders, four sails of French and one ship which the admiral gave the French captain, 3 sails of Captain Venner's fleet of Plymouth, and 4 sails of our admiral's fleet, all these were laden with merchandises, and that of good worth. We stayed in this harbour to pass all this business but only 31 days, and in this time we were occupied with skirmishes and attempts of the enemy 11 times.

We put to the sea, plying after the rest of our fleet which were gone before, which we never heard of till our arrival in England at the Downs in the month of July. At our setting sail from the Downs, according as the custom is, finding the Queen's ships there, we saluted them with certain ordnance. The gunner being careless, as they are many times of their powder in discharging certain pieces in the gunroom, set a barrel of powder on fire, which blew up the admiral's cabin, slew the gunner with 2 others outright, and hurt 20 more, of which 4 or 5 died. This powder made such a smoke in the ship with the fire that burnt in the gunroom, that no man at the first wist what to do: but recalling back their fear, they began to cast water into the gunroom in such abundance (for the Queen's ships now and also the other ships that were in our company came presently to our help) that (God be praised) we put out the fire and saved all, and no great harm was done to the goods. By this may be seen that there is no sure safety of things in this world. Our fire being well put out, and we taking in fresh men (God be praised) we came to Blackwall in safety.

LXV

The discovery of the large, rich, and beautiful Empire of Guiana, with a relation of the great and golden city of Manoa (which the Spaniards call El Dorado). Performed in the year 1595 by Sir Walter Raleigh Knight.

On Thursday the 6 of February in the year 1595 we departed England, and the Sunday following had sight of the north cape of Spain, the wind for the most part continuing prosperous: and so onwards for the Canaries, we directed our course for Trinidad: we abode there 4 or 5 days, and in all that time we came not to the speech of any Indian or Spaniard.

There is that abundance of stone pitch, that all the ships of the world may be therewith laden from thence, and we made trial of it in trimming our ships to be most excellent good, and melteth not with the sun as the pitch of Norway, and therefore for ships trading the south parts very profitable.

This island of Trinidad hath the form of a sheep hook, and is but narrow, the north part is very mountainous, the soil is very excellent and will bear sugar, ginger, or any other commodity that the Indies yield. It hath store of deer, wild pork, fruits, fish and fowl: it hath also for bread sufficient maize, cassava, and of those roots and fruits which are common everywhere in the West Indies. It hath divers beasts which the Indies have not: the Spaniards confessed that they found grains of gold in some of the rivers, but they having a purpose to enter Guiana (the magazine of all rich metals) cared not to spend time in the search thereof any further.

While we remained at Puerto de los Españoles some Spaniards came aboard us to buy linen of the company, and also to view our ships and company, all which I entertained kindly and feasted after our manner: by means whereof I learned as much of the estate of Guiana as I could, or as they knew, for those poor soldiers having been many years without wine, a few draughts made them merry, in which mood they vaunted of Guiana and of the riches thereof, and all what they knew of the ways and passages, myself seeming to purpose nothing, but bred in them an opinion that I was bound only for the relief of those English which I had planted in Virginia, whereof the bruit was come among them.

I found occasions of staying in this place for two causes: the one was to be revenged of Berreo, who the year before had betrayed eight of Captain Whiddon's men, which arrived at Trinidad from the East Indies: Berreo sent a canoe aboard the pinnace only with Indians and dogs inviting the company to go with them into the woods to kill a deer, who followed the Indians, but were no sooner one arquebus shot from the shore, but Berreo's soldiers lying in ambush had them all, notwithstanding that he had given his word to Captain Whiddon that they should take water and wood safely: the other cause of my stay was, for that by discourse with the Spaniards I daily learned more and more of Guiana.

I was assured by another cacique of the north side of the island, that Berreo had sent for soldiers, meaning to have given me a cassado at parting, if it had been possible. For although he had given order through all the island that no Indian should come aboard to trade with me upon pain of hanging and quartering, (having executed two of them for the same, which I afterwards found) yet every night there came some with most lamentable complaints of his cruelty, how he had divided the island and given to every soldier a part, that he made the ancient caciques which were lords of the country to be their slaves, that he kept them in chains, and dropped their naked bodies with burning bacon, and such other torments, which I found afterwards to be true. So as both to be revenged of the former wrong, as also considering that to enter Guiana by small boats, to depart 400 or 500 miles from my ships, and to leave a garrison in my back, I should have savoured very much of the ass: I set upon the corps du garde in the evening, and having put them to the sword, sent Captain

Calfield onwards with 60 soldiers, and myself followed with 40 more and so took their new city which they called St Joseph by break of day: they abode not any fight after a few shot, and all being dismissed but only Berreo and his companion, I brought them with me aboard, and at the instance of the Indians, I set their new city of St Joseph on fire.

We then hastened away towards our purposed discovery, and first I called all the captains of the island together that were enemies to the Spaniards; by my Indian interpreter, which I carried out of England, I made them understand that I was the servant of a Queen, who was the great cacique of the north, and a virgin: that she was an enemy to the [Spaniards] in respect of their tyranny and oppression, and that she delivered all such nations about her, as were by them oppressed, and having freed all the coast of the northern world from their servitude, had sent me to free them also, and withal to defend the country of Guiana from their invasion and conquest. I showed them Her Majesty's picture which they so admired and honoured, as it had been easy to have brought them idolatrous thereof. Having Berreo my prisoner I gathered from him as much of Guiana as he knew.

This Berreo is a gentleman well descended, and had long served the Spanish King very valiant and liberal, and a gentleman of great assuredness, and of a great heart: I used him according to his estate and worth.

I sent Captain Whiddon the year before to get what knowledge he could of Guiana, but my intelligence was far from truth, for the country is situated above 600 English miles further from the sea, than I was made to believe it had been, which afterwards understanding to be true by Berreo, I kept it from the knowledge of my company, who else would never have been brought to attempt the same: of which 600 miles I passed 400 leaving my ships so far from me at anchor in the sea, in the bottom of a galley, and in one barge, two wherries and a ship boat of the *Lion's Whelp*, we carried 100 persons and their victuals for a month, being all driven to lie in the rain and weather, in the open air, in the burning sun, and upon the hard boards, and to dress our meat, and to carry all manner of furniture in them, wherewith they were so pestered and unsavoury, that what with victuals being most fish, with wet clothes of so many men thrust together, and the heat of the sun, I will undertake there was never any prison in England, that could be found more unsavoury and loathsome, especially to myself, who had for many years before been dieted and cared for in a sort far more differing.

But that it had pleased God, we had entered the country but some ten days sooner ere the rivers were overflown, we had adventured either to have gone to the great city of Manoa,[141] or at least taken so many of the other cities and towns nearer at hand, as would have made a royal return: but it pleased not God so much to favour me at this time: I shall willingly spend my life therein, and if any else shall be enabled thereunto, and conquer the same, I assure him thus much, he shall perform more than ever was done in Mexico by Cortez or in Peru by Pizzaro.

Because there may arise many doubts, and how this empire of Guiana is become so populous, and adorned with so many great cities, I thought good to make it known, that the emperor now reigning is descended from those magnificent princes of Peru: for when Francisco Pizarro, and others conquered Peru, and put to death Atabalipa son to Guaynacapa, one of the younger sons of Guaynacapa fled out of Peru, and took with him many thousands of soldiers, and with those and many others which followed him, he vanquished all that tract and valley of America which is situated between the great river of the Amazon and Orinoco and Marañón.

The empire of Guiana is directly east from Peru towards the sea, and lieth under the equinoctial line, and it hath more abundance of gold than any part of Peru, and as many or more great cities than ever Peru had when it flourished most: I have been assured by such of the Spaniards[142] as have seen Manoa the imperial city of Guiana, which the Spaniards call El Dorado, that for the greatness, for the riches, and for the excellent seat, it far exceedeth any of the world.

Although these reports may seem strange, yet if we consider the many millions which are daily brought out of Peru into

Spain, we may easily believe the same: for we find that by the abundant treasure of that country the Spanish King vexeth all the princes of Europe, and is become, in a few years, from a poor King of Castile, the greatest monarch of this part of the world.

How all these rivers cross and encounter, how the country lieth and is bordered mine own discovery, and the way that I entered, with all the rest, your lordship[143] shall receive in a large chart or map, which I have not yet finished, and which I shall most humbly pray your lordship to secrete and not to suffer it to pass your own hands; for by a draught thereof all may be prevented by other nations: for I know it is this very year sought by the French. It was also told me ere I departed from England, that Villiers the admiral was in preparation for the planting of Amazon, to which river the French have made divers voyages, and returned much gold, and other rarities.

Now Berreo for executing of Morequito, and other cruelties, spoils, and slaughters hath lost the love of all the borderers, and dare not send any of his soldiers any further into land than to Carúpano,[144] which he called the port of Guiana: but from thence he had trade further into the country, and always appointed ten Spaniards to reside in Carúpano town, by whose favour, and being conducted by his people, those ten searched the country thereabouts, as well for mines, as for other trades and commodities.

They also have gotten a nephew of Morequito, whom they have christened, and named Don Juan, of whom they have great hope, endeavouring by all means to establish him in the said province. Among many other trades, those Spaniards used canoes to pass to the south side of the mouth of Orinoco, and there buy women and children from the cannibals, which are of that barbarous nature, as they will for three or four hatchets sell the sons and daughters of their own brethren and sisters, and for somewhat more, even their own daughters. Hereof the Spaniards make great profit: for buying a maid of twelve or thirteen years for three or four hatchets, they sell them again at Margarita in the West Indies for fifty and an hundred pesos, which is so many crowns.

The master of my ship, John Dowglas, took one of the canoes which came laden from thence with people to be sold, and the most of them escaped; yet of those he brought, there was one as well favoured, and as well shaped as ever I saw any in England, and afterward I saw many of them, which but for their tawny colour may be compared to any of Europe. They also trade in those rivers for bread of cassava, of which they buy an hundred pound weight for a knife, and sell it at Margarita for ten pesos. They also recover great store of cotton, brazil wood, and those beds which they call hammocks, wherein in hot countries all the Spaniards use to lie commonly, and in no other, neither did we ourselves while we were there. By means of which trades, for ransom of divers of the Guianians, and for exchange of hatchets and knives, Berreo recovered some store of gold plates, eagles of gold, and images of men and divers birds, and despatched his camp-master for Spain, with all that he had gathered, therewith to levy soldiers, and by the show thereof to draw others to the love of the enterprise. And having sent divers images as well of men as beasts, birds and fishes, so curiously wrought in gold, he doubted not but to persuade the King to yield to him some further help, especially for that this land hath never been sacked.

After I had thus learned of his proceedings past and purposed, I told him that I had resolved to see Guiana. Berreo was stricken into a great melancholy and sadness, and used all the arguments he could to dissuade me, and also assured the gentlemen of my company that it would be labour lost, and that they should suffer many miseries if they proceeded. And first he delivered that I could not enter any of the rivers with any bark or pinnace, or hardly with any ship's boat, it was so low, sandy, and full of flats, and that his companies were daily grounded in their canoes, which drew but twelve inches water. He further said, that none of the country would come to speak with us, but would all fly; and if we followed them to their dwellings, they would burn their own towns: and besides that, the way was long, the winter at hand, and that the rivers beginning once to swell, it was impossible to stem the current, and that we could not in those small boats by any means carry victual for half the time, and that (which indeed most discouraged my company) the kings and lords of all the borders of Guiana had decreed that none of them should trade with any Christians for gold, because the same would be their own overthrow, and that for the love of gold the Christians meant to conquer and dispossess them of all together.

Many and the most of these I found to be true, but yet I resolving to make trial of all. How we must either give over our enterprise, or leaving our ships at adventure four hundred mile behind us, must run up in our ships' boats, one barge, and two wherries. I sent away one King, master of the *Lion's Whelp*, with his shipboat, to try another branch of a river, to prove if there were water to be found for either of the small ships to enter. But when he came to the mouth he found it as the rest, but stayed not to discover it thoroughly, because he was assured by an Indian, his guide, that the cannibals would assail them with many canoes, and that they shot poisoned arrows; so as if he hasted not back, they should all be lost.

John Dowglas searched those rivers, and found four goodly entrances, whereof the least was as big as the Thames at Woolwich; but in the bay thitherward it was shoaled, and but six foot water: so we therefore resolved to go on with the boats: we had as much sea to cross over in our wherries, as between Dover and Calais, and in a great billow, the wind and current being both very strong, so as we were driven to go in those small boats directly before the wind into the bottom of the bay and from thence to enter the mouth of some one of those rivers which John Dowglas had last discovered, and had with us for pilot an Indian of Barema, a river to the south of Orinoco, between that and Amazon. This Arawak promised to bring me into the great river of Orinoco, but indeed of that which he entered he was utterly ignorant, for he had not seen it in twelve years before; at which time he was very young, and of no judgement: and if God had not sent us another help, we might have wandered a whole year in that labyrinth of rivers, all the earth doth not yield the like confluence of streams and branches, the one crossing the other so many times, and all so fair and large, and so like one another, as no man can tell which to take.

The great river of Orinoco hath nine branches which fall out on the north side of his own main mouth: on the south side it hath seven other fallings into the sea, so it disembogueth by sixteen arms in all, between islands and broken ground, but the islands are very great, many of them as big as the Isle of Wight. The river's mouth is 300 miles wide at his entrance into the sea.

When three days more were overgone, our companies began to despair, the weather being extremely hot, the river bordered with very high trees, that kept away the air, and the current against us every day stronger than other: so long we laboured, that many days were spent, and we driven to draw ourselves to harder allowance, our bread even at the last, and no drink at all; and our men and ourselves so wearied and scorched, and doubtful withal, whether we should ever perform it or no.

The further we went on (our victual decreasing and the air breeding great faintness) we grew weaker and weaker, when we had most need of strength and ability; for hourly the river ran more violently than other against us, and the barge, wherries, and ships' boats of Captain Gifford and Captain Calfield, had spent all their provisions. On the banks of these rivers were divers sorts of fruits good to eat, flowers and trees of such variety, as were sufficient to make ten volumes of herbals: we relieved ourselves many times with the fruits of the country, and sometimes with fowl and fish. We saw birds of all colours, some carnation, some crimson, orange-tawny, purple, watchet, and of all other sorts both simple and mixed, and it was unto us a great good passing of the time to behold them, besides the relief we found by killing some store of them with our fowling pieces.

Our old pilot told us, that if we would enter a branch of a river on the right hand with our barge and wherries, and leave the galley at anchor the while in the great river, he would bring us to a town of the Arawaks, where we should find store of bread, hens, fish, and of the country wine; and persuaded us, that departing from the galley at noon, we might return ere night. I presently took my bark, with eight musketeers, Captain Gifford's wherry, with himself and four musketeers, and Captain Calfield with his wherry, and as many; and so we entered the mouth of this river: and because we were persuaded that it was so near, we took no victual with us at all. When we had rowed three hours, we marvelled we saw no sign of any dwelling, and asked the pilot where the town was: he told us a little further. After three hours more, the sun being almost set, we began to suspect that he led us that way to betray us. When it grew towards night; and we demanded where the place was; he told us but four reaches more. When we had rowed four and four; we saw no sign; and our poor water-men, even heart-broken, and tired, were ready to give up the ghost: for we had now come from the galley near forty miles.

At the last we determined to hang the pilot; and if we had well known the way back again by night, he had surely gone; but our own necessities pleaded sufficiently for his safety: for it was as dark as pitch, and the river began so to narrow itself, and the trees to hang over from side to side, as we were driven with arming swords to cut a passage through those branches that covered the water. We were very desirous to find this town, hoping of a feast, because we made but a short breakfast aboard the galley in the morning and it was now eight a clock at night, and our stomachs began to gnaw apace: but whether it was best to return or go on, we began to doubt, suspecting treason in the pilot more and more; but the poor old Indian ever assured us that it was but a little further: at the last about one a clock after midnight we saw a light; and rowing towards it, we heard the dogs of the village. When we landed we found few people; for the lord of that place was gone with divers canoes above four hundred miles off, upon a journey towards the head of Orinoco to trade for gold, and to buy women of the cannibals. In his house we had good store of bread, fish, hens, and Indian drink, and so rested that night, and in the morning after we had traded with such of his people as came down, we returned towards our galley, and brought with us some quantity of bread, fish, and hens.

On both sides of this river, we passed the most beautiful country that ever mine eyes beheld: and whereas all that we had seen before was nothing but woods, prickles, bushes, and thorns, here we beheld plains of twenty miles in length, the grass short and green, and in divers parts groves of trees by themselves. In the mean while our companies in the galley thought we had been all lost, (for we promised to return before night) and sent the *Lions Whelp*'s ship's boat with Captain Whiddon to follow us up the river; but the next day, after we had rowed up and down some fourscore miles, we returned, and went on our way, up the great river; Captain Gifford being before the galley and the rest of the boats, seeking out some place to land upon the banks to make fire, espied four canoes coming down the river; after a while two of the four gave over, and ran themselves ashore, every man betaking himself to the fastness of the woods. Those canoes that were taken were laden with bread: but in the lesser there were three Spaniards, who having heard of the defeat of their governor in Trinidad, and that we purposed to enter Guiana, came away in those canoes: one of them was a caballero, another a soldier, and the third a refiner.

In the meantime, nothing on the earth could have been more welcome to us, next unto gold, than the great store of very excellent bread which we found in these canoes; for now our men cried, Let us go on, we care not how far. I took my barge, and went to the bank's side with a dozen shot, where the canoes first ran themselves ashore and landed there: as I was creeping through the bushes, I saw an Indian basket hidden, which was the refiner's basket; for I found in it his quick-silver, saltpetre, and divers things for the trial of metals, and also the dust of such ore as he had refined, but in those canoes which had escaped there was a good quantity of ore and gold. I then landed more men, and offered five hundred pound to what soldier soever could take one of those three Spaniards that we thought were landed. But our labours were in vain; for they put themselves into one of the small canoes: and so while the greater canoes were in taking they escaped. But seeking after the Spaniards, we found the Arawaks hidden in the woods, which were pilots for the Spaniards, and rowed their canoes; of which I kept the chiefest for a pilot, and carried him with me to Guiana, by whom I understood where and in what countries the Spaniards had laboured for gold.

This Arawakan pilot with the rest, feared that we would have eaten them, or otherwise have put them to some cruel death (for the Spaniards, to the end that none of the people in the passage towards Guiana or in Guiana itself might come to speech with us, persuaded all the nations, that we were cannibals) but when the poor men and women had seen us, and that we gave them meat, and to every one something or other, which was rare and strange to them, they began to conceive the deceit and purpose of the Spaniards, who indeed (as they confessed) took from them both their wives and daughters daily, and used them for the satisfying of their own lusts. But I protest before the majesty of the living God, that I neither know nor believe, that any of our company, by violence or otherwise, ever knew any of their women, and yet we saw many hundreds, and had many in our power, and of those very young, and excellently favoured, which came among us without deceit, stark naked.

Nothing got us more love amongst them than this usage: for I suffered not any man to take from any of the nations so much as a pina, or a potato root, without giving them contentment, nor any man so much as to offer to touch any of their wives or daughters: which course so contrary to the Spaniards drew them to admire Her Majesty, whose commandment I told them it was, and also wonderfully to honour our nation.

I caused my Indian interpreter at every place when we departed, to know of the loss or wrong done, and if ought were stolen or taken by violence, either the same was restored, and the party punished in their sight, or else was paid for to their uttermost demand.

They also much wondered at us, after they heard that we had slain the Spaniards at Trinidad, and they wondered even more when I had made them know of that great overthrow that Her Majesty's army and fleet had given them of late years in their own countries.

After we had taken this supply of bread, I gave one of the canoes to the Arawaks, which belonged to the Spaniards that were escaped, I sent back by the same canoe, Ferdinando my first pilot, and gave sufficient victual to carry them back, and by them wrote a letter to the ships, which they promised to deliver, and performed it; the fifteenth day we discovered afar off the mountains of Guiana to our great joy, and towards the evening had a slant of a northerly wind that blew very strong, which brought us in sight of the great river Orinoco; out of which this river descended wherein we were.

That night we came to an anchor at the parting of the three goodly rivers, and landed upon a fair sand, where we found thousands of tortugas' eggs, which are very wholesome meat, and greatly restoring, so as our men were now well filled and highly contented both with the fare, and nearness of the land of Guiana which appeared in sight.

In the morning there came down according to promise the lord of that border called Toparimaca, with some thirty or forty followers, and brought us divers sorts of fruits, and of his wine, bread, fish, and flesh, whom we also feasted. I conferred with this Toparimaca of the next way to Guiana, who conducted us from thence some mile and a half to his town, where some of our captains caroused of his wine till they were reasonably pleasant, they keep it in great earthen pots of ten or twelve gallons very clean and sweet, and are themselves at their meetings and feasts the greatest carousers and drunkards of the world.

A stranger had his wife staying at the port where we anchored, and in all my life I have seldom seen a better favoured woman: she was of good stature, with black eyes, fat of body, of an excellent countenance, her hair almost as long as herself, tied up again in pretty knots, and it seemed she stood not in that awe of her husband, as the rest, for she spake and discoursed, and drank among the gentlemen and captains, and was very pleasant, knowing her own comeliness, and taking great pride therein. I have seen a lady in England so like to her, as but for the difference of colour, I would have sworn might have been the same.

The next day we hasted thence, and having an easterly wind to help us, we spared our arms from rowing. This river is navigable with barks, little less than a thousand miles, and from the place where we entered, it may be sailed up in small pinnaces. I judge the river in this place to be at least thirty miles broad, reckoning the islands which divide the branches in it.

The next morning towards nine of the clock, we weighed anchor, and the breeze increasing, we sailed always west up the river, and after a while opening the land on the right side, the country appeared to be champaign, and the banks showed very perfect red. I therefore sent two of the little barges with some few soldiers to march over the banks of that red land and to discover what manner of country it was on the other side: my old pilot, a man of great travel, told me that those were called the plains of the Sayma, and that there inhabited four principal nations. The fourth are called Aroras, and are as black as negroes, but have smooth hair, and these are very valiant, or rather desperate people, and have the most strong poison on their arrows, of which poison I will speak somewhat being a digression not unnecessary.

There was nothing whereof I was more curious, than to find out the true remedy of these poisoned arrows: for besides the mortality of the wound they make, the party shot endureth the most insufferable torment in the world and abideth a most ugly and lamentable death, sometimes dying stark mad, sometimes their bowels breaking out of their bellies. It is more strange to know, that in all this time there was never Spaniard either by gift or torment that could attain to the true knowledge of the cure, although they have martyred and put to invented torture I know not how many of them. But every one of these Indians know it not, but their soothsayers and priests, who do conceal it, and only teach it but from the father to the son.

Those medicines which are vulgar, and serve for the ordinary poison, are made of the juice of a root called tupara: the same also quencheth marvellously the heat of burning fevers, and healeth inward wounds, and broken veins, that bleed within the body. They taught me the best way of healing as well thereof, as of all other poisons. Some of the Spaniards have been cured in ordinary wounds, of the common poisoned arrows with the juice of garlic: but this is a general rule for all men that shall hereafter travel the Indies where poisoned arrows are used, that they must abstain from drink, for if they take any liquor into their body, as they shall be marvellously provoked thereunto by drought, I say, if they drink before the wound be dressed, or soon upon it, there is no way with them but present death.

The next day we arrived at the port of Morequito, and anchored there, sending away one of our pilots to seek the king of Aromaia, uncle to Morequito slain by Berreo as aforesaid. He came to us on foot from his house, which was fourteen English miles, and with him many of the borderers, with many women and children, that came to wonder at our nation, and to bring us down victual, which they did in great plenty, as venison, pork, hens, chickens, fowl, fish, with divers sorts of excellent fruits and roots, and great abundance of pinas, the princess of fruits, that grow under the sun. They brought us also store of bread, and of their wine, and a sort of paraquitos, no bigger than wrens, and of all other sorts both small and great; one of them gave me a beast called by the Spaniards armadillo, which seemeth to be all barred over with small plates somewhat like to a rhinoceros.

After this old king had rested a while in a little tent, that I caused to be set up, I began by my interpreter to discourse with him of the death of Morequito his predecessor, and afterward of the Spaniards, and ere I went any farther I made him know the cause of my coming thither, whose servant I was, and that the Queen's pleasure was, I should undertake the voyage for their defence, and to deliver them from the tyranny of the Spaniards, dilating at large Her Majesty's greatness, her justice, her charity to all oppressed nations, with as many of the rest of her beauties and virtues, as either I could express, or they conceive: I began to sound the old man as touching Guiana, and the state thereof, what sort of common wealth it was, how governed, of what strength and policy, how far it extended, and what nations were friends or enemies adjoining.

I asked what nations those were which inhabited on the farther side of those mountains, beyond the valley of Amariocapana: he answered with a great sigh (as a man which had inward feeling of the loss of his country and liberty) that he remembered in his father's lifetime when he was very old, and himself a young man, that there came down into that large valley of Guiana, a nation from so far off as the sun slept, (for such were his own words) with so great a multitude as they could not be numbered nor resisted, and that they wore large coats and hats of crimson colour, which colour he expressed by showing a piece of red wood, wherewith my tent was supported, and had now made themselves lords of all, even to that mountain foot.

I desired him to rest with us that night, but I could not entreat him, but he told me that at my return from the country above, he would again come to us, and in the mean time provide for us the best he could, of all that his country yielded.

The next morning we also left the port, and sailed westward up to the river, for that I understood it led to the strongest nations of all the frontiers, and that night we anchored at another island. When we were short of it as low or further down as the port of Morequito we heard the great roar and fall of the river, but when we came to enter with our barge and wherries thinking to have gone up some forty miles, we were not able with a barge of eight oars to row one stone's cast in an hour, and yet the river is as broad as the Thames at Woolwich, and we tried both sides, and the middle, and every part of the river, so as we encamped upon the banks adjoining, (it was on this river side that Morequito slew the friar, and those nine Spaniards which came from Manoa, the city of Inca, and took from them forty thousand pesos of gold).

Upon this river one Captain George, that I took with Berreo told me there was a great silver mine. But by this time, all the rest of the rivers were risen four or five foot in height, so that it was not possible by the strength of any men, or with any boat whatsoever to row into the river against the stream. I therefore sent Captain Thyn, and some thirty shot more to coast the river by land, and to go to a town some twenty miles over the valley called Amnatapoi, and the mean while myself, and some half dozen shot marched over land to view the strange overfalls of the river which roared so far off: I sent also Captain Whiddon, and some eight shot to see if they could find any mineral stone alongst the river side. When we were come to the tops of the first hills of the plains adjoining to the river, we might from that mountain see the river how it ran in three parts, above twenty miles off, and there appeared some ten or twelve overfalls[145] in sight, every one as high over the other as a church-tower, which fell with that fury, that the rebound of water made it seem, as if it had been all covered over with a great shower of rain: and in some places we took it at the first for a smoke that had risen over some great town. For mine own part I was well persuaded from thence to have returned, being a very ill footman, but the rest were all so desirous to go near the said strange thunder of waters, as they drew me on by little and little, till we came into the next valley where we might better discern the same. I never saw a more beautiful country, nor more lively prospects, hills so raised here and there over the valleys, the river winding into divers branches, the plains adjoining without bush or stubble, all fair green grass, the ground of hard sand easy to march on, either for horse or foot, the deer crossing in every path, the birds towards the evening singing on every tree with a thousand several tunes, cranes and herons of white, crimson, and carnation perching in the river's side, the air fresh with a gentle easterly wind, and every stone that we stooped to take up, promised either gold or silver by his complexion. Your lordship shall see of my sorts, and I hope some of them cannot be bettered under the sun, and yet we had no means but with our daggers and fingers to tear them out here and there, the rocks being most hard, and besides the veins lie a fathom or two deep in the rocks. Some of these stones I showed afterwards to a Spaniard of the Caracas, who told me that it was *el Madre del oro*, that is the mother of gold, and that the mine was farther in the ground. But it shall be found a weak policy in me, either to betray myself, or my country with imaginations, neither am I so far in love with that lodging, watching, care, peril, diseases, ill savours, bad fare, and many other mischiefs that accompany these voyages, as to woo myself again into any of them, were I not assured that the sun covereth not so much riches in any part of the earth. Captain Whiddon, brought me a kind of stones like sapphires, what they may prove I know not. I showed them to some of the Orinoco [Indians] and they promised to bring me to a mountain, that had of them very large pieces growing diamond wise: whether it be crystal of the mountain, Bristol-diamond,[146] or sapphire I do not yet know.

There is another goodly river beyond Caroni which is called Aro. Next unto Aro there are two rivers Atoica and Caura, and on that branch which is called Caura, are a nation of people, whose heads appear not above their shoulders; which though it may be thought a mere fable, yet for mine own part I am resolved it is true, because every child in the provinces of Arromaia and Canuri affirm the same: they are reported to have their eyes in their shoulders, and their mouths in the middle of their breasts, and that a long train of hair groweth backward between their shoulders. Such a nation was written of by Mandeville,[147] whose reports were held for fables for many years, and yet since the East Indies were discovered, we find his relations true of such things as heretofore were held incredible: for mine own part I saw them not, but I am resolved that so many people did not all combine, or forethink to make the report.

While we lay at anchor on the coast of Caroni, and had taken knowledge of all the nations upon the head and branches of this river, and had found out so many several people, which were enemies to the new conquerors: I thought it time lost to linger any longer in that place, especially for that the fury of Orinoco began daily to threaten us with dangers in our return: for no half day passed but the river began to rage and overflow very fearfully, and the rains came down in terrible showers, and gusts in great abundance: and withal, our men began to cry out for want of shift, for no man had place to bestow any other apparel than that which he wore upon his back, and that was thoroughly washed on his body for the most part ten times in one day: and we had now been well near a month, every day passing to the westward farther and farther from our ships.

The next day we left the mouth of Caroni, and arrived again at the port of Morequito where we were before: for passing down the stream we went without labour, and against the wind little less than a hundred miles a day. As soon as I came to anchor, I sent away one for old Topiawari, with whom I much desired to have further conference, and also to deal with him for some one of his country, to bring with us into England, as well to learn the language, as to confer withal by the way. I desired him to instruct me what he could, both of the passage into the golden parts of Guiana, and to the civil towns and apparelled people of Inca. He gave me an answer to this effect: first that he could not perceive that I meant to go onward towards the city of Manoa, for neither the time of the year served, neither could he perceive any sufficient numbers for such an enterprise: and if I did, I was sure with all my company to be buried there, for the emperor was of that strength, as that many times so many men more were too few: besides he gave me this good counsel and advised me to hold it in mind that I should not offer by any means hereafter to invade the strong parts of Guiana without the help of all those nations which were also their enemies: for that it was impossible without those, either to be conducted, to be victualled, or to have aught carried with us our people not being able to endure the march in so great heat, unless the borderers gave them help. He told me further that 4 days' journey from his town, were the next and nearest subjects of Inca, and the first town of apparelled and rich people, and that all those plates of gold which were scattered among the borderers and carried to other nations far and near, were there made, but that those of the land within were far finer, and were fashioned after the images of men, beasts, birds, and fishes. I asked him whether he thought that those companies that I had there with me, were sufficient to take that town or no? He told me that he thought they were. I then asked him whether he would assist me with guides, and some companies of his people to join with us? He answered that he would go himself with all the borderers, if the rivers did remain fordable, upon this condition that I would leave with him till my return again fifty soldiers, which he undertook to victual: I answered that I had not above fifty good men in all there, the rest were labourers and rowers, and that I had no provision to leave with them of powder, shot, apparel, or aught else, and that without those things necessary for their defence, they should be in danger of the Spaniards in my absence, who I knew would use the same measure towards mine, that I offered them at Trinidad.

He further alleged, that the Spaniards sought his death, they had him seventeen days in a chain before he was king of the country, and led him like a dog from place to place, until he had paid an hundred plates of gold for his ransom: and now since he became owner of that province, that they had many times laid wait to take him, and that they would be now more vehement, when they should understand of his conference with the English: he therefore prayed us to defer it till next year, when he would undertake to draw in all the borderers to serve us, and then also it would be more seasonable to travel, for at this time of the year, we should not be able to pass any river, the waters were and would be so grown ere our return.

I thought it were evil counsel to have attempted it at that time, although the desire of gold will answer many objections: but it would have been in mine opinion an utter overthrow to the enterprise, if the same should be hereafter by Her Majesty attempted: as yet our desire of gold, or our purpose of invasion is not known to them: and it is likely that if Her Majesty undertake the enterprise, they will rather submit themselves to her obedience than to the Spaniards, of whose cruelty both themselves and the borderers have already tasted: and therefore till I had known Her Majesty's pleasure, I would rather have lost the sack of one or two towns (although they might have been very profitable) than to have defaced or endangered the future hope of so many millions, and the great good, and rich trade which England may be possessed of thereby. I am assured now that they will all die even to the last man against the Spaniards in hope of our succour and return.

After that I had resolved Topiawari, Lord of Aromaia, that I could not at this time leave with him the companies he desired, he freely gave me his only son to take with me into England, and hoped, that though he himself had but a short time to live, yet that by our means his son should be established after his death: and I left with him one Francis Sparrow, a servant of Captain Gifford, (who was desirous to tarry, and could describe a country with his pen) and a boy of mine called Hugh Goodwin,[148] to learn the language. I after asked the manner how the Epuremei wrought those plates of gold, and how they could melt it out of the stone; he told me that the most of the gold which they made in plates and images, was not severed from the stone, but that on the lake of Manoa, and in a multitude of other rivers they gathered it in grains of perfect gold and in pieces as big as small stones, and that they put it to a part of copper, otherwise they could not work it, and that they used a great earthen pot with holes round about it, and when they had mingled the gold with the copper together, they fastened canes to the holes, and so with the breath of men they increased the fire till the metal ran, and then they cast it into moulds of stone and clay, and so to make those plates and images. I have sent your honours of two sorts such as I could by chance recover, more to show the manner of them, than for the value: for I did not in any sort make my desire of gold known, because I had neither time, nor power to have a greater quantity. I gave among them many more pieces of gold, than I received, of the new money of 20 shillings with Her Majesty's picture to wear, with promise that they would become her servants henceforth.

I have also sent the ore, whereof I know some is as rich as the earth yieldeth any. But besides that we were not able to tarry and search the hills, so we had neither pioneers, bars, ledges, nor wedges of iron to break the ground, without which there is no working in mines: but we saw all the hills with stones of the colour of gold and silver. I then parted from old Topiawari, and received his son for a pledge between us, and left him with two of ours. To Francis Sparrow I gave instructions if it were possible, to go on to the great city of Manoa: which being done, we weighed anchor.

We rowed down the stream. I sent away Captain Henry Thyn with the galley, the nearest way, and took with me Captain Gifford, with mine own barge, and the two wherries, and went down that branch of Orinoco, which is called Cararoopana.

In Cararoopana were also many goodly islands, some of six miles long, some of ten, and some of twenty. When it grew towards sunset, we entered a branch of a river that fell into the Orinoco called Winicapora: where I was informed of the mountain of crystal, to which in truth for the length of the way, and the evil season of the year, I was not able to march, nor abide any longer upon the journey: we saw it afar off and it appeared like a white church tower of an exceeding height. There falleth over it a mighty river which toucheth no part of the side of the mountain, but rusheth over the top of it, and falleth to the ground with so terrible a noise and clamour, as if a thousand great bells were knocked against one another. Berreo told me that there were diamonds and other precious stones on it, and that they shined very far off: but what it hath I know not, neither durst he or any of his men ascend to the top of the said mountain, those people adjoining being his enemies.

We landed on the island of Assapano, and there feasted ourselves with that beast which is called armadillo. The day following we recovered the galley at anchor, and the same evening departed with very foul weather and terrible thunder and showers, for the winter was come on very far: the best was, we went no less than 100 miles a day, down the river; but by the way we entered, it was impossible to return, both the breeze and the current of the sea were so forcible.

I protest before God, that we were in a most desperate estate: for the same night which we anchored in the mouth of the river, where it falleth into the sea, there arose a mighty storm, and the river's mouth was at least a league broad, so as we ran before night close under the land with our small boats, and brought the galley as near as we could, but she had as much ado to live as could be, and there wanted little of her sinking, and all those in her; for mine own part I confess, I was very doubtful which way to take, either to go over in the pestered galley, there being but six foot water over the sands, for two leagues together, and that also in the channel, and she drew five: or to adventure in so great a billow, and in so doubtful weather, to cross the seas in my barge. After it cleared up, about midnight, we put ourselves to God's keeping, and thrust out into the sea, leaving the galley at anchor, who durst not adventure but by daylight: and so being all very sober, and melancholy, one faintly cheering another to show courage, it pleased God that the next day about nine of the clock, we descried the island of Trinidad. We kept the shore till we came to where we found our ships at anchor, than which there was never to us a more joyful sight.

Now that it hath pleased God to send us safe to our ships, it is time to leave Guiana to the sun, whom they worship, and steer away towards the north.

The religion of the Epuremei is the same which the Incas, Emperors of Peru used, they believe in the immortality of the soul, worship the sun, and bury with them alive their best beloved wives and treasure, as they likewise do in Pegu in the East Indies, and other places. The Orinoco [Indians] bury not their wives with them, but their jewels, hoping to enjoy them again. The Arawaks dry the bones of their lords, and their wives and friends drink them in powder. They have all many wives, and the lords five-fold to the common sort: their wives never eat with their husbands, nor among the men, but serve their husbands at meals, and afterwards feed by themselves. Those that are past their younger years, make all their bread and drink, and work their cotton beds, and do all else of service and labour, for the men do nothing but hunt, fish, play, and drink, when they are out of the wars.

I was informed of one of the caciques which had buried with him a little before our arrival, a chair of gold most curiously wrought: but if we should have grieved them in their religion at the first, before they had been taught better, and have digged up their graves, we had lost them all.

I will promise these things that follow, which I know to be true. Those that are desirous to discover may be satisfied with this river, above 2000 miles east and west, and 800 miles south and north, and of these, the most either rich in gold, or in other merchandises. The common soldier shall here fight for gold, and pay himself instead of pence, with plates of half a foot broad, whereas he breaketh his bones in other wars for provender and penury. Those commanders and chieftains that shoot at honour and abundance, shall find there more rich and beautiful cities, more temples adorned with golden images, more sepulchres filled with treasure, than either Cortez found in Mexico, or Pizarro in Peru: and the shining glory of this conquest will eclipse all those so far extended beams of the Spanish nation. There is no country which yieldeth more pleasure to the inhabitants, either for the common delights of hunting, hawking, fishing, fowling, or the rest, than Guiana doth.

Both for health, good air, pleasure, and riches I am resolved it cannot be equalled by any region either in the east or the west. Moreover the country is so healthful, as of an hundred persons and more (which lay without shift most sluttishly, and were every day almost melted with heat in rowing and marching, and suddenly wet again with great showers, and did eat of all sorts of corrupt fruits, and made meals of fresh fish without seasoning, of tortugas, of crocodiles, and besides lodged in the open air every night) we lost not any one, nor had one ill disposed to my knowledge, nor found any calentura, or other of those pestilent diseases which dwell in all hot regions.

Where there is store of gold, it is in effect needless to remember other commodities for trade: but it hath towards the south part of the river, great quantities of brazil-wood, and diverse berries that dye a most perfect crimson and carnation. All places yield abundance of cotton, of silk, of balsam, and of those kinds most excellent and never known in Europe, of all sorts of gums, of Indian pepper: and what else the countries may afford within the land we know not, neither had we time to abide the trial, and search. The soil besides is so excellent and so full of rivers, as it will carry sugar, ginger, and all those other commodities, which the West Indies have.

The navigation is short, for it may be sailed with an ordinary wind in six weeks, and in the like time back again.

Guiana is a country that hath yet her maidenhead, never sacked, turned, nor wrought, the face of the earth hath not been torn, nor the virtue and salt of the soil spent by manurance, the graves have not been opened for gold, the mines not broken with sledges, nor their images pulled down out of their temples. It hath never been entered by any army of strength, and never conquered by any Christian prince. It is besides so defensible, that if two forts be built in one of the provinces which I have seen, the flood setteth in so near the bank, where the channel also lieth, that no ship can pass up but within a pike's length of the artillery, first of the one, and afterwards of the other.

Guiana hath but one entrance by the sea (if it hath that) for any vessels of burden: so as whosoever shall first possess it, he shall be found unaccessible for any enemy, except he come in wherries, barges, or canoes, or else in flat bottomed boats, and if he do offer to enter it in that manner, the woods are so thick two hundred miles together upon the rivers of such entrance, as a mouse cannot sit in a boat unhit from the bank. By land it is more impossible to approach, for it hath the strongest situation of any region under the sun, and is so environed with impassable mountains on every side, as it is impossible to victual any company in the passage: which hath been well proved by the Spanish nation, who since the conquest of Peru have never left five years free from attempting this empire, or discovering some way into it, and yet of three and twenty several gentlemen, knights, and noblemen, there was never any that knew which way to lead an army by land, or to conduct ships by sea, anything near the said country. Don Antonio de Berreo (whom we displanted) the last: and I doubt much, whether he himself or any of his yet know the best way into the said empire.

The West Indies were first offered Her Majesty's grandfather by Columbus a stranger, in whom there might be doubt of deceit, and besides it was then thought incredible that there were such and so many lands and regions never written of before. This empire is made known to Her Majesty by her own vassal, and by him that oweth to her more duty than an ordinary subject, so that it shall ill sort with the many graces and benefits which I have received to abuse Her Highness, either with fables or imaginations. The country is already discovered, many nations won to Her Majesty's love and obedience, and those Spaniards which have latest and longest laboured about the conquest, beaten out, discouraged and disgraced, which among these nations were thought invincible. Her Majesty may in this enterprise employ all those soldiers and gentlemen that are younger brethren, and all captains and chieftains that want employment, and the charge will be only the first setting out in victualling and arming them: for after the first or second year I doubt not but to see in London a Contractation House of more receipt for Guiana, than there is now in Seville for the West Indies.

I am resolved that if there were but a small army a foot in Guiana, marching towards Manoa the chief city of Inca, he would yield to Her Majesty by composition so many hundred thousand pounds yearly, as should both defend all enemies abroad, and defray all expenses at home, and that he would besides pay a garrison of three or four thousand soldiers very royally to defend him against other nations.

For whatsoever prince shall possess it, shall be greatest, and if the King of Spain shall enjoy it, he will become unresistible. Her Majesty hereby shall confirm and strengthen the opinions of all nations, as touching her great and princely actions.

I trust in God, this being true, will suffice, and that he which is King of all Kings and Lord of Lords, will put it into her heart which is Lady of Ladies to possess it.

LXVI

A voyage of Master William Parker of Plymouth gentleman, to Jamaica, Puerto de Cavallos situated within the bay of Honduras, and taken by Sir Anthony Sherley and him, and his valiant and happy enterprise upon Campeche the chief town of Yucatan, which he took and sacked with six and fifty men.

In the year 1596, Master William Parker of Plymouth gentleman being furnished with a tall ship called the *Prudence* of one hundred and twenty tons, wherein himself went captain, and the bark called the *Adventure* of five and twenty tons, departed from Plymouth in the month of November, having one hundred men in his company.

The first place where we touched in the West Indies was the isle of Margarita,[149] where we took a Spanish gentleman and others, who for his ransom set at liberty Master James Willis, and five other Englishmen which were prisoners in Cumana, who otherwise were never like to have come from thence. We sailed over to the isle of Jamaica, where the second of March we met with Sir Anthony Sherley, who before our coming had taken the chief town in the island. Here consorting ourselves with him, we departed from Jamaica the sixth of March, and resolved to set upon the strong town of Trujillo near the mouth of the bay of the Honduras. We purposed to entrap the watch, but the watch discovering us, made great fires, and the town presently shot off a great piece, and answered with fires. Notwithstanding, the next day being the one and thirtieth of March, we brought our ships under the fort, and landed our men: but it was a vain purpose, for the town is invincible by nature, and standeth upon the top of a very steep hill joining close to the sea, environed with woods of such exceeding thickness, that there is no passage through the trees: there is also but one very narrow and steep lane to go into the town, at the end whereof is a gate very strongly fortified: so that it is not to be approached unto, unless it be upon the sudden, and with surprise of the watch: wherefore with the loss of some few men we retired from this enterprise.

From hence we passed up farther into the gulf the second of April, to invade the town of Puerto de Cavallos, finding it well fortified, but nothing answering our expectation for wealth. Whereupon Sir Anthony Sherley and I being hitherto frustrate of our hopes, resolved here to enter up to the bottom of Rio Dulce,[150] and to pass overland unto the South Sea. Wherefore we set forward, and entered above thirty leagues up the said Rio Dulce, thinking to have passed overland with two companies of men, and to have carried a pinnace in six quarters to be set together with screws, and therein to have embarked ourselves in the South Sea, and there for a time to have tried our fortune. But this our diligence took no effect, because of the huge highness of the mountains, and the length of the way, being more than was given out at the first. Then with much grief we returned out to Trujillo, where I departed from Sir Anthony Sherley.

After my departure from this worthy knight, I set my course for the east part of Yucatan from whence I ranged all the north coast of the said promontory, until I came unto Cape Desconoscido, where I put 56 of my men into a paragua, or long Indian canoe; leaving my ship six leagues from the town of Campeche at three of the clock in the morning I landed hard by the monastery of San Francisco, and took the said town of Campeche, with the captain and alcalde, finding therein five hundred Spaniards, and in two towns close adjoining to the same eight thousand Indians. The multitude of the Spaniards which fled upon my first assault by ten of the clock in the morning assembling together renewed their strength, and set furiously upon me and my small company. In which assault I lost some six of my men, and myself was shot under the left breast with a bullet, which bullet lieth still in the chine of my back. Being thus put unto our shifts we devised a stratagem: for having divers of the townsmen prisoners, we tied them arm in arm together, and placed them instead of a barricade to defend us from the fury of the enemies' shot. And so with ensign displayed, taking with us our six dead men, we retired with more safety to the haven, where we took a frigate which rode ready freighted with the King's tribute in silver and other good commodities, and brought the same and our paragua to my ship, which lay in two fathom water six leagues from the town, not being able to come any nearer for the shoals upon that coast. Over against the place where our ship rode, stood a town of 300 or 400 Indians called Sebo, which we likewise took, where we found Campeche wood good to dye withall, with wax, and honey. This done we left the coast and turned up to Cape de Catoche again, and in turning up I lost my bark called the *Adventure*, which was taken by two frigates of war, manned out from Campeche: wherein Captain Hen and thirteen of my men were taken, and afterwards executed, as since we understand. After we had stayed five weeks on this coast, we shaped our course for Havana, where finding nothing, we disembarked, and came along by the isle of Bermuda, and from thence sailing for England, we fell with Scilly about the first of July, and within two days after arrived at Plymouth, where we found the Right Honourable the Earl of Essex setting forth with a great fleet for the isles of the Azores.

BIOGRAPHICAL REPERTORY

ACHIM: Sultan Achim (Abu Ali al-Mansur al-Hakim) went counter to the Islamic rule of toleration by forcing Jews and Christians to wear humiliating tokens of identity. In 1009 he demolished several Christian churches including the Holy Sepulchre in Jerusalem, thus helping to provoke the First Crusade. At last he declared himself an incarnation of God, a claim still believed by the Druses. He died by violence.

ALEXIUS: Alexius Comnenus (1048–1118) was Byzantine Emperor (1081–1118). An officer who had served with distinction against the Seljuk Turks, he ousted Nicephorus III and made a brave attempt to preserve and extend the borders of Byzantium. He halted the attack made by the Normans under Robert Guiscard through Epirus into Thessaly and Macedonia. On his northern borders he held off Slav raiders and their heretic allies; and he achieved a treaty with the Turks. He stabilized the currency and was ready to take the offensive against the enemies of Byzantium when the First Crusade arrived. In the long run Christian Byzantium was to have no worse enemies than the crusaders.

BURROUGH: Stephen Burrough (often Borough) (1525–84), born at Northam in Devon, was one of the 1553 Russian expedition's twelve 'counsellors'. He served under Chancellor (q.v.) in *Bonaventure* and brought the ship back the only survivor of the fleet. He was usually pilot for the Russia Company's voyages; in 1556 in *Searchthrift* he found the strait giving entrance to the Kara Sea. About 1558 he was sent by the Crown to Seville, whence he brought back and had translated, 'Arte of Navigation' (1561).

BURROUGH: William Burrough (often Borough) (1536–90) served as a seaman in the first Russian voyage under his brother Stephen (q.v.), and continued sailing round North Cape for the next ten years. In 1570 as captain-general of a fleet of thirteen armed ships he disposed of Danish and other pirates who threatened the route. In

1574–5 he was the Company's agent in Russia. He made charts of the Polar seas, and in *Discourse of the Variation of the Compass*, pointed out the uselessness to navigators of charts which made no allowance for the magnetic variation. In 1583 as Comptroller of the Queen's Navy he managed to catch and hang ten pirates more. He commanded *Lion* in Drake's 1587 raid on Cadiz, when 100 sail were destroyed; but disagreed with Drake on tactics and was put under arrest. In the Armada fight he captained the small ship *Bonavolia*.

CABOT: Sebastian Cabot (1474–1557) was second son of a Venetian pilot John Cabot, who had settled in Bristol about 1472. In 1497 Sebastian sailed with his father to Cape Breton Island and Nova Scotia, and in 1499 explored the American coast from Labrador to latitude 30° N. More talented as a cartographer than as a leader of men, he went into the Spanish service, commanded an expedition to Brazil in 1526, not altogether successfully, and was for eleven years an examiner of Spanish pilots. The world map he drew in 1544 served Ortelius (q.v.) as a basis. Invited back to England in 1548, he was granted a royal pension and became governor for life of the Company of Merchant Venturers formed in 1551 which, when continental war caused trade to stagnate, sent out Chancellor and Willoughby to open up a new market in Russia. When Philip of Spain arrived in London as Mary's consort, Sebastian Cabot's pension was cut.

CAVENDISH: Thomas Cavendish (sometimes Candish) (1560–92) was born near Harwich, and sailed his own ship in the squadron which under Grenville (q.v.) planted Raleigh's colony in Virginia, 1585. In 1586 he sailed round the world, being told by a Flemish ship on his arrival in September of 1588 that the Armada was defeated. He reputedly squandered his wealth at court – hence the second voyage which left Plymouth in 1591, Captain John Davis (q.v.) commanding in *Desire*. Encountering a difficult passage at Magellan's Straits, Cavendish separated from Davis and died at sea near Ascension Island.

CECIL: Sir Robert Cecil (1563?–1612), a son of Lord Burghley, left Cambridge to serve as a diplomat in France, at the time when Hakluyt was ambassador's chaplain there. Francis Bacon was his cousin. Frail, short and bent, Queen Elizabeth referred to him as 'my little elf'. In 1591 she knighted him and brought him into the Privy Council.

In 1596 he was given the title of Secretary of State having long done the work of the office. He began sagaciously to prepare for the accession of James I, who in 1605 made him Earl of Salisbury. His craze was for building, and Hatfield is his monument.

CHANCELLOR: Richard Chancellor (d. 1556) made a voyage with Roger Bodenham to Crete and Chios in 1550. His patron was Sir Henry, father of Sir Philip Sidney. He was chosen pilot-general of Willoughby's expedition in 1553, and captain of *Edward Bonaventure*. In 1556 when making the return journey in November, he was cast away in *Bonaventure* in Aberdour Bay on the Aberdeenshire coast. He left two sons.

CHARLES V: Charles V (1500–1558), became King of Spain in 1516, and was the last Holy Roman Emperor (1519–56) to be crowned by the Pope. He had two life ambitions: to restore church unity, and to resist the Turks. This was asking too much. As he sought to crush the Lutherans, the Turks besieged Vienna (1529), obliging him to buy Protestant support by acknowledging the Augsburg Confession. In 1534 the Turkish admiral Barbarossa took Tunis; in 1535 Charles recaptured it; but in 1541 his attack on Algiers was less successful. Francis I of France was his former rival for the Empire and once had been taken prisoner by Charles after the Battle of Pavia. Francis always played the Turks off against the Holy Roman Emperor. In 1536 when French troops entered Turin, Charles suggested to Francis – but in vain – that the pair of them settle the question by personal combat, with Burgundy and Milan as the stakes, winner to receive command in a crusade against the Turks. After 1556 having allocated Germany to his brother Ferdinand and married Philip off to Mary of England, the Emperor Charles abdicated and retired to a monastery.

CONRAD: King Conrad of Germany (*c.* 990–1039) was the first Western Emperor of the Salian dynasty. Tight-fisted and acquisitive, he was elected to the throne in 1024, and crowned in Rome in 1027. In 1032 he succeeded to Burgundy, to which he had a claim through his wife Gisela.

CUMBERLAND: George Clifford, third Earl of Cumberland, (1558–1605) was born at Brougham Castle in Westmorland. When at

Trinity College, Cambridge, he visited Oxford to study geography, possibly under Hakluyt. He married unhappily, ran through his property, and in 1586, fitting out three ships, he sailed from Plymouth to the River Plate, returning in September 1587, though with not much plunder. From Armada year on, he made privateering raids on Spanish shipping, all adventurous, not all profitable. In 1597–8, with a flotilla of twenty ships fitted out at his own cost, he took Porto Rico in Dominica. Most of his expeditions were mismanaged. A handsome, strong and courageous man, with three large moles on his left cheek, he wore in his hat the Queen's glove, set in diamonds. In Armada year he commanded a Queen's ship, *Elizabeth Bonaventure*, in the action off Gravelines, and was chosen to take the news of success to the royal camp at Tilbury. In 1592, Oxford made him M.A., and the Queen, a Knight of the Garter. He ran through his money fast, at horses and dice, and at his death in London, owed a thousand pounds.

DAVIS: John Davis (1550?–1605) was born at Stoke Gabriel near Dartmouth, a neighbour and acquaintance of those other poor but ambitious West Country gentry, the Gilberts and Raleighs. He made several voyages when young with Adrian Gilbert, and was described, when given command of the 1585 voyage to find the Northwest Passage, as 'very well grounded ... in the art of navigation'. He made further Arctic voyages in 1586 and 1587, sailing through Davis Strait into Baffin Bay, and reaching latitude 73° N. in a ship of not more than 20 tons. His *World's Hydrographical Description* (1595) seeks to prove that the sea is everywhere navigable, and the Northwest Passage therefore feasible. He invented a navigational instrument, the backstaff and double quadrant, and his *Seaman's Secrets* (1594) ran through eight editions to 1657. In 1589 he served with Cumberland (q.v.) off the Azores, and in 1591 commanded *Desire* in the squadron led by Cavendish (q.v.). He served in the Essex expedition to Cadiz and the Azores (1596) and two years later took service as a pilot on a Dutch ship to the Indies, where he saw much fighting. He went out with Lancaster's (q.v.) fleet (1600–1603) as pilot-major. In 1604, when pilot of *Tiger* (240 tons) on a voyage to Sumatra, he was killed by Japanese pirates off Borneo, leaving three sons.

DON (or *DOM*) *ANTONIO* (1531–95). When King Sebastian of Portugal (1557–78) was killed in battle against the Moors, his elderly great-uncle, Cardinal Henry, was next in line. Philip II of Spain had a claim to the Portuguese throne as a nephew of Sebastian's father. When, in 1580, he sent a Spanish army under the Duke of Alva, the illegitimate son of a former Portuguese king's brother was proclaimed King Antonio I, but his supporters were crushed, the Azores island of Terceira holding out longest. Antonio sought alliances with France and England, but in 1582 a French expedition to establish him in the Azores was defeated, and in 1589 an attempt on Lisbon, though led by Drake and Sir John Norris, failed. The Portuguese for their part continued to believe that King Sebastian was still alive. Four pretenders cropped up in turn, the last in Venice in 1600.

DRAKE: Sir Francis Drake (1540?–96) was born near Tavistock in Devon, with only distant claim to gentle blood, of a father said to have suffered for his Protestant convictions, who apprenticed him to the master of a coasting vessel. When this man died, he left Francis his ship. In 1565–7 Drake made voyages to Guinea and the Spanish Main, commanding *Judith* during the disaster at San Juan de Ulloa. In 1572 at Nombre Dios Bay, Cartagena and Porto Bello, he started to equal the account. In 1577 he took *Golden Hind* into the Pacific, captured the treasure ship *Cacafuego*, sailed north as far as the Golden Gate, and home to Deptford (where Elizabeth knighted him) via Celebes, the Cape of Good Hope, and Sierra Leone. In 1582 Drake was mayor of Plymouth, and in 1584–5 M.P. for Bossiney. In 1585, after Spain imposed an embargo on English ships, he raided Vigo, San Domingo in Hispaniola, and Cartagena, bringing back the beleaguered colonists from Virginia, and with them in all probability tobacco and the potato. In 1587 Drake successfully pillaged Armada supplies in Cadiz; he was second-in-command against the Armada. Short, red-haired, broad chested, Drake was energetic, vastly capable and notoriously intolerant. In 1595 he and John Hawkins took a fleet of twenty-seven sail with 2,500 men to raid the West Indies. Hawkins died in Porto Rico. Drake died of dysentery off Porto Bello, and was buried at sea in a lead coffin.

FITCH: Ralph Fitch (1550?–1611) travelled to Aleppo in 1583 with merchants of the Levant Company and went the overland route down the Euphrates Valley to India. On arrival in Goa he was imprisoned; his release was achieved by Thomas Stevens, a Jesuit, formerly of New College, Oxford, the first Englishman known to have reached India by the Cape route. In 1586 Fitch sailed for Burma, and in 1588 reached Malacca, returning via Ormus and Basra and up the Tigris to Aleppo, reaching London in 1591. He was later brought in to advise the newly organized East India Company, their court minutes of 2 October 1600 reading 'ordered that Captain Lancaster together with [. . .] Mr Fitch shall in the meeting tomorrow . . . confer of the merchandise to be provided for the [first] voyage.'

FROBISHER: Sir Martin Frobisher (1535?–94). Yorkshire born, became an orphan ward of Sir John York, Master of the Mint and a merchant venturer of the Russia Company who was later imprisoned and stripped of his wealth for his part in the Lady Jane Grey conspiracy. Frobisher in consequence at the age of eleven was sent to sea, and traded along North Africa to the Levant. In 1566 he was examined 'on suspicion of his having fitted out a vessel as a pirate'. From 1571 he was at sea in the government service off the Irish coast. In 1575 he was granted a licence to go on an expedition for a Northwest Passage. He sailed in two vessels of 25 tons and a pinnace of 10 tons, returning to Harwich in October 1576 with an Eskimo and a quantity of black pyrites which the Italian alchemist Agnello, running counter to the opinions of some London goldsmiths, pronounced to be gold. Sent a second time, in 1577, for gold rather than on a voyage of exploration, Frobisher sighted West Greenland, and returned with 200 tons of ore. All England rejoiced, but no furnace could be got hot enough to 'bring the work to the desired perfection'. After having again returned loaded with pyrites on his third voyage made with fifteen vessels, popular feeling turned against Frobisher. The next Northwest Passage discovery in 1582 was commanded by his lieutenant, Fenton. In 1585 Frobisher captained *Primrose* in Drake's assault on Cartagena. He commanded *Triumph* against the Armada, and was knighted at sea by the Lord High Admiral. In 1593 he was made Justice of the Peace for the West Riding. Frobisher was wounded in the hip in 1594 during an amphibious operation when commanding *Dreadnought* in support of Sir John Norris's relief of Brest from attack by the Spaniards. On return to Plymouth he died.

GILBERT: Sir Humphrey Gilbert (1539?–83), born at Compton near Dartmouth, was step-brother of Sir Walter Raleigh (q.v.) and educated at Eton and Oxford. Gilbert served as captain in Ireland under Sir Henry Sidney. On being sent home with dispatches, he petitioned the Queen for a privilege to discover the Northwest Passage, but was sent back to Ireland, where in 1570 he was knighted. M.P. for Plymouth in 1571, in 1572 he was sent with 1,500 volunteers to help the Dutch, but did not distinguish himself. In retirement at Limehouse from 1573–8, he wrote A *New Passage to Cathaia*, but the licence had been granted to Frobisher (q.v.) in 1575. In 1578 Gilbert received a charter for colonial discovery, but his voyage of that year failed. He returned in debt and went to serve in Ireland. In 1583 he sailed to settle Newfoundland, and on the return trip died, memorably if unnecessarily. A scholar of vision, he was more soldier than seaman, and not always successful even as a soldier.

GILBERT: William Gilbert (1540–1603) was born in Colchester and in 1560 took his B.A. at St John's College, Cambridge. In 1569 he was M.D. and senior fellow. In 1573 he took practice in London becoming physician to Queen Elizabeth, and was President in 1600 of the College of Physicians. His work in Latin on magnetism summarizes previous knowledge in detail, and goes on to investigate the properties of the magnet, including the magnetic polar variation, and its importance for navigators. Gilbert saw the globe as one vast spherical magnet, a poetic insight justifying John Dryden's line, 'Gilbert shall live till loadstones cease to draw.'

GRENVILLE: Sir Richard Grenville (1541?–91) was the son of the Sir Roger Greynville who commanded in *Mary Rose* and was lost when she went down. As a youth he served in Hungary against the Turks, under the Emperor Maximilian. He was M.P. for Cornwall in 1571 and 1584. As representative of his cousin Sir Walter Raleigh, he commanded in 1585 the fleet of five ships sailing from England to Virginia. In Armada year he concerted the land defence of the west coasts. In 1591, under Sir Thomas Howard (q.v.), he was vice-admiral of the mixed fleet of privateers and Queen's ships hovering off the Azores to catch the Spanish treasure fleet. *Revenge* (500 tons, 250 men) had been Drake's ship in the Armada fight; Grenville, who was no seaman, sailed her directly through the midst of the Spanish fleet, suicidally engaging about fifteen Spanish ships and 5,000 men.

After fifteen hours fighting, only twenty of his men were left alive. Described as a man of 'intolerable pride, insatiable ambition', Grenville's death was later written up by Sir Walter Raleigh in a manner to relieve his dead cousin of an accusation of wilful rashness.

HARRIOT: Thomas Harriot (often Hariot) (1560–1621) left Oxford, where he had graduated B.A. in 1580, to become mathematical tutor to Sir Walter Raleigh. He went out to Virginia as surveyor in the expedition commanded by Sir Richard Grenville in 1585, and in 1588 published *A Brief and True Report of the new-found Land of Virginia.* Henry, Earl of Northumberland (who later shared wth Raleigh imprisonment in the Tower) gave Harriot a life pension, and called him one of his 'three magi'. Harriot lived and worked in Sion House (1607–21), but procrastinated about publishing his discoveries. *Praxis*, which contained his decisively important discoveries in algebra, came out ten years after his death. It is evident that he was observing sunspots as early as 1610–13, and using an astronomical telescope of ×50 magnification at the same time as Galileo. Probably the first Englishman to smoke pipe tobacco, he died of cancer of the nose.

HAWKINS: Sir John Hawkins (1532–95) was born in Plymouth, second son of William (d. 1553) and younger brother of William (d. 1589). In 1559 John Hawkins married the daughter of the Treasurer of the Navy. In 1561 he was trading to the Canaries. The year following, backed by a syndicate, he made the first slave-trading voyage. In the second and very profitable slaving voyage in 1564, for which Hawkins was lent the Queen's ship, *Jesus*, the Earls of Pembroke, and Leicester invested. Having subsequently lost about £100,000 of treasure in the disaster at San Juan de Ulloa, Hawkins pretended, with Lord Burleigh's connivance, to be a devout Catholic. He offered the Spaniards to desert to them with the best of the Queen's ships – and by this sleight got £40,000 in redress of his losses, as well as the promised liberation of English prisoners in Seville and a patent as grandee of Spain. In 1572 Hawkins became M.P. for Plymouth. As Treasurer of the Navy he improved ship design and introduced chain pumps and boarding netting, but notoriously made a personal profit from the yard until regulations were tightened up. In financial matters Hawkins, a man spoken of as malicious and covetous, was capable of sailing close to the wind. He was third in command of the Armada fleet, in *Victory*, and was knighted at sea. After building the

Sir John Hawkins Hospital at Chatham, he died at sea off Porto Rico during Drake's West Indies raid in 1595.

HOWARD: Lord Charles Howard (1536–1624), a cousin once removed of Queen Elizabeth and an independent and prudent commander, served at sea during Queen Mary's reign under his father, Lord Howard of Effingham. In 1559 he was a diplomat in France. In 1573 he succeeded to his father's title, and in 1585 became Lord High Admiral. In *Ark Royal* (800 tons) with Drake his second-in-command, he led the English fleet against the Armada. He shared authority with the Earl of Essex in the Cadiz raid of 1596, and the following year was made Earl of Nottingham. He was a commissioner at the trial of Essex, and later, in the reign of King James, helped to negotiate a peace with Spain.

HUNSDON: Lord Hunsdon (1547–1603), as George Carey, matriculated at Trinity College, Cambridge at the age of thirteen. and in 1569 served with his father against the northern rebels. In 1570 he was knighted on the field for his prowess against the Scots, and in 1582 made captain-general of the Isle of Wight: '. . . an attorney coming to settle in the island was, by his command, with a pound of candles at his breach lighted, with bells about his legs, hunted out of the island.' (Oglander: *Memoirs*). In 1588 he fortified the Isle of Wight against the Armada. Succeeding to his peerage in 1596, he became Lord Chamberlain, and his company of players included William Shakespeare.

IVAN THE TERRIBLE: Tsar Ivan (1530–84) was proclaimed Grand Duke of Muscovy at the age of three. His mother died when he was seven, and he grew up in a brutal and degrading environment. At sixteen he took the government into his own hands. The aristocratic boyars having treated him badly, he made common cause with the merchant class and threw the last of the tyrannical boyars to his dogs. He assumed the title of Tsar, and was crowned in 1547 by the Metropolitan of Moscow. In 1550 he summoned a national assembly. Five years later he defeated the Tartars in the Crimea, and in 1556 conquered Kazan and Astrakhan. He began to import craftsmen and welcomed the first English merchants; but from about 1560 persecution mania set in, and by 1570 he mistrusted everyone. In 1569 he had the saintly Metropolitan of Moscow strangled. When the inhabitants of Novgorod were denounced to him for treason, he had them massacred and their city demolished. Finally, in 1580 he struck and killed his favourite son. In 1584, known as the monk Jonah and wearing a hermit's habit, he died, remorseful. He was a big man, highly intelligent and a hard worker, with a sinister face and an enigmatic smile.

JENKINSON: Anthony Jenkinson (d. 1611) was sent in 1546 to the Levant to train as a merchant. He travelled around the Mediterranean. During 1553 he saw the entry of Soleiman the Magnificent into Aleppo, and got from him a safe conduct and a trading permit. Jenkinson entered the Mercers' Company in 1555. Two years later, for a salary of £40 per annum, he was made captain-general of the Russia Company's fleet sailing for Moscow, and agent there. In 1558 he travelled down the Volga to Astrakhan and crossed the Caspian Sea to Khiva and Bokhara. In 1562–3 he reached Persia. In 1565 he addressed a petition to the Queen favouring the Northeast Passage. He was sent to Russia in 1571–2 to sort out difficulties that had arisen with Tsar Ivan. He had a share in Frobisher's second voyage of 1576, and sat on the commission in 1578 which reported on the ore brought back from the Arctic. His descendant was Charles Jenkinson, first Earl of Liverpool.

JOHN XII: Pope John XII (born Octavian) (d. 964) was 'prince and senator of the Romans' at the age of sixteen, and elected Pope in 955 as candidate of the Roman nobility. 'The Lateran Palace was turned into a school for prostitution' (Gibbon). He brought in Otto I to protect Rome from the threats of Berengar II, King of Italy. Pope John crowned Otto Emperor on 2 February 962. Otto then called a council which deposed him as Pope. On 4 December 963 Leo VIII was elected in his place. John, despite being welcomed back by the Roman nobility, died the same year.

LANCASTER: Sir James Lancaster (d. 1618) was brought up among the Portuguese. He commanded *Edward Bonaventure* (300 tons) in Armada year, and in 1591 sailed the same ship to India in company with *Penelope* and *Merchant Royal*. He reached Penang and Ceylon, returning in 1594 with only twenty-five out of his 198 men alive, despite his daily habit of issuing three spoonfuls of lemon juice against scurvy. He had acquired plunder and the news that the

Portuguese monopoly of the Cape route was broken. In the summer of 1594 Lancaster led the fleet which looted Pernambuco. In 1600 he was appointed to command the first East India Company fleet. In 1603 he returned again from the East Indies with a rich cargo of pepper and was knighted. Afterwards he helped organize the East India Company.

LODGE: Sir Thomas Lodge (d. 1584) was born in Shropshire, entered the Grocers' Company, was made alderman in 1553, sheriff in 1556, and Lord Mayor in 1562. A governor of the Russia Company, he was trading to the Barbary and African coasts in 1562, and became a shareholder in Hawkins's first slave-trading voyage. He cut his second son Thomas Lodge, the celebrated poet (who later sailed with Cavendish in his 1591 voyage), out of his will.

LOK: John Lok (sometimes Lock) (1533?–1615?) was sent to Flanders and France when he was thirteen by his father, a London merchant and alderman. In 1552 in Lisbon Lok observed 'the marvellous great trade of the Spanish West Indies'. He was himself a trader for twenty-four years, captaining a 1000 ton ship to the Levant, and spending more than £500 on books and instruments. He helped finance Frobisher's 1576 Northwest Passage attempt, and was a governor of the Cathay Company. The Northwest Passage ruined him. In 1579 he petitioned the Privy Council for relief, claiming that his voyages of exploration had cost him £7,500. The Privy Council allowed him £430. By 1581, when imprisoned by William Burrough (q.v.) for £200 outstanding in respect of one of Frobisher's ships, he was still £3,000 in debt, and was being sued for £200 due for stores to Frobisher as late as 1614. In 1592 he went to Aleppo as the Levant Company's consul. Late in life he translated into English a portion of Hakluyt's Latin edition of *Peter Martyr*. (*Historie of the West Indies*, 1612.) He had fifteen children.

MERCATOR: Gerardus Mercator (1512–94) was the latinized name of Gerhard Kremer. Born in East Flanders, he matriculated to the University of Louvain in 1530 and studied mathematics, astronomy and cosmography. He became a maker of scientific instruments – constructing a set of field instruments for Charles V. Mercator also compiled, drew and engraved: he produced maps of the Holy Land (1537), of the World (1538), and of Flanders (1540). Arrested in 1544 on suspicion of the Protestant heresy, he emigrated in 1552 to

Duisberg in the German Rhineland. Two years later he published there his great map of Europe, in fifteen sheets, reducing therein Ptolemy's length for the Mediterranean from 62° to 53°. He published a critical edition of Ptolemy's maps. In 1564 he produced an eight-sheet wall map of the British Isles, and in 1569 his world chart, using the useful projection (which takes Mercator's name, though he did not invent it) whereby a pilot can lay off a compass course, on a chart, as a straight line making the same angle with all meridians.

ORTELIUS: Abraham Ortelius (1527–98) was the latinized name of Wortels, born in Antwerp and trained as an engraver. He began selling antiquities and maps, and about 1560, under Mercator's influence, commenced map-making. in 1564 he produced a map of the world on a heart-shaped projection. After publishing maps of Egypt and Asia he issued (1567) the first edition of his masterpiece, *Theatrum Orbis Terrarum*, an atlas of seventy uniform maps (in 1573 increased to eighty-seven) all given regular critical revision. An abridgement of his atlas, *Epitome theatri orteliani*, appeared in the last year of his life and *Theatrum* was constantly reissued thereafter.

OSBORNE: Sir Edward Osborne (1530?–91) was apprenticed as a clothworker to William Hewett, who later became Lord Mayor of London. A careless nursemaid having dropped Hewett's infant daughter off London Bridge, the young apprentice dived in and saved her. When she was eighteen and an heiress, Osborne married her, had five children by her, and succeeded in due course to his father-in-law's property. He traded with Spain and Turkey, and was a governor of the Turkey Company. Alderman in 1573, Lord Mayor in 1583, and M.P. for the City in 1586, he was renowned for having sent Irish beggars all the way back to Ireland and for having asked Mr Secretary Walsingham to stop pack-horses from travelling through the London suburbs on the Sabbath. He was progenitor of the Dukes of Leeds.

RALEIGH: Sir Walter Raleigh (1552?–1618) was born at Budleigh Salterton in Devon, one of a clan of poor country gentry of Protestant persuasion. Raleigh always spoke with a broad Devon accent. At seventeen as a volunteer with the Huguenots he fought at Moncontour. About 1572 he was at Oriel College, and in 1578 joined Gilbert's (q.v.) first voyage. Befriended at court by Leicester, he got involved in affrays, departed to Ireland as a captain, and was there involved in a cold-blooded massacre, Edmund Spenser being also present, of 600 Spanish prisoners. Returning to England in 1581 with dispatches he caught the Queen's eye, was granted a monopoly of wine licences in 1583, in 1584 was knighted, and in 1586 received a substantial grant of land in Ireland. He became Captain of the Queen's Guard. Raleigh is said to have spent £40,000 in colonizing – he sent Amadas and Barlow to explore Virginia in 1584, and an expedition to colonize it in 1585 which was taken off by Drake. He introduced the potato to Ireland and popularized tobacco. An admirer of *The Faerie Queen*, in 1589 he brought Spenser to court. He associated not only with fellow poets, but with sceptics like Harriot (q.v.) and Christopher Marlowe; the warrant for Marlowe implicated him. Of his Guiana project, his wife wrote to Cecil imploring the Secretary of State 'rather to stay him than further him'. But in 1595 Raleigh sailed with a royal commission to Guiana, brought back some evidence of gold, and introduced mahogany. In the 1596 expedition to Cadiz he was wounded. King James supposed he had reason to believe that Raleigh opposed his succession, and in 1604 sent him to the Tower as 'guilty of compassing the death of the King' by intriguing to set Arabella Stuart on the throne. Reprieved of the death sentence, but stripped of his property, Raleigh studied science and literature during his imprisonment, wrote his *History of the World*, and condensed fresh water from salt. Let go to Guiana again by James in 1616 on a promise to find gold, he was given a ship's crew who were 'the world's scum'. The enterprise failed; Raleigh returned to England in 1618, penniless. His last words as he knelt to the block were, 'What matter how the head lie, so the heart be right?'

RANDOLFE: Thomas Randolfe (sometimes Randolph) (1523–90) was Principal of Broadgates Hall (now Pembroke College), Oxford until 1553 when he escaped Queen Mary's persecution of the Protestants by retiring to France. In 1558–9 after Elizabeth's succession he became the English government's agent in Germany. In 1560 he was sent to Scotland as a secret agent with the Protestant party there. He failed however to prevent the marriage between Mary Queen of Scots and Darnley, and was recalled to England in 1566. He married a sister of Francis Walsingham (q.v.). Sent in 1568 on a special embassy to Russia for the English merchants, he got privileges for them from Ivan the Terrible which led to the formation of the Russia Company.

He went on numerous special embassies to France and Scotland, on the last of which he was nearly assassinated. In 1585–6 he achieved the signature of a treaty with Scotland, and was Chancellor of the Exchequer and Postmaster-General until his death.

TAMASSO: Tamasso (sometimes Thomas, more correctly Tahmasp) (1524–76) was Shah of Persia, succeeding to the throne when a boy. His country, then weak, was looking for counterweights to the Turks, who under Soleiman the Magnificent had invaded Persia as far as Tabriz. Despite this, when Jenkinson (q.v.) at Qazvīn presented to him the letters of Queen Elizabeth, the Shah replied, 'O, thou unbeliever, we have no need to have friendship with the unbelievers', and proposed sending Jenkinson's head to Sultan Soleiman as a personal gift.

SHERLEY: Anthony Sherley (sometimes Shirley) (1565–1635?) was the son of a Sussex squire who had served in the Low Countries as Treasurer at War, emerging hopelessly in debt. Sherley took his B.A. at Oxford in 1581, saw service in the Low Countries under Leicester in 1586, and was present at the Battle of Zutphen. In 1591 he joined Essex in the expedition to Normandy in support of Henry of Navarre. Henry dubbed him knight; which displeased Queen Elizabeth since Sherley had accepted this honour without her consent. Unhappily married to a cousin of Essex, he went on the Islands Voyage in 1597, and in 1596 made an expedition to Dominica and Venezuela reputedly to get away from her. He got home 'alive but poor'. In 1598 he was sent by Essex quite unofficially as ambassador to Persia to make an alliance against the Turks and open up trade. Essex, soon to fall into disgrace, had omitted to consult the government. Sherley raised money in Istanbul and Aleppo by bills, and went overland to Isfahan where he got a firman from the Shah granting Christian merchants religious liberty, freedom from customs, and the right to trade. He was sent north into Russia on an embassy from the Shah to achieve a Christian alliance against the Turk. He went on to visit Moscow, Prague and Rome. Barred from England, he retired to Venice where he was arrested for debt, and on his release became a spy for Robert Cecil. In 1605 the Emperor sent him to Morocco to negotiate the release of prisoners. He afterwards made his way to Madrid and was appointed Admiral of the Levant Seas. Sailing from Sicily in 1609 to attack the Turks, he made a futile landing on

Mytilene and was dismissed. 1611 saw him in poverty in Madrid, 'making himself believe that he shall one day be a great prince, when for the present he wants shoes to wear'. The King of Spain pensioned him. In 1613 he published in London an account of his embassy to Persia.

STAFFORD: Sir Edward Stafford (1552?–1605) was sent in 1578, his mother being mistress-of-the-robes to Elizabeth, on a minor diplomatic mission to Catherine de Medici. For the next few years he was involved in the negotiations for the abortive marriage between Elizabeth and Anjou. In 1583 he was knighted and sent to Paris as ambassador where he served seven years: flaunting his Protestantism, openly helping Henry of Navarre and, when he besieged Paris, even going out to serve in the trenches with him.

TURBERVILLE: George Turberville (1540?–1610?), of an old Dorset family, was at Winchester and New College, Oxford but left before taking his degree and entered the Inns of Court. When serving as secretary to Thomas Randolfe (q.v.) on his Russian embassy, Turberville wrote a book, now lost, *Poems describing the Places and Manners of the Country and People of Russia* (1568), which included the metrical epistle preserved by Hakluyt. An early practitioner in blank verse, Turberville translated Ovid and later published *Epitaphs, Epigrams, Songs and Sonnets* (1567), *The Book of Faulconrie* (1575) and other works about country sports.

WALSINGHAM: Sir Francis Walsingham (1530?–90) was the son of a lawyer prominent in the City who had bought Foots Cray Manor in Kent. After King's College, Cambridge (1548) and Grays Inn (1552), Walsingham, as a convinced Protestant, went abroad on Queen Mary's accession, and studied languages and foreign manners. In Elizabeth's reign, having become M.P. for Banbury, he was employed by Cecil (later Lord Burghley) to procure secret intelligence from foreign agents. Ambassador in Paris at the time of the Anjou marriage negotiations, he managed to hide his close friend Sir Philip Sidney in the embassy during the Saint Bartholomew's Day massacre. From 1573 until his death Walsingham was joint Secretary of State, controlling foreign policy, though his advice – too radically Protestant for Elizabeth – was often not acted upon. He used not only public funds but his private fortune for espionage – 'knowledge is never too dear' – employing fifty-three private agents in foreign courts, and eighteen others, including Thomas Phelippes an expert in deciphering. He gained excellent advanced information of, for example, the Armada. In 1583 his daughter married Sir Philip Sidney whose death involved Walsingham in heavy debt. He had been a member of the Russia Company in 1562 and of the Levant Company with Sir Edward Osborne (q.v.). He invested in Frobisher's second voyage and in Drake's circumnavigation, and encouraged American colonization, planning to kill two birds with one stone by sending English Catholics out there. A literary patron, he encouraged writers about the New World, among them Hakluyt.

WILLOUGHBY: Hugh Willoughby (d. 1554) came of a Nottinghamshire family. He served in the 1544 expedition against Scotland, and was knighted there by the Earl of Hertford who later became Duke of Somerset. In 1548–9 he was Captain of Lowther Castle. Appointed captain-general of the Cathay Fleet, commanding in *Bona Esperanza*, he had such experienced pilots as Burrough (q.v.) and Chancellor (q.v.) serving under him. Separated from his fleet he was blown to the coast of Novaya Zemlya and driven back to harbour in Lapland, where unprepared for wintering he perished on some date subsequent to January 1554. A few years later the ship and bodies were found together with Willoughby's journal.

GLOSSARY

Ambergris: A wax-like substance secreted by the sperm whale, found floating in tropical seas and used in perfumery.
Aquavitae: Any form of ardent spirits taken as a drink.
Armada: (Sp=armed) A fleet of ships of war; sometimes: a single armed ship.
Arquebus: An early type of portable gun fired usually from a tripod or forked rest; a name applied loosely to any portable firearm of the time.
Arras: A hanging screen of tapestry placed around the walls of a room.
Baricoes: (Fr. *baricaut*) Kegs.
Bastinado: A blow with a stick, especially upon the sole of the foot.
Beetle: A ram or mall, usually of wood, for driving in wedges.
Bigg: The four-rowed barley.
Bloody flux: Dysentery.
Brage: A drink of honey or spice with ale.
Bombasine: Raw cotton; cotton wool.
Bristow friezes: A coarse woollen cloth with nap on one side only, made near Bristol.
Broadcloth: A fine, plain-woven, double-width, black cloth used mostly for men's garments.
Brown bills: A kind of halberd, painted or burnished brown, used by foot-soldiers.
Caballero: (Sp.) Man on horseback; hence, gentleman.
Cable (Height of): Cable's length; about 100 fathoms or 200 yards.
Cade: Cask.
Calentura: Disease to which sailors were liable in the tropics; in their delirium they fancied the blue sea to be green fields, and wanted to jump in.
Calicut cloth: Cotton cloth imported from Calicut on the Malibar coast; hence later: calico.
Caliver: A light arquebus (q.v.) fired without a rest.
Canhook: A short rope with a hook at each end used for slinging casks.

Carrack: The large merchant ship, also fitted with guns as a ship of war, used by the Portuguese for their East Indies trade; sometimes used loosely for galleon.

Caravel: Can mean either a small fast light ship (from Port. *caravela*), or a Turkish war frigate (from It. *caravella*).

Card: The circular piece of stiff paper on which the 32 points of the compass are marked; sometimes: a map.

Carmosell: A Turkish or Moorish merchant ship.

Cassava: A fleshy root used as food in tropical America; also called manioc.

Cayro: (Malayālam *kāyar*) Coco-nut fibre; coir.

Chamlet (sometimes: *Camlet*): Originally a costly eastern fabric; substitutes were later made of various combinations of wool, silk and hair.

Charter party: The deed made between merchants and shipowners for the hire of a ship and safe delivery of its cargo.

Cod and yard: The genitals (cod signifying scrotum; yard: penis). Cod is sometimes used improperly for the testicles.

Cosmographer: Geographer.

Crasko: Crash (?): a coarse linen used for towels.

Culver (short for *Culverin*): A large cannon, very long in proportion to its bore.

Cypress: A kind of satin originally brought from Cyprus and used for hat bands.

Dag: An early and heavy kind of pistol.

Demi-culverin: A cannon of about 4½ inches bore.

Denizen: One who lives habitually in a country but is not native born.

Estacha: (Sp. *estaca*) Clamp-nail.

Falcon: A light cannon.

Fights: A kind of screen used to hide and protect fighting men on board ship.

Flux: Dysentery.

Foist: A light galley propelled by oars and sail.

Fore-course: The sail attached to the lower yard of the foremast.

Frieze: (see Bristow friezes).

Frisadoes: (Sp. *frisado*) Silk plush.

Furmenty (sometimes: *Frumenty*): Hulled wheat boiled in milk and seasoned with cinnamon, sugar &c.

Galliass: A heavy, low-built vessel of war, larger than a galley, using oars and sail.

Garbanzo: Chick pea.

Gillyflower: A clove; clove-scented pink; or wallflower.

Groat: An old coin, equal to four pence.

Grograins (sometimes: *Grogran* or *Grogram*): (Fr. *gros grain*) A coarse silk fabric, sometimes with an admixture of mohair and wool, and stiffened with gum.

Gullet: A water channel.

Halberd: A combined spear and battle axe, five to seven feet long.

Hogshead: A large cask of a definite capacity, which varied for different liquids and commodities from about 50 to about 140 gallons.

Jacynth: In antiquity, a jewel of a blue colour, probably the sapphire; later, a reddish-orange gem, a variety of zircon; the name is also used for varieties of topaz and garnet.

Kern: To make salt into grains.

Machetos: (Sp. *machete*) A broad knife like a cutlass, but used as a tool.

Machico: (Sp. *machado* (?)) Hatchet.

Manurance: Tenure, occupation (of land &c), cultivation, tillage.

Minion: A small gun, of about 3-inch bore.

Montego de porco: Lard.

Motley: A cloth of mixed colour.

Neap: The tide shortly after the first and third quarters of the moon, when the highwater level is at its lowest point.

Noble: A gold coin worth 6s. 8d. in Elizabethan times.

Outlandish: Foreign, alien.

Parcel gilt: Partly gilded, especially when silver cups and bowls have their inner surface gilt.

Pina: Pineapple.

Pinnace: A small light vessel, rigged usually as a two-masted schooner, and used as an auxiliary.

Pioneer: An advance guard which prepares with spade and pickaxe the way for the main body of troops.

Pintado: A kind of eastern chintz.

Plantain: A tropical plant similar to a banana.

Plat: Plan or sketch.

Pomecitron: Used for our citron: a thick-skinned, juicy fruit resembling an orange; citron then included lemon and lime.

Pompion: Pumpkin.
Port-base: port-piece (?): an old type of ship's gun.
Provedor: (Sp. *proveedor*) Purveyor, furnisher, contractor.
Quarter: 28 lb.
Quern: A small hand mill for grinding pepper, mustard &c.
Quintal: A hundredweight.
Recusant: One, especially a Roman Catholic, who refused to attend the services of the Church of England.
Rove: (Sp. *arroba*) Roughly 25 lb., used as a measure of sugar.
Saker: Cannon smaller than a demi-culverin (q.v.), much used at sea.
Shuba: A Russian fur gown or greatcoat.
Snaphance: An early form of flint-lock for musket and pistol.
Spermaceti: A fatty substance found in sperm and other whales; used in medicine and for making candles.
Spinel: A red gem like a ruby.
Spodium: A fine powder compounded of ashes.
Sucket: Sweetmeat of fruit, candied or preserved in syrup.
Sumach: A small tree; its chopped leaves and shoots are used in tanning, and for dyeing and staining leather black.
Tierce: A third part.
Tolmach: Servant.
Tortuga: (Sp. *tortuga*) Tortoise or turtle.
Trencher: A platter.
Tun: A large cask containing four hogsheads or 252 old wine-gallons.
Turkish roll: Turban.
Twist: The junction of the thighs.
Watchet: A light blue colour.
Writing tables: Writing tablets.
Yard: (see *Cod and yard.*)

NOTES

1. *Nova Hispania:* Mexico.
2. *Luzones:* Luzon is the principal island of the Philippines.
3. *Novaya Zemlya:* Two large islands off the Arctic coast of European Russia, snow-covered October to May and colder than Spitzbergen, with no trees, but with many fur-bearing animals. Stephen Burrough reached their southern extremity in 1556. Barents, in 1596 after his discovery of Spitzbergen, wintered in Ice Haven in 76° 12′ N.
4. *Dvina:* A river, 466 miles long, flowing northward into the Dvina Gulf of the White Sea; frozen from October to April, it then floods, and is navigable thereafter for 342 miles southward.
5. *Kazan:* City on the upper Volga, capital of a kingdom founded in 1438 by Ulu Mohammed, Khan of the Golden Horde; conquered in 1552 by Tsar Ivan IV who thus opened up the Volga to trade.
6. *Astrakhan:* City on the Volga delta on the north Caspian shore, and capital of the Khanate of the Golden Horde until captured in 1556 by Tsar Ivan IV. In 1558 the Tsar moved it to a more defensible island site on the left bank and Russian colonists were brought in to push out the Kalmucks.
7. *Candia:* Port on the north shore of Crete, now known as Iraklion, which then gave its name to the island; Crete was still a Venetian possession.
8. *Chios:* Now a Greek island off the Turkish coast, but at that time Genoese since 1346. From 1415 the Genoese on Chios were tributaries of the Ottoman Turks; in 1566 the rule of the Genoese company which dominated the economy of the island was abolished.
9. *Ormuz:* Now Hormoz, a waterless island in the mouth of the Persian Gulf offshore from Bandar Abbas; then used as an entrepôt by the Portuguese.
10. *Heligoland:* An island 38 miles off the German coast opposite the mouth of the River Elbe, once the shrine of the heathen goddess Hertha; St Willibrord, in the seventh century, first preached Christianity there.

11. *Lycia:* Region of Asia Minor now in southwest Turkey, extending inland to the Taurus Mountains.
12. *Joppa:* Modern Jaffa.
13. *Machico:* Traditionally the man was named Robert Machim or à Machin or Macham, and his mistress was called Anna d'Arfet.
14. *Prussia:* The order of Teutonic Knights Hospitaller, limited to Germans of noble birth, was founded in Jerusalem during the Third Crusade. In 1226 the order was invited by the Bishop of Prussia to subdue the heathen Prussians; by 1234 Prussia was a fief of the Holy See. Once its connection with the Holy Land was lost, the order became a governing aristocracy closely connected with the Hans towns. In the fourteenth century the knights fought the heathen Lithuanians; in 1410 they were defeated by the Polish King Ladislaus at Tannenberg.
15. *Hans:* The Hanseatic League, an association of Baltic towns including Lübeck, Hamburg, Wismar, Rostock and Stralsund, was formed in the mid thirteenth century; they set up trading posts in common, helped each other against pirates, and won special trading rights in London (where their establishment was sometimes called the Steelyard). Their motto was, 'whoever touches one, touches all', but there was no unified German political power to back them and their importance declined. Under Elizabeth the Steelyard's privileges were withdrawn, to the advantage of the English factory in Hamburg.
16. *Seventh clime:* The geography of antiquity conceived of seven climates in the known world, presided over by the seven planets and indicated by lines drawn through the following latitudes: Meroe, 17°; Syene, 24°; Alexandria, 31°; Rhodes, 36°; Borysthenes, 45°; and the Riphaean Mountains, 48°.
17. *Sestos:* A river – sometimes known as Cestos, sometimes as Cess – on the Grain Coast in present-day Liberia.
18. *Malta:* In 1530 the Emperor Charles V granted Malta (with Gozo, and Tripoli) to the Knights of St John who had been driven by the Turks from Rhodes. In anticipation of Turkish attack, the knights began to fortify. They held off a raid by Arabs from Africa in 1551; and in 1565 though outnumbered fourfold, withstood an epic siege by the Turks under Soleiman the Magnificent.
19. *Wardhouse:* Vardö, a port on a small island off the north Norway coast used by whalers.

20. *Shama:* A town between Takoradi and Accra in Ghana, where in 1471 the Portuguese first landed on the Gold Coast. Fort St Sebastian was begun by them in 1526. Cape Tres Puntas is now Cape Three Points.
21. *Barnacles:* Wild geese breeding in Arctic seas; but traditionally supposed to be produced out of shellfish – by roundabout confusion with a feathery seaweed which attaches itself to ships' bottoms.
22. *Kholmogory:* Port on the River Dvina, near where it enters the White Sea; replaced in 1583 by Archangel.
23. *Pleskau:* Now Pskov, formerly a sister republic of Novgorod and one of the last free cities to be brought under by Moscow (in 1510, by Basil Ivanovich, Prince of Moscow).
24. *Warped in:* To move a ship along by hauling on a rope (or 'warp').
25. *Samoyed:* A people with a distinctive language, Samoyedic, who were found eastward from the White Sea to the Yenisei River. They had domesticated the reindeer, and lived by fishing.
26. *Easterlings:* A name applied to natives of east Germany or the Baltic coasts, especially from the Hans towns; and to Baltic and German ships.
27. *Nijni Novgorod:* A city at the confluence of the Volga and Oka rivers; once a free city, it was annexed to Moscow in 1392. It repelled Tartar attacks in 1513, 1520 and 1536 and thereafter became a commercial depot for the Volga basin, building ships with timber from the local forests and sending a caravan yearly down the Volga. Its importance increased after the conquest of Kazan and Astrakhan by Ivan the Terrible.
28. *Carbuse:* (Modern Gr. *Καρπουζι*) Water melon.
29. *Urgenj:* City on the River Oxus (now the River Amu Darya) near the Sea of Kharazm (now Aral Sea). Urgenj was the capital of an Uzbek dynasty until 1573 when the Amu Darya changed its course and the capital was moved to Khiva.
30. *Yaik:* Now the Ural River.
31. *Media:* Ancient name of the northwestern part of Iran – corresponding to the provinces of Azerbaijan, Kurdistan, and parts of Kermanshah – extending eastward to the Caspian Sea and westward to the Zagros Mountains; Shamakha is a town in Shirvan, west of Baku.
32. *Scrachick:* Serakhs, a town between mountains and desert to the southeast of the Caspian Sea.

33. *Occient:* Khokand, junction of caravan routes arriving from Samarkand and from Tashkent.
34. *Cascar:* Kashgar.
35. *Sowchick:* Sukchur (now Su-chow) at the western end of the Great Wall of China, not far from the junction of caravan routes coming eastward from Kashgar and southward from Karakorum.
36. *Cherkesy:* A people living between the Kuban River and the Black Sea who became Russian vassals about 1556.
37. *Challica Ostriva:* the Tyulen'i Islands; to the south and east is Turkmenia (Tumen).
38. *Hircania:* 'At length we came upon the mighty river of ... Volga ... it runneth into a certain lake or sea which of late they called the Hircan sea, according to the name of a certain city in Persia, standing upon the shore thereof.' (Sir John Mandeville: *Travels.*)
39. *Arrash:* Possibly Agdash, a town on the Tiflis–Baku road about 60 miles west of Shamakha in the direction of Georgia.
40. *Qazvīn*: A city of 60 miles southwest of the Caspian shore and east of Tabriz.
41. *Yaroslavl:* A Russian city on the right upper bank of the Volga at its confluence with the Kotorosl': though northerly, these rivers are navigable for half the year or more.
42. *Gilan:* The city is now Bandar Pahlavi, a Persian port on the southwest Caspian coast; the province (Gilyan) is the westernmost of present-day Iran's three Caspian provinces and was incorporated by Shah Tahmasp in 1567.
43. *Cabo Blanco:* A Portuguese post in West Africa about midway between the Canary Islands and the Cape Verde Islands.
44. *Callowsa:* Probably the Gambia River.
45. *Cumaná:* This port on the Spanish Main, east of the Pearl Coast and a mile from the mouth of the Manzanares River in what is now Venezuela, claims to be the oldest European settlement on the South American mainland.
46. *Tortuga:* Not the better-known Tortuga Island, later a buccaneering base off the north shore of Hispaniola, but Isla la Tortuga, west of Margarita off the Venezuelan coast.
47. *Nizovaya:* A small port on the west Caspian shore between Derbent and Baku near the mouth of the River Kudial.
48. *Shahi:* Formerly a small silver coin of Persia.
49. *Kersey:* A kind of coarse narrow cloth woven from long wool and usually ribbed.

50. *Galls:* An excrescence caused by an insect, found particularly on oak trees; once used in the manufacture of ink and for dyeing.
51. *Lahilan:* A town, now Persian, east of Gilan near the Caspian.
52. *Hallap:* A town on the Persian-Turkish border on the Tabriz–Erzerum road northeast of Lake Van.
53. *Ducat:* A gold coin of varying value, formerly current in most European countries (also, a silver coin current in Italy).
54. *St Nicholas Road:* In Dvina Bay in the White Sea.
55. *Vologda:* Russian trading city on the route between the headwaters of the River Dvina and the upper Volga.
56. *Panuco:* River coming to the sea at Tampico, north of Vera Cruz (San Juan de Ulloa) in the Gulf of Mexico.
57. *Cavallos:* A port in the Bay of Honduras, now Puerto Cortés.
58. *St Mary Port:* Puerto de Santa Maria, on the north shore of the River Guadalete near its mouth, about 5 nautical miles across the bay from Cadiz.
59. *Samogitia:* A former Baltic province.
60. *Nogaia:* The territory of the Great Nogai Horde, between the Don and the Volga.
61. *Turcoman:* The Turkomans now live mainly in Soviet Turkmenistan, but are scattered across Persia and north Afghanistan. Jenkinson would perhaps have encountered them on the southeast shore of the Caspian, north of the River Atrek.
62. *Zagatay:* The Jaghatai Khanate was centered on Kashgar, north of the Karakorum Mountains; Jenkinson can hardly have got this far.
63. *Parthia:* The mountainous country southeast of the Caspian Sea extending from the Elburz Mountains eastward towards Herat.
64. *Baiben:* (Sp. *vaivén*) Line or cord.
65. *Gallipoli:* Ag. Gallini on the south coast of Crete is a possibility; Gallipoli in the Dardanelles is out of the question.
66. *Geese:* Penguins; what else?
67. *Tarapacá:* Name of an inland city and of a province in northern Chile, of which Iquique is now the port.
68. *Patia:* A good natural harbour and port town in northern Peru, now a centre for the export of Panama hats.
69. *Buena Esperanza:* Cape of Good Hope.
70. *Bay:* There is evidence to suppose that this was a small creek or bay on the north side of the Golden Gate near present-day San

Francisco; Cape Reyes is white, somewhat like the cliffs of Dover, and gophers are to be found.

71. *Islands:* Most probably the Marshall Islands northeast of New Guinea.

72. *Islands:* Not readily identifiable, but in the Moluccas or Spice Islands.

73. *Fastings:* Ramadan, the ninth month of the Moslem year, is a strict 29-day fast during the hours of daylight.

74. *Barateve:* This mountainous island in the Moluccas is now called Batjan; the fort built there by the Portuguese was captured by the Dutch in 1609.

75. *Sea unicorn:* The narwhal, smaller than most other members of the whale family, has one or both its two teeth developed as a spirally-twisted horn; unicorn's horn was held to be potent against poison.

76. *Ingenio:* (Sp. *Ingenio de azúcar*) Sugar mill, and by extension, a sugar factory.

77. *West Friesland:* Friesland was a name applied in this case to Greenland, but sometimes to an imaginary land in the Arctic.

78. *Anil:* The indigo shrub; or the dye derived from it.

79. *Logwood:* A dyestuff made from vegetable fibres derived from the heartwood of *Haematoxylon Campechianum* – a tree imported into Europe soon after the discovery of South America; logwood was used for black and compound shades in woollens, and for dyeing silks black.

80. *Tabin:* Yamal peninsula, which with the large island of Novaya Zemlya encloses the Kara Sea – into which flow the Rivers Ob and Yenisei, after crossing the Siberian Plain. Mercator then lacked accurate information about the vast extent of Siberia and Central Asia.

81. *Loadstone:* Magnetic oxide of iron; hence the pole of the loadstone is magnetic north.

82. *Great Khan:* Properly, the title given to Genghis Khan and those who succeeded to his rule over Turkish, Tartar and Mongol tribes; but sometimes in the Middle Ages, as also here, applied to the Emperor of China.

83. *Cambalu:* Pekin; Mercator's estimate of 300 German miles from the mouth of the River Ob to Pekin is of course extremely misleading.

84. *Cleveland:* The county of Cleves.

85. *Uvek:* A town on the Volga south of Saratov.
86. *Crims:* Nogaians and Crims were Tartars; Crims were found in or near the Crimea, Nogaians between the Caspian Sea and the Sea of Azov.
87. *Derbent:* A port on the western shore of the Caspian Sea on the eastern flank of the Caucasus, founded by the Persians in antiquity to guard the Caspian or 'Iron' Gates against northern invaders. Its monuments include a wall built by Alexander the Great. Derbent was captured by the Arabs in 728; by the Mongols in 1220; and was included in the Russian Empire in 1813.
88. *Bildih:* The context suggests that this possibly may be the Russian Caspian port of Makhachkala.
89. *Pasha:* (Turkish *padisah*: ruler) Formerly the highest title (borne after the name) of Turkish civil and military officials, but at this time of military commanders.
90. *Baku:* The best harbour on the Caspian, lying in a gulf on the western coast, sheltered by offshore islands; in the Russian Empire since 1806.
91. *Stove:* The *Hamam*, or Turkish Bath.
92. *Mark:* A coin worth 160 silver pennies (8 oz. silver) or two-thirds of a pound sterling.
93. *Cochineal:* A natural dyestuff, used for scarlet, crimson and orange tints, made of the dried and powdered bodies of the cochineal insect which lives on cactus; the dye was introduced from Mexico and was in use there long before the Spaniards.
94. *Shaffe:* A bundle of iron or steel containing a definite number of *gads*, or iron bars.
95. *Batman:* An oriental weight, varying greatly according to locality: between 16 and 55 lb.
96. *Pieces of brass:* Here means brazen cannon.
97. *Goletta in Barbary:* A fortress on the Gulf of Tunis, south of the ruins of Carthage. Tunis had been captured by Charles V in 1535; the Spaniards were driven out in 1569, retook it in 1573, and ceded it to the Turks in 1574.
98. *Golconda:* A kingdom comprising what later became part of Hyderabad and northern Madras; south of the River Penner was the Coromandel Coast; north, the Golconda Coast.
99. *Pegu:* A city and kingdom in what is, now, lower Burma.
100. *Cabie:* (*Cabaan*) A white cloth, worn by Arabs over their shoulders.

101. *Hing:* The drug asafoetida.
102. *Negrais:* Cape Negrais, one of the mouths of the Irrawaddy.
103. *Caplan:* This could hardly be a mangled hearsay version of Mogok, the ruby-mining centre in the mountains north of Mandalay – but what else might it be?
104. *Cochin:* Port in southwest India on the Malabar Coast.
105. *Banda:* Group of small islands, then Portuguese-controlled, in Moluccas, 66 miles east of Ceram.
106. *Occam:* Roanoke Island stands offshore of Albemarle Sound, a large estuary in what is now North Carolina, where the Rivers Roanoke and Chowan reach the sea.
107. *Mountain:* Possibly Sugar Loaf mountain, now in the mouth of Rio de Janeiro harbour, with Ilha Cotunduba to the seaward.
108. *São Sebastião:* An island off the coast of Brazil, 60 miles east of Santos.
109. *Santa Maria:* In Arauco Bay between Valparaíso and Valdivia.
110. *Arauco:* A mountainous coastal province of southern Chile which includes Santa Maria Island; skilfully defended at the time of the Spanish invasion by the Araucanian Indians, who by 1662 had overwhelmed the fort at Arauco three times.
111. *Morro Moreno:* Bahia Moreno, 12 miles north of Antofagasta in Chile.
112. *Puna:* An island in the Gulf of Guayaquil off the southern coast of Ecuador.
113. *Capul:* A small island in the San Bernadino Strait between the Philippine islands of Luzon and Samar.
114. *Light horseman:* An old name for the light boat, since called a gig.
115. *Brazil:* Originally the name of a dye-wood; a tree yielding a similar dye was found in South America, so the place was named 'terra de brasil' – or, 'red-dye-wood land'. Brazil-wood produced tints of red, orange and peach.
116. *Nombre de Dios:* A port on the Spanish Main east of Portobello, to which bullion, shipped along the Pacific coast from Peru to the Isthmus of Panama, was transported overland for eventual convoy to Seville.
117. *Puerto Real:* A port on the mainland side of the roadstead formed by the island of Leon, on the tip of which Cadiz stands; about 7 nautical miles from Cadiz.

118. *Cape Sagres:* Punta de Sagres, east-southeast of Cape St Vincent on the southwest extremity of Portugal.

119. *Newhaven:* Le Havre in Normandy.

120. *Master:* The officer who then navigated a ship of war: a competent seaman, as distinct from the commander who took her into action, who might on occasion be a gentleman with military experience and court connections only.

121. *Main trestle-trees:* Two strong pieces of timber fixed horizontally fore-and-aft on the opposite sides of a masthead to support the cross-trees, the top, and the fid of the mast above.

122. *Rosa solis:* Originally a cordial or liqueur flavoured with sundew; but often the name signified brandy or other spirits flavoured with essences or spices and sugar.

123. *Scurvy:* A deficiency disease caused by shortage of vitamin C (found in green and root vegetables and fresh fruits) resulting in spongy gums, haemorrhages in body tissues, muscular pains, sunken eyes, sallow complexions, mental depression and failure of strength.

124. *Match:* Their weapons were fired by matchlocks, in which a glowing match had to be placed to ignite the powder.

125. *Bahía:* A port in Brazil with a deep natural harbour; Bahia de Todos os Santos was founded in 1549 and was the capital until 1763. Now known as Salvador.

126. *Bear-haven:* Castletown Bearehaven in Bantry Bay, County Cork.

127. *Comoro Islands:* A group of volcanic islands in latitude 11° to 13°S. in the strait between northern Madagascar and northern Mozambique.

128. *Pulo Pinaou:* In the Mergui archipelago.

129. *Junsalaom:* In the Malay peninsula in what is now south Thailand, nowadays called Phuket.

130. *Abath:* The Asian rhinoceros, now rare, the powdered horn of which is still in demand as an aphrodisiac.

131. *Nieblas:* Possibly the Virgin Islands.

132. *Mona:* A small island in the Mona Passage between Porto Rico and Hispaniola.

133. *Chiego:* Probably a phonetic rendering of the peak 832 metres high, now called Bujeo, west of Algeciras.

134. *Cadi:* A civil judge among the Turks, Arabs, Persians &c.; usually the judge of a town or village.

135. *Morocco:* The city of Marrakesh.

136. *Santa Cruz:* English ships might then have called at three nearby ports called Santa Cruz – in Madeira, La Palma and Tenerife; the last is the likeliest.

137. *Maio:* Probably Maceió, a harbour 125 miles south-southwest of Pernambuco, where a stone reef offshore gives a safe anchorage.

138. *Murdering piece:* A small cannon or mortar, used in close quarters combat, particularly against enemy boarding the ship.

139. *Hail shot:* Small shot which scatters like hail when fired, as distinct from ball or bullet which is aimed at a mark.

140. *Pernambuco-wood:* Brazil-wood, from which a dye was extracted.

141. *Manoa:* In 1586 a privateer sent out by Raleigh captured for ransom Don Pedro Sarmiento de Gamboa, who told Raleigh about an unexplored Indian kingdom between the Orinoco and Amazon rivers, 'and of that great and golden city, which the Spaniards call El Dorado, and the naturals Manoa'. Berreo later confirmed the legendary origin of Manoa – that Guaynacapa, Emperor of Peru, had fled after the Spanish invasion with several thousand warriors and founded a new capital, Manoa, on a large inland lake called the Lake of Parima – possibly the flooded headwaters of the Caroni River, upstream from the falls. A Spaniard called Juan Martínez, sole survivor of an expedition up the Orinoco, claimed to have been taken by Indians to Manoa and to have seen the treasure there; on his deathbed he swore to the truth of all this in the presence of a priest. Thereafter came successive Spanish expeditions to find Manoa: several by Berreo, who told Raleigh he had spent 300,000 ducats on the search.

142. *Spaniards:* This is almost certainly disingenuous; Raleigh could hardly have met personally any Spaniard claiming to have seen Manoa, but he would have heard or read stories at second-hand; including an account of the plausible-sounding sojourn there of Juan Martínez.

143. *Your lordship:* Raleigh's account of Guiana was presented to Lord Charles Howard, Lord High Admiral and to Sir Robert Cecil, later Secretary of State.

144. *Carúpano:* A port on the Venezuelan coast between Trinidad and Cumana.

145. *Overfalls:* The cataract in the gorge of the Caroní River, a south-

ward tributary of the Orinoco. Berreo too had failed to pass the cataract on his final expedition. He believed that Manoa lay upstream.

146. *Bristol diamond:* A kind of transparent rock crystal resembling a diamond in brilliancy, found near Bristol in Clifton limestone.

147. *Mandeville:* 'And in another isle toward the south dwell folk of foul stature and of cursed kind that have no heads. And their eyes be in their shoulders.' Sir John Mandeville: *Travels.*

148. *Hugh Goodwin:* Raleigh's men who returned to Guiana 22 years later found Hugh Goodwin still there; he had almost forgotten how to speak English.

149. *Margarita:* An island off the Venezuelan shore and facing Cumana, 180 miles west of Trinidad.

150. *Río Dulce:* A short navigable river running inland from the Caribbean coast, and giving such easy access to the old city of Guatemala, from which all central America was in those days governed, that it had to be protected against buccaneers by a fort.

THOMAS NASHE

THE UNFORTUNATE TRAVELLER AND OTHER WORKS

Edited by J. B. Steane

Thomas Nashe, a contemporary of Shakespeare, was a pamphleteer, poet, story-teller, satirist, scholar, moralist and jester. His work epitomizes everything that comes to mind when we think of the character of the Elizabethans: their shameless minglings of devoutness and bawdy, scholarship and slang, the inexhaustible fluency of their language, their strictly *ad personam* controversies, their occasional brutality, their relish for life and their constant awareness of the immanence of death. Nashe himself wrote: 'I have written in all sorts of humours . . . more than any young man of my age in England'. (He died in his early thirties.)

This volume offers the modern reader a selection of those of his works whose interest is perennial. As well as *The Unfortunate Traveller* it contains *Piers Penniless*, *Terrors of the Night*, *Lenten Stuff* and *A Choice of Valentines*, and extracts from *Christ's Tears over Jerusalem*, *The Anatomy of Absurdity* and other works.

LAURENCE STERNE

A SENTIMENTAL JOURNEY

With an Introduction by A. Alvarez

Owing, perhaps, to his Irish blood, Laurence Sterne is one of the most engaging buttonholers in literature. He launches into conversation with no story to tell, little plan of narration, and a habit of slipping down every side-turning . . . but there is no getting away from him. *A Sentimental Journey* began as an account of a tour by coach through France and Italy: it ends as a treasury of dramatic sketches, pathetic and ironic incidents, philosophical musings, reminiscences, and anecdotes. 'It is perhaps the most bodiless novel ever written', as Mr Alvarez remarks in his introduction. Nevertheless the studied artlessness of a work which was written by the dying author of *Tristram Shandy* forestalled by nearly two centuries those modern writers who in some ways resemble him – Joyce, Beckett and Virginia Woolf.

The Penguin English Library

'It seems certain that the P.E.L. will continue to offer the best supply of well-edited literary texts for academic and general use in paperback form, and that the series will survive and grow' – John Sutherland

A selection

Edward Gibbon

THE DECLINE AND FALL OF THE ROMAN EMPIRE

A one-volume abridgement by Dero A. Saunders

D. H. Lawrence

WOMEN IN LOVE

Edited by Charles Ross

Edward Thomas

SELECTED POEMS AND PROSE

Edited by David Wright

Adam Smith

THE WEALTH OF NATIONS

Edited by Andrew Skinner

Oliver Goldsmith

THE VICAR OF WAKEFIELD

Edited by Stephen Coote

Daniel Defoe

ROXANA

Edited by David Blewett

SELECTIONS FROM THE TATLER AND THE SPECTATOR

Edited by Angus Ross